Jerry Muldoon 3/03

Fundamentals of Chemical Reaction Engineering

Fundamentals of Chemical Reaction Engineering

Mark E. Davis
California Institute of Technology

Robert J. Davis
University of Virginia

Boston   Burr Ridge, IL   Dubuque, IA   Madison, WI   New York   San Francisco   St. Louis
Bangkok   Bogotá   Caracas   Kuala Lumpur   Lisbon   London   Madrid   Mexico City
Milan   Montreal   New Delhi   Santiago   Seoul   Singapore   Sydney   Taipei   Toronto

McGraw-Hill Higher Education

A Division of The **McGraw-Hill** *Companies*

FUNDAMENTALS OF CHEMICAL REACTION ENGINEERING

Published by McGraw-Hill, a business unit of The McGraw-Hill Companies, Inc., 1221 Avenue of the Americas, New York, NY 10020. Copyright © 2003 by The McGraw-Hill Companies, Inc. All rights reserved. No part of this publication may be reproduced or distributed in any form or by any means, or stored in a database or retrieval system, without the prior written consent of The McGraw-Hill Companies, Inc., including, but not limited to, in any network or other electronic storage or transmission, or broadcast for distance learning.

Some ancillaries, including electronic and print components, may not be available to customers outside the United States.

This book is printed on acid-free paper.

International   1 2 3 4 5 6 7 8 9 0 DOC/DOC 0 9 8 7 6 5 4 3 2
Domestic   1 2 3 4 5 6 7 8 9 0 DOC/DOC 0 9 8 7 6 5 4 3 2

ISBN   0–07–245007–X
ISBN   0–07–119260–3 (ISE)

Publisher: *Elizabeth A. Jones*
Sponsoring editor: *Suzanne Jeans*
Developmental editor: *Maja Lorkovic*
Marketing manager: *Sarah Martin*
Project manager: *Jane Mohr*
Production supervisor: *Sherry L. Kane*
Senior media project manager: *Tammy Juran*
Coordinator of freelance design: *Rick D. Noel*
Cover designer: *Maureen McCutcheon*
Compositor: *TECHBOOKS*
Typeface: *10/12 Times Roman*
Printer: *R. R. Donnelley/Crawfordsville, IN*

Cover image: *Adapted from artwork provided courtesy of Professor Ahmed Zewail's group at Caltech. In 1999, Professor Zewail received the Nobel Prize in Chemistry for studies on the transition states of chemical reactions using femtosecond spectroscopy.*

Library of Congress Cataloging-in-Publication Data

Davis, Mark E.
    Fundamentals of chemical reaction engineering / Mark E. Davis, Robert J. Davis. — 1st ed.
        p.   cm. — (McGraw-Hill chemical engineering series)
    Includes index.
    ISBN 0–07–245007–X (acid-free paper) — ISBN 0–07–119260–3 (acid-free paper : ISE)
    1. Chemical processes.   I. Davis, Robert J.   II. Title.   III. Series.

TP155.7 .D38   2003
660'.28—dc21   2002025525
                                                    CIP

INTERNATIONAL EDITION ISBN 0–07–119260–3
Copyright © 2003. Exclusive rights by The McGraw-Hill Companies, Inc., for manufacture and export. This book cannot be re-exported from the country to which it is sold by McGraw-Hill. The International Edition is not available in North America.

www.mhhe.com

CONTENTS

T his book is an introduction to the quantitative treatment of chemical reaction engineering. The level of the presentation is what we consider appropriate for a one-semester course. The text provides a *balanced* approach to the understanding of: (1) *both* homogeneous and heterogeneous reacting systems and (2) *both* chemical reaction engineering and chemical reactor engineering. We have emulated the teachings of Prof. Michel Boudart in numerous sections of this text. For example, much of Chapters 1 and 4 are modeled after his superb text that is now out of print (*Kinetics of Chemical Processes*), but they have been expanded and updated. Each chapter contains numerous worked problems and vignettes. We use the vignettes to provide the reader with discussions on real, commercial processes and/or uses of the molecules and/or analyses described in the text. Thus, the vignettes relate the material presented to what happens in the world around us so that the reader gains appreciation for how chemical reaction engineering and its principles affect everyday life. Many problems in this text require numerical solution. The reader should seek appropriate software for proper solution of these problems. Since this software is abundant and continually improving, the reader should be able to easily find the necessary software. This exercise is useful for students since they will need to do this upon leaving their academic institutions. Completion of the entire text will give the reader a good introduction to the fundamentals of chemical reaction engineering and provide a basis for extensions into other nontraditional uses of these analyses, for example, behavior of biological systems, processing of electronic materials, and prediction of global atmospheric phenomena. We believe that the emphasis on chemical *reaction* engineering as opposed to chemical *reactor* engineering is the appropriate context for training future chemical engineers who will confront issues in diverse sectors of employment.

We gratefully acknowledge Prof. Michel Boudart who encouraged us to write this text and who has provided intellectual guidance to both of us. MED also thanks Martha Hepworth for her efforts in converting a pile of handwritten notes into a final product. In addition, Stacey Siporin, John Murphy, and Kyle Bishop are acknowledged for their excellent assistance in compiling the solutions manual. The cover artwork was provided courtesy of Professor Ahmed Zewail's group at Caltech, and we gratefully thank them for their contribution. We acknowledge with appreciation the people who reviewed our project, especially A. Brad Anton of Cornell University, who provided extensive comments on content and accuracy. Finally, we thank and apologize to the many students who suffered through the early drafts as course notes.

We dedicate this book to our wives and to our parents for their constant support.

Mark E. Davis
Pasadena, CA

Robert J. Davis
Charlottesville, VA

Nomenclature

a_i	activity of species i
a_v	external catalyst particle surface area per unit reactor volume
A_i	representation of species i
A_C	cross sectional area of tubular reactor
A_C^P	cross sectional area of a pore
A_H	heat transfer area
$\overline{A}$	pre-exponential factor
Bo_a	dimensionless group analogous to the axial Peclet number for the energy balance
C_i or $[A_i]$	concentration of species i
C_{iB}	concentration of species i in the bulk fluid
C_{iS}	concentration of species i at the solid surface
C_p	heat capacity per mole
$\overline{C}_p$	heat capacity per unit mass
d_e	effective diameter
d_p	particle diameter
d_t	diameter of tube
D_a	axial dispersion coefficient
D^e	effective diffusivity
D_{ij}	molecular diffusion coefficient
D_{Ki}	Knudsen diffusivity of species i
D_r	radial dispersion coefficient
D_{TA}	transition diffusivity from the Bosanquet equation
Da	Damkohler number
$\overline{Da}$	dimensionless group
E	activation energy
E_D	activation energy for diffusion
$E(t)$	$E(t)$-curve; residence time distribution
$\overline{E}$	total energy in closed system
f_f	friction factor in Ergun equation and modified Ergun equation
f_i	fractional conversion based on species i
f_i^{eq}	fractional conversion at equilibrium

$\bar{f}_i$	fugacity of species i
$\bar{f}_i^0$	fugacity at standard state of pure species i
ff	frictional force
F_i	molar flow rate of species i
g	gravitational acceleration
$(g/g_c)\bar{z}$	gravitational potential energy per unit mass
g_c	gravitational constant
gm	mass of catalyst
ΔG	change in Gibbs function ("free energy")
h	Planck's constant
h_i	enthalpy per mass of stream i
h_t	heat transfer coefficient
H	enthalpy
ΔH	change in enthalpy
ΔH_r	enthalpy of the reaction (often called heat of reaction)
H_w	dimensionless group
$\bar{H}_w$	dimensionless group
$\bar{I}$	ionic strength
J	Colburn J factor
$\bar{J}_i$	flux of species i with respect to a coordinate system
k	rate constant
$\bar{k}$	Boltzmann's constant
$\bar{k}_c$	mass transfer coefficient
K_a	equilibrium constant expressed in terms of activities
K_C	portion of equilibrium constant involving concentration
K_P	portion of equilibrium constant involving total pressure
K_X	portion of equilibrium constant involving mole fractions
$K_{\bar{\phi}}$	portion of equilibrium constant involving activity coefficients
L	length of tubular reactor
L_c	length of microcavity in Vignette 6.4.2
L_p	generalized length parameter
$\bar{L}$	length in a catalyst particle
$\overline{m}_i$	mass of stream i
$\dot{m}_i$	mass flow rate of stream i
M_i	molecular weight of species i
$\overline{M}$	ratio of concentrations or moles of two species
$\overline{MS}$	total mass of system
n_i	number of moles of species i

N_i	flux of species i
$NCOMP$	number of components
$NRXN$	number of independent reactions
P	pressure
Pe_a	axial Peclet number
Pe_r	radial Peclet number
PP	probability
q	heat flux
Q	heat transferred
$\dot{Q}$	rate of heat transfer
r	reaction rate
r_t	turnover frequency or rate of turnover
$\bar{r}$	radial coordinate
$\bar{r}_t$	radius of tubular reactor
R	recycle ratio
R_g	universal gas constant
R_p	radius of pellet
R_{pore}	radius of pore
$\bar{R}$	dimensionless radial coordinate in tubular reactor
$\bar{R}_{cc}$	correlation coefficient
Re	Reynolds number
s_i	instantaneous selectivity to species i
ΔS	change in entropy
S_c	sticking coefficient
S_i	overall selectivity to species i
S_p	surface area of catalyst particle
$\bar{S}$	number of active sites on catalyst
SA	surface area
Sc	Schmidt number
SE	standard error on parameters
Sh	Sherwood number
t	time
$\langle t \rangle$	mean residence time
$\bar{t}^*$	student t-test value
T	temperature
T_B	temperature of bulk fluid
T_S	temperature of solid surface
TB	third body in a collision process
u	linear fluid velocity (superficial velocity)

$\bar{u}(\bar{r})$	laminar flow velocity profile
U	overall heat transfer coefficient
$\bar{U}_i$	internal energy
v	volumetric flow rate
V	volume
V_i	mean velocity of gas-phase species i
V_p	volume of catalyst particle
V_R	volume of reactor
V_{total}	average velocity of all gas-phase species
W_c	width of microcavity in Vignette 6.4.2
x	length variable
x_p	half the thickness of a slab catalyst particle
X_i	mole fraction of species i
$\bar{X}^2$	defined by Equation (B.1.5)
y	dimensionless concentration
Y_i	yield of species i
z	axial coordinate
$\bar{z}$	height above a reference point
Z	dimensionless axial coordinate
$\bar{Z}_i$	charge of species i
α_i	when used as a superscript is the order of reaction with respect to species i
$\bar{\alpha}_i$	coefficients; from linear regression analysis, from integration, etc.
$\alpha\alpha_1$	parameter groupings in Section 9.6
$\alpha\alpha_2$	parameter groupings in Section 9.6
β	Prater number
β_T	dimensionless group
$\beta\beta_i$	dimensionless groups
γ	Arrhenius number
$\bar{\gamma}_i$	activity coefficient of species i
Γ	dimensionless temperature in catalyst particle
$\bar{\Gamma}$	dimensionless temperature
$\delta(t)$	Dirac delta function
$\bar{\delta}$	thickness of boundary layer
ε_i	molar expansion factor based on species i
$\bar{\varepsilon}$	deviation of concentration from steady-state value
$\bar{\varepsilon}_B$	porosity of bed
$\bar{\varepsilon}_p$	porosity of catalyst pellet

η	intraphase effectiveness factor
η_o	overall effectiveness factor
$\overline{\eta}$	interphase effectiveness factor
θ	dimensionless time
θ_i	fractional surface coverage of species i
$\overline{\theta}$	dimensionless temperature
λ	universal frequency factor
λ^e	effective thermal conductivity in catalyst particle
λ_λ	parameter groupings in Section 9.6
$\overline{\lambda}_r$	effective thermal conductivity in the radial direction
μ_i	chemical potential of species i
$\overline{\mu}$	viscosity
ξ	number of moles of species reacted
ρ	density (either mass or mole basis)
ρ_B	bed density
ρ_p	density of catalyst pellet
σ_i	standard deviation
$\overline{\sigma}_i$	stoichiometric number of elementary step i
τ	space time
$\overline{\tau}$	tortuosity
v_i	stoichiometric coefficient of species i
ϕ	Thiele modulus
ϕ_0	Thiele modulus based on generalized length parameter
$\overline{\phi}_i$	fugacity coefficient of species i
Φ	extent of reaction
χ	dimensionless length variable in catalyst particle
ψ	dimensionless concentration in catalyst particle for irreversible reaction
$\overline{\psi}$	dimensionless concentration in catalyst particle for reversible reaction
Ψ	dimensionless concentration
ω	dimensionless distance in catalyst particle

Notation used for stoichiometric reactions and elementary steps

	Stoichiometric reaction	Elementary step
Irreversible (one-way)	$\Longrightarrow$	$\longrightarrow$
Reversible (two-way)	$=$	$\rightleftarrows$
Equilibrated	$\ominus$	$\ominus$
Rate-determining		$\nrightarrow$ or $\nleftrightarrow$

The Basics of Reaction Kinetics for Chemical Reaction Engineering

1.1 | The Scope of Chemical Reaction Engineering

The subject of chemical reaction engineering initiated and evolved primarily to accomplish the task of describing how to choose, size, and determine the optimal operating conditions for a reactor whose purpose is to produce a given set of chemicals in a petrochemical application. However, the principles developed for chemical reactors can be applied to most if not all chemically reacting systems (e.g., atmospheric chemistry, metabolic processes in living organisms, etc.). In this text, the principles of chemical reaction engineering are presented in such rigor to make possible a comprehensive understanding of the subject. Mastery of these concepts will allow for generalizations to reacting systems independent of their origin and will furnish strategies for attacking such problems.

The two questions that must be answered for a chemically reacting system are: (1) what changes are expected to occur and (2) how fast will they occur? The initial task in approaching the description of a chemically reacting system is to understand the answer to the first question by elucidating the thermodynamics of the process. For example, dinitrogen (N_2) and dihydrogen (H_2) are reacted over an iron catalyst to produce ammonia (NH_3):

$$N_2 + 3H_2 = 2NH_3, \quad -\Delta H_r = 109 \text{ kJ/mol (at 773 K)}$$

where ΔH_r is the enthalpy of the reaction (normally referred to as the heat of reaction). This reaction proceeds in an industrial ammonia synthesis reactor such that at the reactor exit approximately 50 percent of the dinitrogen is converted to ammonia. At first glance, one might expect to make dramatic improvements on the production of ammonia if, for example, a new catalyst (a substance that increases the rate of reaction without being consumed) could be developed. However, a quick inspection of the thermodynamics of this process reveals that significant enhancements in the production of ammonia are not possible unless the temperature and pressure of the reaction are altered. Thus, the constraints placed on a reacting system by thermodynamics should always be identified first.

VIGNETTE 1.1.1

The initial success of a large-scale catalytic technology began in 1913 when the first industrial chemical reactor to synthesize ammonia from dinitrogen and dihydrogen began operation in Germany. Most of the ammonia manufactured today is used to produce nitrogen-rich fertilizers that have an enormous impact on meeting worldwide food demands. According to figures for U.S. agriculture, the 800,000 tons of dinitrogen converted to ammonia in the first Haber reactor (ammonia synthesis over an iron catalyst is called the Haber process after the inventor F. Haber) could grow 700 million additional bushels of corn, enough to nourish 50 million man-years [P. B. Weisz, *CHEMTECH*, **14** (1984) 354]. Because ammonia is a low-priced commodity chemical, the catalyst must be cheap and durable, and its activity must be high so that temperature and pressure can be maintained as low as possible to minimize the size and cost of the huge industrial reactors. The catalyst is essentially iron and the reaction is now well understood. Certain groups of iron atoms on the surface of the catalyst dissociate dinitrogen and dihydrogen into atoms of nitrogen and hydrogen and combine them into ammonia. The catalyst operates at high temperature to increase the rate of the reaction and at high pressure to increase the thermodynamic yield of ammonia (see Example 1.1.1). In a commercial reactor, the catalyst can run for a long time before it is replaced. Thus, the high productivity leads to low cost: the catalyst can give products worth approximately 2000 times its value.

EXAMPLE 1.1.1

In order to obtain a reasonable level of conversion at a commercially acceptable rate, ammonia synthesis reactors operate at pressures of 150 to 300 atm and temperatures of 700 to 750 K. Calculate the equilibrium mole fraction of dinitrogen at 300 atm and 723 K starting from an initial composition of $X_{N_2} = 0.25$, $X_{H_2} = 0.75$ (X_i is the mole fraction of species i). At 300 atm and 723 K, the equilibrium constant, K_a, is 6.6×10^{-3}. (K. Denbigh, *The Principles of Chemical Equilibrium*, Cambridge Press, 1971, p. 153).

■ **Answer**

(See Appendix A for a brief overview of equilibria involving chemical reactions):

$$\tfrac{1}{2} N_2 + \tfrac{3}{2} H_2 = NH_3$$

$$K_a = \left[\frac{a_{NH_3}}{a_{N_2}^{1/2} a_{H_2}^{3/2}} \right], \quad \text{where } a = \text{activity}$$

The definition of the activity of species i is:

$$a_i = \bar{f}_i/\bar{f}_i^0, \quad \bar{f}_i^0 = \text{fugacity at the standard state, that is, 1 atm for gases}$$

and thus

$$K_a = \left[\frac{\bar{f}_{NH_3}}{\bar{f}_{N_2}^{1/2} \bar{f}_{H_2}^{3/2}} \right] \left[\frac{(\bar{f}_{N_2}^0)^{1/2} (\bar{f}_{H_2}^0)^{3/2}}{(\bar{f}_{NH_3}^0)} \right] = \left[\frac{\bar{f}_{NH_3}}{\bar{f}_{N_2}^{1/2} \bar{f}_{H_2}^{3/2}} \right] [1 \text{ atm}]$$

Use of the Lewis and Randall rule gives:

$$\bar{f}_i = X_i \bar{\phi}_i P, \quad \bar{\phi}_i = \text{fugacity coefficient of pure component } i \text{ at } T \text{ and } P \text{ of system}$$

then

$$K_a = K_X K_{\bar{\phi}} K_P = \left[\frac{X_{NH_3}}{X_{N_2}^{1/2} X_{H_2}^{3/2}} \right] \left[\frac{\bar{\phi}_{NH_3}}{\bar{\phi}_{N_2}^{1/2} \bar{\phi}_{H_2}^{3/2}} \right] [P^{-1}] [1 \text{ atm}]$$

Upon obtaining each $\bar{\phi}_i$ from correlations or tables of data (available in numerous references that contain thermodynamic information):

$$\left[\frac{X_{NH_3}}{X_{N_2}^{1/2} X_{H_2}^{3/2}} \right] = \left[\frac{(1.14)^{1/2}(1.09)^{3/2}}{(0.91)} \right] (6.6 \times 10^{-3})(300 \text{ atm})(1 \text{ atm})^{-1}$$

$$\left[\frac{X_{NH_3}}{X_{N_2}^{1/2} X_{H_2}^{3/2}} \right] = 2.64$$

If a basis of 100 mol is used (ξ is the number of moles of N_2 reacted):

Species	Initial	At equilibrium
N_2	25	$25 - \xi$
H_2	75	$75 - 3\xi$
NH_3	0	2ξ
total	100	$100 - 2\xi$

then

$$\frac{2\xi/(100 - 2\xi)}{(\frac{25 - \xi}{100 - 2\xi})^{1/2} (\frac{75 - 3\xi}{100 - 2\xi})^{3/2}} = \frac{(2\xi)(100 - 2\xi)}{(25 - \xi)^{1/2}(75 - 3\xi)^{3/2}} = 2.64$$

Thus, $\xi = 13.1$ and $X_{N_2} = (25 - 13.1)/(100 - 26.2) = 0.16$. At 300 atm, the equilibrium mole fraction of ammonia is 0.36 while at 100 atm it falls to approximately 0.16. Thus, the equilibrium amount of ammonia increases with the total pressure of the system at a constant temperature.

The next task in describing a chemically reacting system is the identification of the reactions and their arrangement in a *network*. The kinetic analysis of the network is then necessary for obtaining information on the rates of individual *reactions* and answering the question of how fast the chemical conversions occur. Each reaction of the network is stoichiometrically simple in the sense that it can be described by the single parameter called the *extent of reaction* (see Section 1.2). Here, a stoichiometrically simple reaction will just be called a reaction for short. The expression "simple reaction" should be avoided since a stoichiometrically simple reaction does not occur in a simple manner. In fact, most chemical reactions proceed through complicated sequences of *steps* involving reactive intermediates that do not appear in the stoichiometries of the reactions. The identification of these intermediates and the sequence of steps are the core problems of the kinetic analysis.

If a step of the sequence can be written as it proceeds at the molecular level, it is denoted as an *elementary step* (or an *elementary reaction*), and it represents an irreducible molecular event. Here, elementary steps will be called *steps* for short. The hydrogenation of dibromine is an example of a stoichiometrically simple reaction:

$$H_2 + Br_2 \implies 2HBr$$

If this reaction would occur by H_2 interacting directly with Br_2 to yield two molecules of HBr, the step would be elementary. However, it does not proceed as written. It is known that the hydrogenation of dibromine takes place in a sequence of two steps involving hydrogen and bromine atoms that do not appear in the stoichiometry of the reaction but exist in the reacting system in very small concentrations as shown below (an initiator is necessary to start the reaction, for example, a photon: $Br_2 + light \rightarrow 2Br$, and the reaction is terminated by $Br + Br + TB \rightarrow Br_2$ where TB is a third body that is involved in the recombination process—see below for further examples):

$$Br + H_2 \rightarrow HBr + H$$
$$H + Br_2 \rightarrow HBr + Br$$

In this text, stoichiometric reactions and elementary steps are distinguished by the notation provided in Table 1.1.1.

Table 1.1.1 | Notation used for stoichiometric reactions and elementary steps.

	Stoichiometric reaction	Elementary step
Irreversible (one-way)	$\Longrightarrow$	$\longrightarrow$
Reversible (two-way)	$=\!\!=$	$\rightleftharpoons$
Equilibrated	$\ominus$	$\ominus$
Rate-determining		$\nrightarrow$ or $\nleftrightarrow$

In discussions on chemical kinetics, the terms *mechanism* or *model* frequently appear and are used to mean an assumed reaction network or a plausible sequence of steps for a given reaction. Since the levels of detail in investigating reaction networks, sequences and steps are so different, the words *mechanism* and *model* have to date largely acquired bad connotations because they have been associated with much speculation. Thus, they will be used carefully in this text.

As a chemically reacting system proceeds from reactants to products, a number of species called *intermediates* appear, reach a certain concentration, and ultimately vanish. Three different types of intermediates can be identified that correspond to the distinction among networks, reactions, and steps. The first type of intermediates has reactivity, concentration, and lifetime comparable to those of stable reactants and products. These intermediates are the ones that appear in the reactions of the network. For example, consider the following proposal for how the oxidation of methane at conditions near 700 K and atmospheric pressure may proceed (see Scheme 1.1.1). The reacting system may evolve from two stable reactants, CH_4 and O_2, to two stable products, CO_2 and H_2O, through a network of four reactions. The intermediates are formaldehyde, CH_2O; hydrogen peroxide, H_2O_2; and carbon monoxide, CO. The second type of intermediate appears in the sequence of steps for an individual reaction of the network. These species (e.g., free radicals in the gas phase) are usually present in very small concentrations and have short lifetimes when compared to those of reactants and products. These intermediates will be called *reactive intermediates* to distinguish them from the more stable species that are the ones that appear in the reactions of the network. Referring to Scheme 1.1.1, for the oxidation of CH_2O to give CO and H_2O_2, the reaction may proceed through a postulated sequence of two steps that involve two reactive intermediates, CHO and HO_2. The third type of intermediate is called a *transition state,* which by definition cannot be isolated and is considered a species in transit. Each elementary step proceeds from reactants to products through a transition state. Thus, for each of the two elementary steps in the oxidation of CH_2O, there is a transition state. Although the nature of the transition state for the elementary step involving CHO, O_2, CO, and HO_2 is unknown, other elementary steps have transition states that have been elucidated in greater detail. For example, the configuration shown in Fig. 1.1.1 is reached for an instant in the transition state of the step:

$$OH^- + C_2H_5Br \rightarrow HOC_2H_5 + Br^-$$

The study of elementary steps focuses on transition states, and the kinetics of these steps represent the foundation of chemical kinetics and the highest level of understanding of chemical reactivity. In fact, the use of lasers that can generate femtosecond pulses has now allowed for the "viewing" of the real-time transition from reactants through the transition-state to products (A. Zewail, *The*

SCHEME 1.1.1 | ILLUSTRATION OF THREE TYPES OF INTERMEDIATES

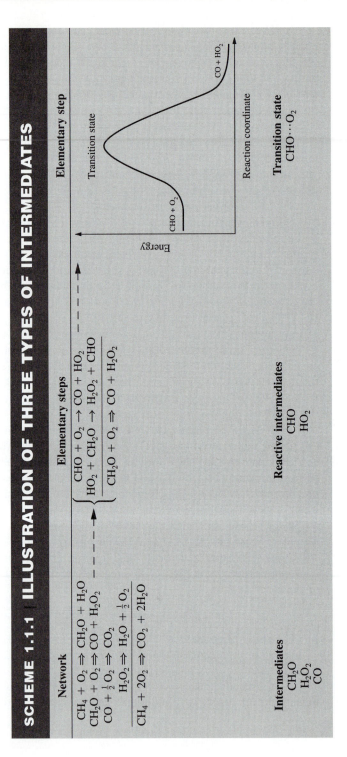

Network

$$CH_4 + O_2 \rightleftharpoons CH_2O + H_2O$$
$$CH_2O + O_2 \rightleftharpoons CO + H_2O_2$$
$$CO + \tfrac{1}{2}O_2 \rightleftharpoons CO_2$$
$$H_2O_2 \rightleftharpoons H_2O + \tfrac{1}{2}O_2$$
$$\overline{CH_4 + 2O_2 \rightleftharpoons CO_2 + 2H_2O}$$

Elementary steps

$$CHO + O_2 \rightarrow CO + HO_2$$
$$HO_2 + CH_2O \rightarrow H_2O_2 + CHO$$
$$\overline{CH_2O + O_2 \rightleftharpoons CO + H_2O_2}$$

Elementary step

Transition state

CHO + O₂

CO + HO₂

Energy

Reaction coordinate

Intermediates
CH₂O
H₂O₂
CO

Reactive intermediates
CHO
HO₂

Transition state
CHO···O₂

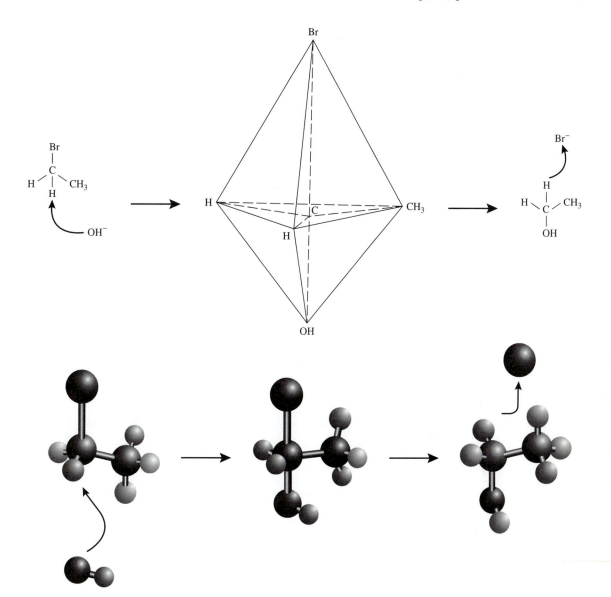

Figure 1.1.1 |

The transition state (trigonal bipyramid) of the elementary step:

$$OH^- + C_2H_5Br \longrightarrow HOC_2H_5 + Br^-$$

The nucleophilic substituent OH^- displaces the leaving group Br^-.

Chemical Bond: Structure and Dynamics, Academic Press, 1992). However, in the vast majority of cases, chemically reacting systems are investigated in much less detail. The level of sophistication that is conducted is normally dictated by the purpose of the work and the state of development of the system.

1.2 | The Extent of Reaction

The changes in a chemically reacting system can frequently, but not always (e.g., complex fermentation reactions), be characterized by a stoichiometric equation. The stoichiometric equation for a simple reaction can be written as:

$$0 = \sum_{i=1}^{NCOMP} v_i A_i \tag{1.2.1}$$

where *NCOMP* is the number of components, A_i, of the system. The stoichiometric coefficients, v_i, are positive for products, negative for reactants, and zero for inert components that do not participate in the reaction. For example, many gas-phase oxidation reactions use air as the oxidant and the dinitrogen in the air does not participate in the reaction (serves only as a diluent). In the case of ammonia synthesis the stoichiometric relationship is:

$$N_2 + 3H_2 = 2NH_3$$

Application of Equation (1.2.1) to the ammonia synthesis, stoichiometric relationship gives:

$$0 = 2NH_3 - N_2 - 3H_2$$

For stoichiometric relationships, the coefficients can be ratioed differently, e.g., the relationship:

$$0 = 2NH_3 - N_2 - 3H_2$$

can be written also as:

$$0 = NH_3 - \tfrac{1}{2}N_2 - \tfrac{3}{2}H_2$$

since they are just mole balances. However, for an elementary reaction, the stoichiometry is written as the reaction should proceed. Therefore, an elementary reaction such as:

$$2NO + O_2 \rightarrow 2NO_2 \qquad \text{(correct)}$$

CANNOT be written as:

$$NO + \tfrac{1}{2}O_2 \rightarrow NO_2 \qquad \text{(not correct)}$$

EXAMPLE 1.2.1

If there are several simultaneous reactions taking place, generalize Equation (1.2.1) to a system of $NRXN$ different reactions. For the methane oxidation network shown in Scheme 1.1.1, write out the relationships from the generalized equation.

■ Answer

If there are $NRXN$ reactions and $NCOMP$ species in the system, the generalized form of Equation (1.2.1) is:

$$0 = \sum_i^{NCOMP} v_{i,j} A_i, \quad j = 1, \cdots, NRXN \tag{1.2.2}$$

For the methane oxidation network shown in Scheme 1.1.1:

$$0 = 0CO_2 + 1H_2O - 1O_2 + 0CO + 0H_2O_2 + 1CH_2O - 1CH_4$$

$$0 = 0CO_2 + 0H_2O - 1O_2 + 1CO + 1H_2O_2 - 1CH_2O + 0CH_4$$

$$0 = 1CO_2 + 0H_2O - \tfrac{1}{2}O_2 - 1CO + 0H_2O_2 + 0CH_2O + 0CH_4$$

$$0 = 0CO_2 + 1H_2O + \tfrac{1}{2}O_2 + 0CO - 1H_2O_2 + 0CH_2O + 0CH_4$$

or in matrix form:

$$
\begin{bmatrix} 0 \\ 0 \\ 0 \\ 0 \end{bmatrix} =
\begin{bmatrix}
0 & 1 & -1 & 0 & 0 & 1 & -1 \\
0 & 0 & -1 & 1 & 1 & -1 & 0 \\
1 & 0 & -\tfrac{1}{2} & -1 & 0 & 0 & 0 \\
0 & 1 & \tfrac{1}{2} & 0 & -1 & 0 & 0
\end{bmatrix}
\begin{bmatrix} CO_2 \\ H_2O \\ O_2 \\ CO \\ H_2O_2 \\ CH_2O \\ CH_4 \end{bmatrix}
$$

Note that the sum of the coefficients of a column in the matrix is zero if the component is an intermediate.

Consider a closed system, that is, a system that exchanges no mass with its surroundings. Initially, there are n_i^0 moles of component A_i present in the system. If a single reaction takes place that can be described by a relationship defined by Equation (1.2.1), then the number of moles of component A_i at any time t will be given by the equation:

$$n_i(t) = n_i^0 + v_i \Phi(t) \tag{1.2.3}$$

that is an expression of the *Law of Definitive Proportions* (or more simply, a mole balance) and defines the parameter, Φ, called the *extent of reaction*. The extent of reaction is a function of time and is a natural reaction variable.

Equation (1.2.3) can be written as:

$$\Phi(t) = \frac{n_i(t) - n_i^0}{v_i} \tag{1.2.4}$$

Since there is only one Φ for each reaction:

$$\frac{n_i(t) - n_i^0}{v_i} = \frac{n_j(t) - n_j^0}{v_j} \qquad (1.2.5)$$

or

$$n_j(t) = n_j^0 + \left(\frac{v_j}{v_i}\right)[n_i(t) - n_i^0] \qquad (1.2.6)$$

Thus, if n_i is known or measured as a function of time, then the number of moles of all of the other reacting components can be calculated using Equation (1.2.6).

EXAMPLE 1.2.2

If there are numerous, simultaneous reactions occurring in a closed system, each one has an extent of reaction. Generalize Equation (1.2.3) to a system with *NRXN* reactions.

■ **Answer**

$$n_i = n_i^0 + \sum_{j=1}^{NRXN} v_{i,j}\Phi_j \qquad (1.2.7)$$

EXAMPLE 1.2.3

Carbon monoxide is oxidized with the stoichiometric amount of air. Because of the high temperature, the equilibrium:

$$N_2 + O_2 \; \rightleftharpoons \; 2NO \qquad (1)$$

has to be taken into account in addition to:

$$CO + \tfrac{1}{2}O_2 \; \rightleftharpoons \; CO_2 \qquad (2)$$

The total pressure is one atmosphere and the equilibrium constants of reactions (1) and (2) are:

$$K_{X_1} = \frac{(X_{NO})^2}{(X_{N_2})(X_{O_2})}, \quad K_{X_2} = \frac{(X_{CO_2})}{(X_{CO})(X_{O_2})^{\frac{1}{2}}}$$

where $K_{X_1} = 8.26 \times 10^{-3}$, $K_{X_2} = 0.7$, and X_i is the mole fraction of species i (assuming ideal gas behavior). Calculate the equilibrium composition.

■ **Answer**

Assume a basis of 1 mol of CO with a stoichiometric amount of air (ξ_1 and ξ_2 are the number of moles of N_2 and CO reacted, respectively):

Species	Initial	At equilibrium
N_2	1.88	$1.88 - \xi_1$
O_2	0.5	$0.5 - \frac{1}{2}\xi_2 - \xi_1$
CO	1	$1 - \xi_2$
CO_2	0	ξ_2
NO	0	$2\xi_1$
total	3.38	$3.38 - \frac{1}{2}\xi_2$

$$K_{X_1} = \frac{(2\xi_1)^2}{(1.88 - \xi_1)(0.5 - \frac{1}{2}\xi_2 - \xi_1)} = 8.26 \times 10^{-3}$$

$$K_{X_2} = \frac{(\xi_2)(3.38 - \frac{1}{2}\xi_2)^{\frac{1}{2}}}{(1 - \xi_2)(0.5 - \frac{1}{2}\xi_2 - \xi_1)^{\frac{1}{2}}} = 0.7$$

The simultaneous solution of these two equations gives:

$$\xi_1 = 0.037, \qquad \xi_2 = 0.190$$

Therefore,

Species	Mole fraction at equilibrium
N_2	0.561
O_2	0.112
CO	0.247
CO_2	0.058
NO	0.022
	1.000

EXAMPLE 1.2.4

Using the results from Example 1.2.3, calculate the two equilibrium extents of reaction.

■ **Answer**

$$\Phi_1^{eq} = \xi_1^{eq} = 0.037$$

$$\Phi_2^{eq} = \xi_2^{eq} = 0.190$$

VIGNETTE 1.2.1

Below is shown the pollution standard index (PSI) for Pasadena, California, in July 1995 from 7 A.M. until 6 P.M.:

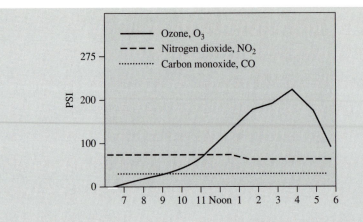

On these scales, 100 is the Federal standard and values above this are unhealthy. Notice that the ozone somewhat follows the intensity of the sunlight. That is because ozone is formed by numerous reactions that involve sunlight, for example,

$$NO_2 + sunlight \rightarrow NO + O$$

$$O + O_2 + TB \rightarrow O_3 + TB$$

where TB represents N_2, O_2 or another third body that absorbs the excess vibrational energy and thereby stabilizes the O_3 formed. One major source of NO_2 is exhaust from vehicles. NO_x formation occurs when N_2 and O_2 are raised to a high temperature (see Example 1.2.3) via the reactions:

$$N_2 + O_2 \Rightarrow 2NO$$

$$NO + \tfrac{1}{2}O_2 \Rightarrow NO_2$$

and is called thermal NO_x. Additionally, fuels possess nitrogen-containing compounds like porphyrins that when combusted in an engine also give NO_x (called fuel NO_x). Since NO_x and O_3 are health hazards, much effort has been extended to develop catalytic converters for automobiles. The main function of the catalytic converter is to accomplish the following reactions:

$$hydrocarbons + O_2 \Rightarrow CO_2 + H_2O$$

$$CO + \tfrac{1}{2}O_2 \Rightarrow CO_2$$

$$2NO + 2CO \Rightarrow N_2 + 2CO_2$$

Catalytic converters (see Figure 1.2.1) contain metal catalysts (Pd, Pt, Rh) that carry out the above reactions and thus significantly reduce pollution. These catalysts have been in use since 1980 and have dramatically aided the reduction of air pollution in Los Angeles and elsewhere. For example, in the 1970s the PSI in Pasadena showed levels over 400 on really smoggy days. In the 1990s, these levels never occurred. Thus, catalytic converters significantly contributed to pollution reduction and are one of the major success stories for chemical reaction engineering.

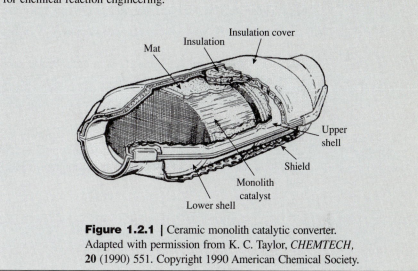

Figure 1.2.1 | Ceramic monolith catalytic converter. Adapted with permission from K. C. Taylor, *CHEMTECH,* **20** (1990) 551. Copyright 1990 American Chemical Society.

The drawback of Φ is that it is an extensive variable, that is, it is dependent upon the mass of the system. The *fractional conversion, f,* does not suffer from this problem and can be related to Φ. In general, reactants are not initially present in stoichiometric amounts and the reactant in the least amount determines the maximum value for the extent of reaction, Φ_{max}. This component, called the *limiting component* (subscript ℓ) can be totally consumed when Φ reaches Φ_{max}. Thus,

$$0 = n_\ell^0 + v_\ell \Phi_{max} \tag{1.2.8}$$

The fractional conversion is defined as:

$$f(t) = \frac{\Phi(t)}{\Phi_{max}} \tag{1.2.9}$$

and can be calculated from Equations (1.2.3) and (1.2.8):

$$f_\ell = (-v_\ell)\frac{\Phi}{n_\ell^0} = 1 - \frac{n_\ell}{n_\ell^0} \tag{1.2.10}$$

Equation (1.2.10) can be rearranged to give:

$$n_\ell = n_\ell^0(1 - f_\ell) \tag{1.2.11}$$

where $0 \leq f_\ell \leq 1$. When the thermodynamics of the system limit Φ such that it cannot reach Φ_{max} (where $n_\ell = 0$), Φ will approach its equilibrium value Φ^{eq} ($n_\ell \neq 0$ value of n_ℓ determined by the equilibrium constant). When a reaction is limited by thermodynamic equilibrium in this fashion, the reaction has historically been called *reversible*. Alternatively, the reaction can be denoted as *two-way*. When Φ^{eq} is equal to Φ_{max} for all practical purposes, the reaction has been denoted *irreversible* or *one-way*. Thus, when writing the fractional conversion for the limiting reactant,

$$0 \leq f_\ell \leq f_\ell^{eq} \leq 1 \tag{1.2.12}$$

where f_ℓ^{eq} is the fractional conversion at equilibrium conditions.

Consider the following reaction:

$$\bar{a}A + \bar{b}B + \cdots = \bar{s}S + \bar{w}W + \cdots \tag{1.2.13}$$

Expressions for the change in the number of moles of each species can be written in terms of the fractional conversion and they are [assume A is the limiting reactant, lump all inert species together as component I and refer to Equations (1.2.6) and (1.2.11)]:

$$n_A = n_A^0 - n_A^0 f_A$$

$$n_B = n_B^0 - \left(\frac{\bar{b}}{\bar{a}}\right) n_A^0 f_A$$

$$\vdots$$

$$n_S = n_S^0 + \left(\frac{\bar{s}}{\bar{a}}\right) n_A^0 f_A$$

$$n_W = n_W^0 + \left(\frac{\bar{w}}{\bar{a}}\right) n_A^0 f_A$$

$$\vdots$$

$$n_I = n_I^0 \qquad\qquad \text{(inerts)}$$

$$n_{\text{TOTAL}} = n_{\text{TOTAL}}^0 + n_A^0 \left[\frac{\bar{s} + \bar{w} + \cdots - \bar{a} - \bar{b}\cdots}{\bar{a}}\right] f_A$$

or

$$\frac{n_{\text{TOTAL}}}{n_{\text{TOTAL}}^0} = 1 + \frac{n_A^0}{n_{\text{TOTAL}}^0}\left[\frac{\bar{s} + \bar{w} + \cdots - \bar{a} - \bar{b}\cdots}{\bar{a}}\right] f_A \tag{1.2.14}$$

By defining ε_A as the *molar expansion factor*, Equation (1.2.14) can be written as:

$$n_{\text{TOTAL}} = n_{\text{TOTAL}}^0(1 + \varepsilon_A f_A) \tag{1.2.15}$$

where

$$\varepsilon_A = \frac{n_A^0}{n_{\text{TOTAL}}^0}\left[\frac{\sum_i \nu_i}{|\nu_A|}\right] = X_A^0\left[\frac{\sum_i \nu_i}{|\nu_A|}\right] \tag{1.2.16}$$

Notice that ε_A contains two terms and they involve stoichiometry and the initial mole fraction of the limiting reactant. The parameter ε_A becomes important if the density of the reacting system is changing as the reaction proceeds.

EXAMPLE 1.2.5

Calculate ε_A for the following reactions:

  (i)   n-butane = isobutane (isomerization)
  (ii) n-hexane $\Rightarrow$ benzene + dihydrogen (aromatization)
  (iii) reaction (ii) where 50 percent of the feed is dinitrogen.

■ **Answer**

  (i) $CH_3CH_2CH_2CH_3 = CH_3CH(CH_3)_2$, pure n-butane feed

$$\varepsilon_A = \frac{n^0_{TOTAL}}{n^0_{TOTAL}}\left[\frac{1-1}{|-1|}\right] = 0$$

  (ii) $CH_3CH_2CH_2CH_2CH_2CH_3 \Longrightarrow \bigcirc + 4H_2$, pure n-hexane feed

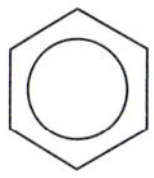

$$\varepsilon_A = \frac{n^0_{TOTAL}}{n^0_{TOTAL}}\left[\frac{4+1-1}{|-1|}\right] = 4$$

  (iii) $CH_3CH_2CH_2CH_2CH_2CH_3 \Longrightarrow \bigcirc + 4H_2$, 50 percent of feed is n-hexane

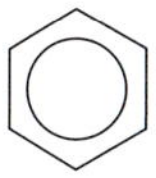

$$\varepsilon_A = \frac{0.5 n^0_{TOTAL}}{n^0_{TOTAL}}\left[\frac{4+1-1}{|-1|}\right] = 2$$

EXAMPLE 1.2.6

If the decomposition of N_2O_5 into N_2O_4 and O_2 were to proceed to completion in a closed volume of size V, what would the pressure rise be if the starting composition is 50 percent N_2O_5 and 50 percent N_2?

■ **Answer**

The ideal gas law is:

$$PV = n_{TOTAL}R_gT \quad (R_g : \text{universal gas constant})$$

At fixed T and V, the ideal gas law gives:

$$P = P^0\left(\frac{n_{TOTAL}}{n^0_{TOTAL}}\right) = P^0(1 + \varepsilon_A f_A)$$

The reaction proceeds to completion so $f_A = 1$ at the end of the reaction. Thus,

$$\frac{P}{P^0} = 1 + \varepsilon_A, \quad A : N_2O_5$$

with

$$N_2O_5 \Rightarrow N_2O_4 + \tfrac{1}{2}O_2,$$

$$\varepsilon_A = \frac{0.5n^0_{TOTAL}}{n^0_{TOTAL}} \left[\frac{1 + 0.5 - 1}{|-1|} \right] = 0.25$$

Therefore,

$$\frac{P}{P^0} = 1.25$$

1.3 | The Rate of Reaction

For a homogeneous, closed system at uniform pressure, temperature, and composition in which a single chemical reaction occurs and is represented by the stoichiometric Equation (1.2.1), the extent of reaction as given in Equation (1.2.3) increases with time, t. For this situation, the *reaction rate* is defined in the most general way by:

$$\frac{d\Phi}{dt} \quad \left(\frac{mol}{time} \right) \tag{1.3.1}$$

This expression is either positive or equal to zero if the system has reached equilibrium. The reaction rate, like Φ, is an extensive property of the system. A specific rate called the *volumic rate* is obtained by dividing the reaction rate by the volume of the system, V:

$$r = \frac{1}{V}\frac{d\Phi}{dt} \quad \left(\frac{mol}{time \cdot volume} \right) \tag{1.3.2}$$

Differentiation of Equation (1.2.3) gives:

$$dn_i = v_i \, d\Phi \tag{1.3.3}$$

Substitution of Equation (1.3.3) into Equation (1.3.2) yields:

$$r = \frac{1}{v_i V}\frac{dn_i}{dt} \tag{1.3.4}$$

since v_i is not a function of time. Note that the volumic rate as defined is an extensive variable and that the definition is not dependent on a particular reactant or product. If the volumic rate is defined for an individual species, r_i, then:

$$r_i = v_i \, r = \frac{1}{V}\frac{dn_i}{dt} \tag{1.3.5}$$

Since v_i is positive for products and negative for reactants and the reaction rate, $d\Phi/dt$, is always positive or zero, the various r_i will have the same sign as the v_i (dn_i/dt has the same sign as r_i since r is always positive). Often the use of molar concentrations, C_i, is desired. Since $C_i = n_i/V$, Equation (1.3.4) can be written as:

$$r = \frac{1}{v_i V}\frac{d}{dt}(C_i V) = \frac{1}{v_i}\frac{dC_i}{dt} + \frac{C_i}{v_i V}\frac{dV}{dt} \tag{1.3.6}$$

Note that only when the volume of the system is constant that the volumic rate can be written as:

$$r = \frac{1}{v_i}\frac{dC_i}{dt}, \text{ constant } V \tag{1.3.7}$$

When it is not possible to write a stoichiometric equation for the reaction, the rate is normally expressed as:

$$r = \frac{(COEF)}{V}\frac{dn_i}{dt}, (COEF) = \left\{ \begin{array}{l} -, \text{ reactant} \\ +, \text{ product} \end{array} \right. \tag{1.3.8}$$

For example, with certain polymerization reactions for which no unique stoichiometric equation can be written, the rate can be expressed by:

$$r = \frac{-1}{V}\frac{dn}{dt}$$

where n is the number of moles of the monomer.

Thus far, the discussion of reaction rate has been confined to *homogeneous* reactions taking place in a closed system of uniform composition, temperature, and pressure. However, many reactions are *heterogeneous;* they occur at the interface between phases, for example, the interface between two fluid phases (gas-liquid, liquid-liquid), the interface between a fluid and solid phase, and the interface between two solid phases. In order to obtain a convenient, specific rate of reaction it is necessary to normalize the reaction rate by the interfacial surface area available for the reaction. The interfacial area must be of uniform composition, temperature, and pressure. Frequently, the interfacial area is not known and alternative definitions of the specific rate are useful. Some examples of these types of rates are:

$$r = \frac{1}{gm}\frac{d\Phi}{dt} \left(\frac{\text{mol}}{\text{mass·time}} \right) \quad \text{(specific rate)}$$

$$r = \frac{1}{SA}\frac{d\Phi}{dt} \left(\frac{\text{mol}}{\text{area·time}} \right) \quad \text{(areal rate)}$$

where *gm* and *SA* are the mass and surface area of a solid phase (catalyst), respectively. Of course, alternative definitions for specific rates of both homogeneous and heterogeneous reactions are conceivable. For example, numerous rates can be defined for enzymatic reactions, and the choice of the definition of the specific rate is usually adapted to the particular situation.

For heterogeneous reactions involving fluid and solid phases, the areal rate is a good choice. However, the catalysts (solid phase) can have the same surface area but different concentrations of active sites (atomic configuration on the catalyst capable of catalyzing the reaction). Thus, a definition of the rate based on the number of active sites appears to be the best choice. The *turnover frequency* or *rate of turnover* is the number of times the catalytic cycle is completed (or turned-over) per catalytic site (active site) per time for a reaction at a given temperature, pressure, reactant ratio, and extent of reaction. Thus, the turnover frequency is:

$$r_t = \frac{1}{\overline{S}}\frac{dn}{dt} \tag{1.3.9}$$

where $\overline{S}$ is the number of active sites on the catalyst. The problem of the use of r_t is how to count the number of active sites. With metal catalysts, the number of metal atoms exposed to the reaction environment can be determined via techniques such as chemisorption. However, how many of the surface atoms that are grouped into an active site remains difficult to ascertain. Additionally, different types of active sites probably always exist on a real working catalyst; each has a different reaction rate. Thus, r_t is likely to be an average value of the catalytic activity and a lower bound for the true activity since only a fraction of surface atoms may contribute to the activity. Additionally, r_t is a rate and *not* a rate constant so it is always necessary to specify all conditions of the reaction when reporting values of r_t.

The number of turnovers a catalyst makes before it is no longer useful (e.g., due to an induction period or poisoning) is the best definition of the life of the catalyst. In practice, the turnovers can be very high, $\sim 10^6$ or more. The turnover frequency on the other hand is commonly in the range of $r_t = 1\ s^{-1}$ to $r_t = 0.01\ s^{-1}$ for practical applications. Values much smaller than these give rates too slow for practical use while higher values give rates so large that they become influenced by transport phenomena (see Chapter 6).

EXAMPLE 1.3.1

Gonzo and Boudart [*J. Catal.*, **52** (1978) 462] studied the hydrogenation of cyclohexene over Pd supported on Al_2O_3 and SiO_2 at 308 K, atmospheric pressure of dihydrogen and 0.24M cyclohexene in cyclohexane in a stirred flask:

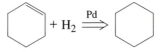

The specific rates for 4.88 wt. % Pd on Al_2O_3 and 3.75 wt. % Pd on SiO_2 were 7.66×10^{-4} and 1.26×10^{-3} mol/(gcat · s), respectively. Using a technique called *titration,* the percentage of Pd metal atoms on the surface of the Pd metal particles on the Al_2O_3 and SiO_2 was 21 percent and 55 percent, respectively. Since the specific rates for Pd on Al_2O_3 and SiO_2 are different, does the metal oxide really influence the reaction rate?

Titration is a technique that can be used to measure the number of surface metal atoms. The procedure involves first chemisorbing (chemical bonds formed between adsorbing species and surface atoms) molecules onto the metal atoms exposed to the reaction environment. Second, the chemisorbed species are reacted with a second component in order to recover and count the number of atoms chemisorbed. By knowing the stoichiometry of these two steps, the number of surface atoms can be calculated from the amount of the recovered chemisorbed atoms. The technique is illustrated for the problem at hand:

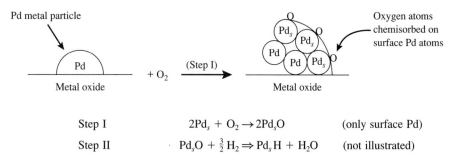

Step I	$2Pd_s + O_2 \rightarrow 2Pd_sO$	(only surface Pd)
Step II	$Pd_sO + \frac{3}{2}H_2 \Rightarrow Pd_sH + H_2O$	(not illustrated)

By counting the number of H_2 molecules consumed in Step II, the number of surface Pd atoms (Pd_s) can be ascertained. Thus, the percentage of Pd atoms on the surface can be calculated since the total number of Pd atoms is known from the mass of Pd.

■ Answer

The best way to determine if the reaction rates are really different for these two catalysts is to compare their values of the turnover frequency. Assume that each surface Pd atom is an active site. Thus, to convert a specific rate to a turnover frequency:

$$r_t(s^{-1}) = r\left(\frac{mol}{gcat \cdot s}\right) \cdot \left(\frac{gcat}{mass\ metal}\right) \cdot \left(\frac{molecular\ weight\ of\ metal}{fraction\ of\ surface\ atoms}\right)$$

$$r_{t_{Pd/Al_2O_3}} = (7.66 \times 10^{-4}) \cdot (\tfrac{1}{0.0488})(106.4)(0.21)^{-1}$$

$$r_{t_{Pd/Al_2O_3}} = 8.0\ s^{-1}$$

Likewise for Pd on SiO_2,

$$r_{t_{Pd/SiO_2}} = 6.5\ s^{-1}$$

Since the turnover frequencies are approximately the same for these two catalysts, the metal oxide support does not appear to influence the reaction rate.

1.4 | General Properties of the Rate Function for a Single Reaction

The rate of reaction is generally a function of temperature and composition, and the development of mathematical models to describe the form of the reaction rate is a central problem of applied chemical kinetics. Once the reaction rate is known, information can often be derived for rates of individual steps, and reactors can be designed for carrying out the reaction at optimum conditions.

Below are listed general rules on the form of the reaction rate function (M. Boudart, *Kinetics of Chemical Processes,* Butterworth-Heinemann, 1991, pp. 13–16). The rules are of an approximate nature but are sufficiently general that exceptions to them usually reveal something of interest. It must be stressed that the utility of these rules is their applicability to many single reactions.

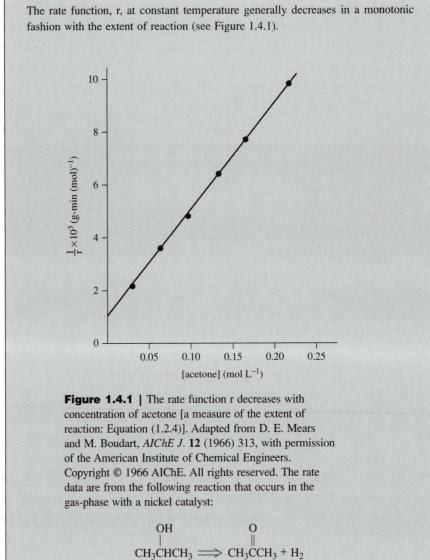

Rule I

The rate function, r, at constant temperature generally decreases in a monotonic fashion with the extent of reaction (see Figure 1.4.1).

Figure 1.4.1 | The rate function r decreases with concentration of acetone [a measure of the extent of reaction: Equation (1.2.4)]. Adapted from D. E. Mears and M. Boudart, *AIChE J.* **12** (1966) 313, with permission of the American Institute of Chemical Engineers. Copyright © 1966 AIChE. All rights reserved. The rate data are from the following reaction that occurs in the gas-phase with a nickel catalyst:

Rule II

The rate of an irreversible (one-way) reaction can generally be written in the form:

$$r = k(T)\overline{F}(C_i, T) \tag{1.4.1}$$

where $\overline{F}(C_i, T)$ is a function that depends on the composition of the system as expressed by the concentrations C_i.

The coefficient k does not depend on the composition of the system or time. For this reason, k is called the *rate constant*. If $\overline{F}$ is not a function of the temperature,

$$r = k(T)\overline{F}(C_i)$$

then the reaction rate is called separable since the temperature and composition dependencies are separated in k and $\overline{F}$, respectively.

Rule III

The rate constant generally depends on the absolute temperature, T, following the law first proposed by Arrhenius in 1889:

$$k = \overline{A} \exp\left[\frac{-E}{R_g T}\right] \tag{1.4.2}$$

In Equation (1.4.2), the *pre-exponential factor, $\overline{A}$,* does not depend appreciably on temperature, and E is called the *activation energy.* Figure 1.4.2 is an example of a typical Arrhenius plot.

Rule IV

Frequently, the function $\overline{F}(C_i)$ in the expression $r = k\overline{F}(C_i)$ can approximately be written as:

$$\overline{F}(C_i) = \prod_i C_i^{\alpha_i}$$

The product $\prod$ is taken over all components of the system. The exponents α_i are small integers or fractions that are positive, negative, or zero and are temperature independent at least over a restricted interval (see Table 1.4.1 for an example).

Consider the general reaction:

$$\overline{a}A + \overline{b}B \implies \overline{w}W \tag{1.4.3}$$

If the reaction rate follows Rule IV then it can be written as:

$$r = kC_A^{\alpha_A}C_B^{\alpha_B}$$

The exponent α_i is called the *order of reaction* with respect to the corresponding component of the system and the sum of the exponents is called the *total order* of

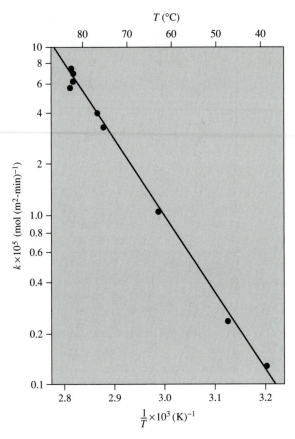

Figure 1.4.2 |

A typical Arrhenius plot, $\ln k$ vs $1/T$. The slope corresponds to $-E/R_g$. Adapted from D. E. Mears and M. Boudart, *AIChE J.* **12** (1966) 313, with permission of the American Institute of Chemical Engineers. Copyright © 1966 AIChE. All rights reserved.

the reaction. In general $\alpha_i \neq |v_i|$ and is rarely larger than two in absolute value. If $\alpha_i = |v_i|$ for reactants and equal to zero for all other components of the system, the expression:

$$\overline{F}(C_i) = \prod_i C_i^{|v_i|} \qquad \text{(for reactants only)}$$

would be of the form first suggested by Guldberg and Waage (1867) in their Law of Mass Action. Thus, a rate function of the form:

$$\mathrm{r} = k \prod_i C_i^{\alpha_i} \qquad\qquad (1.4.4)$$

Table 1.4.1 | Kinetic parameters for the simultaneous hydroformylation[1] and hydrogenation[2] of propene over a rhodium zeolite catalyst at $T = 423$ K and 1 atm total pressure [from Rode et al., *J. Catal.*, **96** (1985) 563].

$$\text{Rate}\left(\frac{\text{mol}}{g\text{Rh} \cdot \text{hr}}\right) = \bar{A} \exp\left[\frac{-E}{(R_g T)}\right] P_{C_3}^{\alpha_1} P_{H_2}^{\alpha_2} P_{CO}^{\alpha_3}$$

	$(C_3H_8)^1$	$(i - C_4H_8O)^2$	$(n - C_4H_8O)^2$
$\bar{A}$	4.12×10^6	1.36×10^3	1.86×10^3
E (kcal/mol), (393–423 K)	15.8	11.0	10.8
α_1	1.01	1.00	0.97
α_2	0.35	0.40	0.48
α_3	−0.71	−0.70	−0.72

1 $CH_2{=}CHCH_3 + H_2 \Longrightarrow CH_3CH_2CH_3$ (C_3H_8)

$$2 \ CH_2{=}CHCH_3 + CO + H_2 \begin{cases} CH_3CH_2CH_2\overset{\overset{\textstyle O}{\|}}{C}{-}H & (\text{n-}C_4H_8O) \\[2mm] CH_3\underset{\underset{\textstyle CH_3}{|}}{\overset{\overset{\textstyle O}{\|}}{C}}H{-}\!C{-}H & (\text{i-}C_4H_8O) \end{cases}$$

Table 1.4.2[1] | Examples of rate functions of the type: $r = k \prod_i C_i^{\alpha_i}$.

Reaction	Rate function
$CH_3CHO \Rightarrow CH_4 + CO$	$k(CH_3CHO)^{1.5}$
$C_2H_6 + H_2 \Rightarrow 2CH_4 \,(\text{catalytic})$	$k(C_2H_6)^{0.9}(H_2)^{-0.7}$
$SbH_3 \Rightarrow Sb + \frac{3}{2}H_2$	$k(SbH_3)^{0.6}$
$N_2 + 3H_2 \Rightarrow 2NH_3 \,(\text{catalytic})$	$k(N_2)(H_2)^{2.25}(NH_3)^{-1.5}$

[1]From M. Boudart, *Kinetics of Chemical Processes*, Butterworth-Heinemann, 1991, p. 17.

which is normally referred to as "pseudo mass action" or "power law" is really the Guldberg-Waage form only when $\alpha_i = |v_i|$ for reactants and zero for all other components (note that orders may be negative for catalytic reactions, as will be discussed in a later chapter). For the reaction described by Equation (1.4.3), the Guldberg-Waage rate expression is:

$$r = kC_A^{\bar{a}}C_B^{\bar{b}}$$

Examples of power law rate expressions are shown in Table 1.4.2.

For elementary steps, $\alpha_i = |v_i|$. Consider again the gas-phase reaction:

$$H_2 + Br_2 \Rightarrow 2HBr$$

If this reaction would occur by H_2 interacting directly with Br_2 to yield two molecules of HBr, the step would be elementary and the rate could be written as:

$$r = kC_{H_2}C_{Br_2}$$

However, it is known that this is not how the reaction proceeds (Section 1.1) and the real rate expression is:

$$r = \frac{k_1 C_{H_2} C_{Br_2}^{1/2}}{k_2 + \dfrac{C_{HBr}}{C_{Br_2}}}$$

For elementary steps the number of molecules that partcipate in the reaction is called the molecularity of the reaction (see Table 1.4.3).

Rule V

When a reaction is two-way (or reversible), its rate can generally be expressed as the difference between a rate in the forward direction r_+ and in the reverse direction r_-

$$r = r_+ - r_- \tag{1.4.5}$$

When Rule II applies to the rate functions r_+ and r_- so that:

$$r_+ = k_+ \overline{F}_+(C_i)$$
$$r_- = k_- \overline{F}_-(C_i)$$

both rate constants k_+ and k_- are related to the equilibrium constant, K_C. For example, the reaction $A \rightleftarrows B$ at ideal conditions gives (see Chapter 5 for a more rigorous explanation of the relationships between rates and equilibrium expressions):

$$K_C = \frac{k_+}{k_-} \tag{1.4.6}$$

1.5 | Examples of Reaction Rates

Consider the unimolecular reaction:

$$A \rightarrow \text{products} \tag{1.5.1}$$

Using the Guldberg-Waage form of the reaction rate to describe this reaction gives:

$$r = kC_A \tag{1.5.2}$$

From Equations (1.3.4) and (1.5.2):

$$r = \frac{1}{v_i V} \frac{dn_i}{dt} = \frac{-1}{V} \frac{dn_A}{dt} = kC_A$$

or

$$\frac{dn_A}{dt} = -kn_A \qquad \text{(variable } V) \tag{1.5.3}$$

$$\frac{df_A}{dt} = k(1 - f_A) \qquad [\text{using Equation (1.2.11)}] \tag{1.5.4}$$

Table 1.4.3 | Molecularity and rates of elementary steps.

Molecularity	Number of reactant molecules	General description	Example (1)	Rate constant (1)
Unimolecular	1	$A \rightarrow$ products	$N_2O_5 \rightarrow NO_2 + NO_3$	$1.96 \times 10^{14} \exp[-10660/T], \, s^{-1}$
Bimolecular	2	$2A \rightarrow$ products $A + B \rightarrow$ products	$\overline{NO + NO_3 \rightarrow 2NO_2}$	$2.0 \times 10^{11}, \, cm^3/s/molecule \,(2)$
Trimolecular (rare)	3	$3A \rightarrow$ products $2A + B \rightarrow$ products $A + B + C \rightarrow$ products	$\overline{2NO + O_2 \rightarrow 2NO_2}$ $NO + NO_2 + H_2O \rightarrow 2HNO_2$	$3.3 \times 10^{-39} \exp(530/T), \, cm^6/s/molecule^2 \,(2)$ $\leq 4.4 \times 10^{-40}, \, cm^6/s/molecule^2 \,(2)$

(1) From J. H. Seinfeld, *Atmospheric Chemistry and Physics of Air Pollution*, Wiley, 1986, p. 175.
(2) Concentrations are in molecules/cm^3.

Table 1.5.1 | Examples of reactions that can be described using first-order reaction rates.

Reactions	Examples
Isomerizations	$CH_3CH_2CH{=}CH_2 \Longrightarrow CH_3{-}\underset{\underset{CH_2}{\|}}{C}{-}CH_3$
Decompositions	$N_2O_5 \Longrightarrow NO_2 + NO_3$
	$CH_2{-}CH_2 \Longrightarrow CH_4 + CO$
Radioactive decay (each decay can be described by a first-order reaction rate)	$^{135}I \xrightarrow{-\beta} {}^{135}Xe \xrightarrow{-\beta} {}^{135}Cs \xrightarrow{-\beta} {}^{135}Ba(stable)$

$$\frac{dC_A}{dt} = -kC_A \qquad (\text{constant } V) \tag{1.5.5}$$

$$\frac{dP_A}{dt} = -kP_A \qquad \left[\text{constant } V: C_i = P_i/(R_gT)\right] \tag{1.5.6}$$

Thus, for first-order systems, the rate, r, is proportional (via k) to the amount present, n_i, in the system at any particular time. Although at first glance, first-order reaction rates may appear too simple to describe real reactions, such is not the case (see Table 1.5.1). Additionally, first-order processes are many times used to approximate complex systems, for example, lumping groups of hydrocarbons into a generic hypothetical component so that phenomenological behavior can be described.

In this text, concentrations will be written in either of two notations. The notations C_i and $[A_i]$ are equivalent in terms of representing the concentration of species i or A_i, respectively. These notations are used widely and the reader should become comfortable with both.

EXAMPLE 1.5.1 |

The natural abundance of ^{235}U in uranium is 0.79 atom %. If a sample of uranium is enriched to 3 at. % and then is stored in salt mines under the ground, how long will it take the sample to reach the natural abundance level of ^{235}U (assuming no other processes form ^{235}U; this is not the case if ^{238}U is present since it can decay to form ^{235}U)? The half-life of ^{235}U is 7.13×10^8 years.

■ **Answer**

Radioactive decay can be described as a first-order process. Thus, for any first-order decay process, the amount of material present declines in an exponential fashion with time. This is easy to see by integrating Equation (1.5.3) to give:

$$n_i = n_i^0 \exp(-kt), \quad \text{where } n_i^0 \text{ is the amount of } n_i \text{ present at } t = 0.$$

The *half-life*, $t_{\frac{1}{2}}$, is defined as the time necessary to reduce the amount of material in half. For a first-order process $t_{\frac{1}{2}}$ can be obtained as follows:

$$\tfrac{1}{2}\, n_i^0 = n_i^0 \exp(-kt_{\frac{1}{2}})$$

or

$$t_{\frac{1}{2}} = \frac{\ln(2)}{k}$$

Given $t_{\frac{1}{2}}$, a value of k can be calculated. Thus, for the radioactive decay of ^{235}U, the first-order rate constant is:

$$k = \frac{\ln(2)}{t_{\frac{1}{2}}} = 9.7 \times 10^{-10} \text{ years}^{-1}$$

To calculate the time required to have 3 at. % ^{235}U decay to 0.79 at. %, the first-order expression:

$$\frac{n_i}{n_i^0} = \exp(-kt) \quad \text{or} \quad t = \frac{\ln\!\left(\dfrac{n_i^0}{n_i}\right)}{k}$$

can be used. Thus,

$$t = \frac{\ln\!\left(\dfrac{3}{0.79}\right)}{9.7 \times 10^{-10}} = 1.4 \times 10^9 \text{ years}$$

or a very long time.

EXAMPLE 1.5.2

N_2O_5 decomposes into NO_2 and NO_3 with a rate constant of $1.96 \times 10^{14} \exp[-10{,}660/T]\,s^{-1}$. At $t = 0$, pure N_2O_5 is admitted into a constant temperature and volume reactor with an initial pressure of 2 atm. After 1 min, what is the total pressure of the reactor? $T = 273 \text{ K}$.

■ Answer
Let n be the number of moles of N_2O_5 such that:

$$\frac{dn}{dt} = -kn$$

Since $n = n^0(1 - f)$:

$$\frac{df}{dt} = k(1 - f), f = 0 @ t = 0$$

Integration of this first-order, initial-value problem yields:

$$\ln\!\left(\frac{1}{1 - f}\right) = kt \quad \text{for } t \geq 0$$

or

$$f = 1 - \exp(-kt) \quad \text{for } t \geq 0$$

At 273 K, $k = 2.16 \times 10^{-3} \, s^{-1}$. After reaction for 1 min:

$$f = 1 - \exp[-(60)(2.16 \times 10^{-3})] = 0.12$$

From the ideal gas law at constant T and V:

$$\frac{P}{P^0} = \frac{n}{n^0} = \frac{n^0(1 + \varepsilon f)}{n^0}$$

For this decomposition reaction:

$$\varepsilon_{N_2O_5} = 1.0 \left[\frac{2 - 1}{|-1|} \right] = 1$$

Thus,

$$P = P^0(1 + f) = 2(1 + 0.12) = 2.24 \text{ atm}$$

EXAMPLE 1.5.3

Often isomerization reactions are highly two-way (reversible). For example, the isomerization of 1-butene to isobutene is an important step in the production of methyl tertiary butyl ether (MTBE), a common oxygenated additive in gasoline used to lower emissions. MTBE is produced by reacting isobutene with methanol:

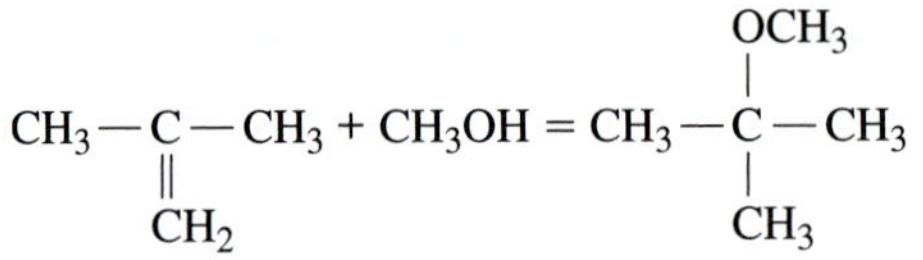

$$CH_3 - \underset{\underset{CH_2}{\|}}{C} - CH_3 + CH_3OH = CH_3 - \underset{\underset{CH_3}{|}}{\overset{\overset{OCH_3}{|}}{C}} - CH_3$$

In order to make isobutene, n-butane (an abundant, cheap C_4 hydrocarbon) can be dehydrogenated to 1-butene then isomerized to isobutene. Derive an expression for the concentration of isobutene formed as a function of time by the isomerization of 1-butene:

$$CH_2 = CHCH_2CH_3 \underset{k_2}{\overset{k_1}{\rightleftharpoons}} \underset{\underset{CH_2}{\|}}{CH_3CCH_3}$$

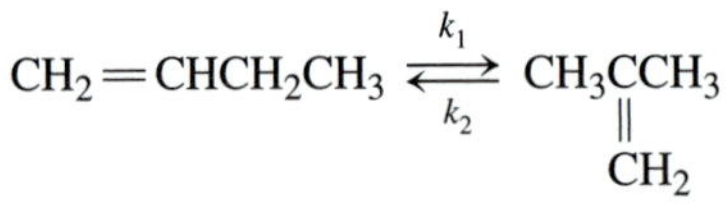

■ **Answer**

Let isobutene be denoted as component I and 1-butene as B. If the system is at constant T and V, then:

$$\frac{dC_B}{dt} = -k_1 C_B + k_2 C_I \quad \text{or} \quad \frac{d[B]}{dt} = -k_1[B] + k_2[I]$$

Since $[B] = [B]^0(1 - f_B)$:

$$[I] = [I]^0 + [B]^0 f_B = [B]^0 (\overline{M} + f_B), \quad \overline{M} = {[I]^0} \Big/ {[B]^0} \neq 0$$

Thus,

$$\frac{df_B}{dt} = k_1(1 - f_B) - k_2(\overline{M} + f_B)$$

At equilibrium $\dfrac{df_B}{dt} = 0$, so:

$$K_C = \frac{k_1}{k_2} = \frac{[I]^{eq}}{[B]^{eq}} = \frac{[B]^0(\overline{M} + f_B^{eq})}{[B]^0(1 - f_B^{eq})}$$

Insertion of the equilibrium relationship into the rate expression yields:

$$\frac{df_B}{dt} = k_1(1 - f_B) - k_1(1 - f_B^{eq})\left[\frac{\overline{M} + f_B}{\overline{M} + f_B^{eq}}\right]$$

or after rearrangement:

$$\frac{df_B}{dt} = \frac{k_1(\overline{M} + 1)}{(\overline{M} + f_B^{eq})}(f_B^{eq} - f_B), \quad f_B = 0 \; @ \; t = 0$$

Integration of this equation gives:

$$\ln\left[\frac{1}{1 - \dfrac{f_B}{f_B^{eq}}}\right] = \left[\frac{k_1(\overline{M} + 1)}{\overline{M} + f_B^{eq}}\right]t, \quad \overline{M} \neq 0$$

or

$$f_B = f_B^{eq}\left\{1 - \exp\left[-\left(\frac{k_1(\overline{M} + 1)}{\overline{M} + f_B^{eq}}\right)t\right]\right\}$$

Using this expression for f_B:

$$[I] = [B]^0(\overline{M} + f_B)$$

Consider the bimolecular reaction:

$$A + B \rightarrow \text{products} \tag{1.5.7}$$

Using the Guldberg-Waage form of the reaction rate to describe this reaction gives:

$$\text{r} = kC_A C_B \tag{1.5.8}$$

From Equations (1.3.4) and (1.5.8):

$$r = \frac{1}{v_i V}\frac{dn_i}{dt} = \frac{-1}{V}\frac{dn_A}{dt} = kC_A C_B$$

or

$$V\frac{dn_A}{dt} = -kn_A n_B \qquad \text{(variable } V) \tag{1.5.9}$$

$$\frac{dC_A}{dt} = -kC_A C_B \qquad \text{(constant } V) \tag{1.5.10}$$

$$\frac{dP_A}{dt} = -\frac{k}{R_g T}P_A P_B \quad [\text{constant } V: C_i = P_i/(R_g T)] \tag{1.5.11}$$

For second-order kinetic processes, the limiting reactant is always the appropriate species to follow (let species denoted as A be the limiting reactant). Equations (1.5.9–1.5.11) cannot be integrated unless C_B is related to C_A. Clearly, this can be done via Equation (1.2.5) or Equation (1.2.6). Thus,

$$n_B = n_B^0 - (n_A^0 - n_A)$$

or if the volume is constant:

$$C_B = C_B^0 - (C_A^0 - C_A) \quad \text{or} \quad P_B = P_B^0 - (P_A^0 - P_A)$$

If $\overline{M} = \dfrac{n_B^0}{n_A^0} = \dfrac{C_B^0}{C_A^0} = \dfrac{P_B^0}{P_A^0}$ then:

$$\left.\begin{aligned}
n_B &= n_A + n_A^0(\overline{M} - 1) \quad \text{(variable } V)\\
C_B &= C_A + C_A^0(\overline{M} - 1) \quad \text{(constant } V)\\
P_B &= P_A + P_A^0(\overline{M} - 1) \quad \text{(constant } V)
\end{aligned}\right\} \tag{1.5.12}$$

Inserting Equation (1.5.12) into Equations (1.5.9–1.5.11) gives:

$$V\frac{dn_A}{dt} = -kn_A[n_A + n_A^0(\overline{M} - 1)] \qquad \text{(variable } V) \tag{1.5.13}$$

$$\frac{dC_A}{dt} = -kC_A[C_A + C_A^0(\overline{M} - 1)] \qquad \text{(constant } V) \tag{1.5.14}$$

$$\frac{dP_A}{dt} = -\frac{k}{R_g T}P_A[P_A + P_A^0(\overline{M} - 1)] \quad \text{(constant } V) \tag{1.5.15}$$

If V is not constant, then $V = V^0(1 + \varepsilon_A f_A)$ by using Equation (1.2.15) and the ideal gas law. Substitution of this expression into Equation (1.5.13) gives:

$$\frac{df_A}{dt} = \frac{k\left(\frac{n_A^0}{V^0}\right)(1 - f_A)[\overline{M} - f_A]}{(1 + \varepsilon_A f_A)} \tag{1.5.16}$$

EXAMPLE 1.5.4

Equal volumes of 0.2 M trimethylamine and 0.2 M n-propylbromine (both in benzene) were mixed, sealed in glass tubes, and placed into a constant temperature bath at 412 K. After various times, the tubes were removed and quickly cooled to room temperature to stop the reaction:

$$N(CH_3)_3 + C_3H_7Br \Rightarrow C_3H_7 \overset{+}{N}(CH_3)_3Br^-$$

The quaternization of a tertiary amine gives a quaternary ammonium salt that is not soluble in nonpolar solvents such as benzene. Thus, the salt can easily be filtered from the remaining reactants and the benzene. From the amount of salt collected, the conversion can be calculated and the data are:

Time at 412 K (min)	Conversion (%)
5	4.9
13	11.2
25	20.4
34	25.6
45	31.6
59	36.7
80	45.3
100	50.7
120	55.2

Are these data consistent with a first- or second-order reaction rate?

■ Answer

The reaction occurs in the liquid phase and the concentrations are dilute. Thus, a good assumption is that the volume of the system is constant. Since $C_A^0 = C_B^0$:

$$\text{(first-order)} \quad \ln\left[\frac{1}{1 - f_A}\right] = kt$$

$$\text{(second-order)} \quad \frac{f_A}{1 - f_A} = kC_A^0 t$$

In order to test the first-order model, the $\ln[1/(1 - f_A)]$ is plotted versus t while for the second-order model, $f_A/(1 - f_A)$ is plotted versus t (see Figures 1.5.1 and 1.5.2). Notice that both models conform to the equation $y = \overline{\alpha}_1 t + \overline{\alpha}_2$. Thus, the data can be fitted via linear

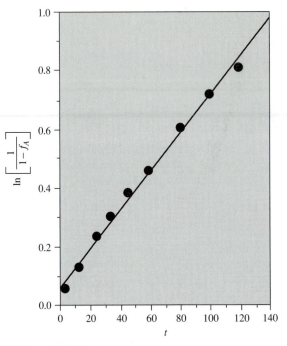

Figure 1.5.1 |
Reaction rate data for first-order kinetic model.

regression to both models (see Appendix B). From visual inspection of Figures 1.5.1 and 1.5.2, the second-order model appears to give a better fit. However, the results from the linear regression are (*SE* is the standard error):

first-order $\bar{\alpha}_1 = 6.54 \times 10^{-3}$ $SE(\bar{\alpha}_1) = 2.51 \times 10^{-4}$

$\bar{\alpha}_2 = 5.55 \times 10^{-2}$ $SE(\bar{\alpha}_2) = 1.63 \times 10^{-2}$

$\bar{R}_{cc} = 0.995$

second-order $\bar{\alpha}_1 = 1.03 \times 10^{-2}$ $SE(\bar{\alpha}_1) = 8.81 \times 10^{-5}$

$\bar{\alpha}_2 = -5.18 \times 10^{-3}$ $SE(\bar{\alpha}_2) = 5.74 \times 10^{-3}$

$\bar{R}_{cc} = 0.999$

Both models give high correlation coefficients $(\bar{R}_{cc})$, and this problem shows how the correlation coefficient may not be useful in determining "goodness of fit." An appropriate way to determine "goodness of fit" is to see if the models give $\bar{\alpha}_2$ that is not statistically different from zero. This is the reason for manipulating the rate expressions into forms that have zero intercepts (i.e., a known point from which to check statistical significance). If a student $\bar{t}^*$-test is used to test significance (see Appendix B), then:

$$\bar{t}^* = \frac{|\bar{\alpha}_2 - 0|}{SE(\bar{\alpha}_2)}$$

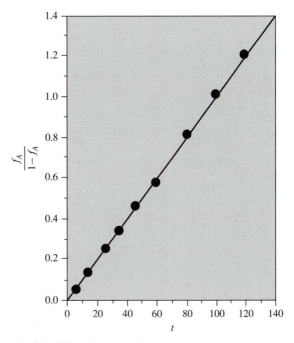

Figure 1.5.2 |
Reaction rate data for second-order kinetic model.

The values of $\bar{t}^*$ for the first- and second-order models are:

$$\bar{t}_1^* = \frac{|5.55 \times 10^{-2} - 0|}{1.63 \times 10^{-2}} = 3.39$$

$$\bar{t}_2^* = \frac{|-5.18 \times 10^{-3} - 0|}{5.74 \times 10^{-3}} = 0.96$$

For 95 percent confidence with 9 data points or 7 degrees of freedom (from table of student $\bar{t}^*$ values):

$$\bar{t}_{\text{exp}}^* = \frac{\text{expected deviation}}{\text{standard error}} = 1.895$$

Since $\bar{t}_1^* > \bar{t}_{\text{exp}}^*$ and $\bar{t}_2^* < \bar{t}_{\text{exp}}^*$, the first-order model is rejected while the second-order model is accepted. Thus,

$$kC_A^0 = 1.030 \times 10^{-2}$$

and

$$k = \frac{1.030 \times 10^{-2}}{0.1\,\text{M}} = 0.1030 \ \frac{1}{\text{M} \cdot \text{min}}$$

When the standard error is known, it is best to report the value of the correlated parameters:

$$r = [0.1030 \pm 0.0009 \text{ M}^{-1} \text{ min}^{-1}]C_A C_B$$

EXAMPLE 1.5.5

The following data were obtained from an initial solution containing methyl iodide (MI) and dimethyl-p-toludine (PT) both in concentrations of 0.050 mol/L. The equilibrium constant for the conditions where the rate data were collected is 1.43. Do second-order kinetics adequately describe the data and if so what are the rate constants?

Data:

t (min)	Fractional conversion of PT
10	0.18
26	0.34
36	0.40
78	0.52

■ **Answer**

$$CH_3I + (CH_3)_2N \overset{}{\underset{}{\longrightarrow}} CH_3 \underset{k_2}{\overset{k_1}{\rightleftarrows}} (CH_3)_3\overset{+}{N} \overset{}{\underset{}{\longrightarrow}} CH_3 + I^-$$

(MI)   (PT)   (NQ)   (I)

At constant volume,

$$\frac{dC_{PT}}{dt} = -k_1 C_{PT} C_{MI} + k_2 C_{NQ} C_I$$

$$C_{PT}^0 = C_{MI}^0 = 0.05, \quad C_{NQ}^0 = C_I^0 = 0,$$

$$C_{PT} = C_{PT}^0(1 - f_{PT}), \quad C_{MI} = C_{PT}^0(1 - f_{PT})$$

$$C_{NQ} = C_I = C_{PT}^0 f_{PT}$$

Therefore,

$$\frac{df_{PT}}{dt} = k_1 C_{PT}^0 (1 - f_{PT})^2 - k_2 C_{PT}^0 f_{PT}^2$$

At equilibrium,

$$K_C = \frac{k_1}{k_2} = \frac{(f_{PT}^{eq})^2}{(1 - f_{PT}^{eq})^2}$$

Substitution of the equilibrium expression into the rate expression gives:

$$\frac{df_{PT}}{dt} = k_1 C_{PT}^0 (1 - f_{PT}^{eq})^2 \left[\left(\frac{1 - f_{PT}}{1 - f_{PT}^{eq}} \right)^2 - \left(\frac{f_{PT}}{f_{PT}^{eq}} \right)^2 \right]$$

Upon integration with $f_{PT} = 0$ at $t = 0$:

$$\ln\left[\frac{f_{PT}^{eq} - (2f_{PT}^{eq} - 1)f_{PT}}{(f_{PT}^{eq} - f_{PT})}\right] = 2k_1\left[\frac{1}{f_{PT}^{eq}} - 1\right]C_{PT}^0 t$$

Note that this equation is again in a form that gives a zero intercept. Thus, a plot and linear least squares analysis of:

$$\ln\left[\frac{f_{PT}^{eq} - (2f_{PT}^{eq} - 1)f_{PT}}{(f_{PT}^{eq} - f_{PT})}\right] \quad \text{versus } t$$

will show if the model can adequately describe the data. To do this, f_{PT}^{eq} is calculated from K_C and it is $f_{PT}^{eq} = 0.545$. Next, from the linear least squares analysis, the model does fit the data and the slope is 0.0415. Thus,

$$0.0415 = 2k_1\left[\frac{1}{f_{PT}^{eq}} - 1\right]C_{PT}^0$$

giving $k_1 = 0.50$ L/mol/min. From K_C and k_1, $k_2 = 0.35$ L/mol/min.

Consider the trimolecular reaction:

$$A + B + C \rightarrow \text{products} \qquad (1.5.17)$$

Using the Guldberg-Waage form of the reaction rate to describe this reaction gives:

$$\text{r} = kC_AC_BC_C \qquad (1.5.18)$$

From Equations (1.3.4) and (1.5.18):

$$\text{r} = \frac{1}{v_iV}\frac{dn_i}{dt} = \frac{-1}{V}\frac{dn_A}{dt} = kC_AC_BC_C$$

$$V^2\frac{dn_A}{dt} = -kn_An_Bn_C \qquad \text{(variable } V) \qquad (1.5.19)$$

$$\frac{dC_A}{dt} = -kC_AC_BC_C \qquad \text{(constant } V) \qquad (1.5.20)$$

$$\frac{dP_A}{dt} = -\frac{k}{(R_gT)^2}P_AP_BP_C \quad \text{(constant } V) \qquad (1.5.21)$$

Trimolecular reactions are very rare. If viewed from the statistics of collisions, the probability of three objects colliding with sufficient energy and in the correct configuration for reaction to occur is very small. Additionally, only a small amount of these collisions would successfully lead to reaction (see Chapter 2, for a detailed discussion). Note the magnitudes of the reaction rates for unimolecular and bimolecular reactions as compared to trimolecular reactions (see Table 1.4.3). However, trimolecular reactions do occur, for example:

$$O + O_2 + TB \rightarrow O_3 + TB$$

where the third body TB is critical to the success of the reaction since it is necessary for it to absorb energy to complete the reaction (see Vignette 1.2.1).

In order to integrate Equation (1.5.20), C_B and C_C must be related to C_A and this can be done by the use of Equation (1.2.5). Therefore, analysis of a trimolecular process is a straightforward extension of bimolecular processes. If trimolecular processes are rare and give slow rates, then the question arises as to how reactions like hydroformylations (Table 1.4.1) can be accomplished on a commercial scale (Vignette 1.5.1). The hydroformylation reaction is for example (see Table 1.4.1):

$$CH_2{=}CH{-}R + CO + H_2 \Longrightarrow R{-}CH_2{-}CH_2{-}\overset{\overset{\displaystyle O}{\displaystyle \|}}{C}{-}H$$

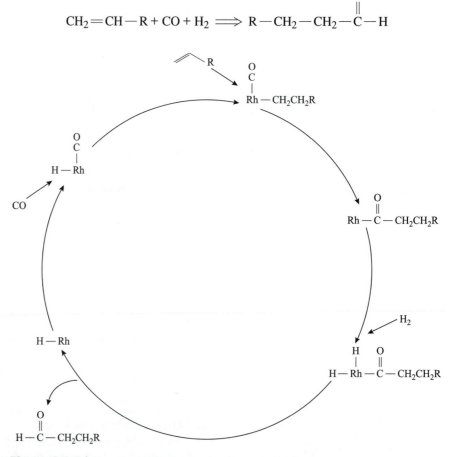

Figure 1.5.3 |
Simplified version of the hydroformylation mechanism. Note that other ligands on the Rh are not shown for ease in illustrating what happens with reactants and products.

This reaction involves three reactants and the reason that it proceeds so efficiently is that a catalyst is used. Referring to Figure 1.5.3, note that the rhodium catalyst coordinates and combines the three reactants in a closed cycle, thus breaking the "statistical odds" of having all three reactants collide together simultaneously. Without a catalyst the reaction proceeds only at nonsignificant rates. This is generally true of reactions where catalysts are used. More about catalysts and their functions will be described later in this text.

VIGNETTE 1.5.1

Ingredients in soap (e.g., a shampoo) contain compounds like sodium lauryl sulfate. Sodium lauryl sulfate is $CH_3(CH_2)_{10}-CH_2-OSO_3^- Na^+$ and is a good detergent because "oily" substances interact with the hydrocarbon chain and the sodium sulfate portion of the molecule makes them water-soluble. Thus, when molecules like sodium lauryl sulfate contact "oily" clothes, dishes, hair, etc., they are able to "solubilize" the "oily" fraction and remove it when combined with water. Sodium lauryl sulfate is manufactured from lauryl alcohol (1-dodecanol) by sulfonation and neutralization with NaOH. Lauryl alcohol is produced by the hydroformylation of 1-undecene:

$$CH_3-(CH_2)_8-CH{=}CH_2 + CO + H_2 \Longrightarrow CH_3-(CH_2)_{10}-\overset{\displaystyle O}{\overset{\|}{C}}-H$$

after which the aldehyde is hydrogenated to the alcohol:

$$CH_3-(CH_2)_{10}-\overset{\displaystyle O}{\overset{\|}{C}}-H + H_2 \Longrightarrow CH_3-(CH_2)_{10}-CH_2OH$$

Thus, large-scale hydroformylation reactors are used worldwide to ultimately produce detergent-grade surfactants.

When conducting a reaction to give a desired product, it is common that other reactions proceed simultaneously. Thus, more than a single reaction must be considered (i.e., a reaction network), and the issue of selectivity becomes important. In order to illustrate the challenges presented by reaction networks, small reaction networks are examined next. Generalizations of these concepts to larger networks are only a matter of patience.

Consider the reaction network of two irreversible (one-way), first-order reactions in series:

$$A \xrightarrow{k_1} B \xrightarrow{k_2} C \tag{1.5.22}$$

This network can represent a wide variety of important classes of reactions. For example, oxidation reactions occurring in excess oxidant adhere to this reaction network, where B represents the partial oxidation product and C denotes the complete oxidation product CO_2:

$$
\text{CH}_3\text{CH}_2\text{OH} \xrightarrow{\text{excess air}} \overset{\overset{\displaystyle O}{\displaystyle \|}}{\text{CH}_3\text{C}} \!-\! \text{H} \xrightarrow{\text{excess air}} 2\text{CO}_2 + 2\text{H}_2\text{O}
$$

$$
\text{ethanol} \qquad\qquad\qquad \text{acetaldehyde}
$$

For this situation the desired product is typically B, and the difficulty arises in how to obtain the maximum concentration of B given a particular k_1 and k_2. Using the Guldberg-Waage form of the reaction rates to describe the network in Equation (1.5.22) gives for constant volume:

$$
\left.
\begin{aligned}
\frac{dC_A}{dt} &= -k_1 C_A \\[2mm]
\frac{dC_B}{dt} &= k_1 C_A - k_2 C_B \\[2mm]
\frac{dC_C}{dt} &= k_2 C_B
\end{aligned}
\right\} \tag{1.5.23}
$$

with

$$
C_A^0 + C_B^0 + C_C^0 = C^0 = C_A + C_B + C_C
$$

Integration of the differential equation for C_A with $C_A = C_A^0$ at $t = 0$ yields:

$$
C_A = C_A^0 \exp[-k_1 t] \tag{1.5.24}
$$

Substitution of Equation (1.5.24) into the differential equation for C_B gives:

$$
\frac{dC_B}{dt} + k_2 C_B = k_1 C_A^0 \exp[-k_1 t]
$$

This equation is in the proper form for solution by the integrating factor method, that is:

$$
\frac{dy}{dt} + p(t)y = g(t), \quad I = \exp\left[\int p(t)dt\right]
$$

$$
\int d(Iy) = \int Ig(t)dt
$$

Now, for the equation concerning C_B, $p(t) = k_2$ so that:

$$
\int d(C_B \exp[k_2 t]) = \int k_1 C_A^0 \exp[(k_2 - k_1)t]dt
$$

Integration of the above equation gives:

$$
C_B \exp[k_2 t] = \left[\frac{k_1 C_A^0}{k_2 - k_1}\right]\exp[(k_2 - k_1)t] + \gamma
$$

or

$$C_B = \left[\frac{k_1 C_A^0}{k_2 - k_1} \right] \exp(-k_1 t) + \gamma \exp(-k_2 t)$$

where γ is the integration constant. Since $C_B = C_B^0$ at $t = 0$, γ can be written as a function of C_A^0, C_B^0, k_1 and k_2. Upon evaluation of γ, the following expression is found for $C_B(t)$:

$$C_B = \frac{k_1 C_A^0}{k_2 - k_1} [\exp(-k_1 t) - \exp(-k_2 t)] + C_B^0 \exp(-k_2 t) \qquad (1.5.25)$$

By knowing $C_B(t)$ and $C_A(t)$, $C_C(t)$ is easily obtained from the equation for the conservation of mass:

$$C_C = C^0 - C_A - C_B \qquad (1.5.26)$$

For $C_B^0 = C_C^0 = 0$ and $k_1 = k_2$, the normalized concentrations of C_A, C_B, and C_C are plotted in Figure 1.5.4.

Notice that the concentration of species B initially increases, reaches a maximum, and then declines. Often it is important to ascertain the maximum amount of species B and at what time of reaction the maximum occurs. To find these quantities, notice that at C_B^{max}, $dC_B/dt = 0$. Thus, if the derivative of Equation (1.5.25) is set equal to zero then t_{max} can be found as:

$$t_{max} = \frac{1}{(k_2 - k_1)} \ln \left[\left(\frac{k_2}{k_1} \right) \left(1 + \frac{C_B^0}{C_A^0} - \left(\frac{k_2}{k_1} \right) \frac{C_B^0}{C_A^0} \right) \right] \qquad (1.5.27)$$

Using the expression for t_{max} (Equation (1.5.27)) in Equation (1.5.25) yields C_B^{max}.

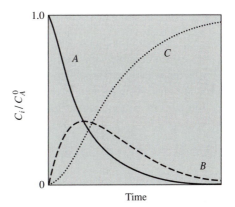

Figure 1.5.4 |

Normalized concentration of species i as a function of time for $k_1 = k_2$.

EXAMPLE 1.5.6

For $C_B^0 = C_C^0 = 0$, find the maximum concentration of C_B for $k_1 = 2k_2$.

■ **Answer**

From Equation (1.5.27) with $C_B^0 = 0$, t_{max} is:

$$t_{max} = \frac{1}{(k_2 - k_1)} \ln\left[\left(\frac{k_2}{k_1}\right)\right]$$

Substitution of t_{max} into Equation (1.5.25) with $C_B^0 = 0$ gives:

$$C_B^{max}/C_A^0 = \frac{k_1}{(k_2 - k_1)} \left[\exp\left[\frac{-k_1}{(k_2 - k_1)} \ln\left(\frac{k_2}{k_1}\right)\right] - \exp\left[\frac{-k_2}{(k_2 - k_1)} \ln\left(\frac{k_2}{k_1}\right)\right] \right]$$

or

$$C_B^{max}/C_A^0 = \frac{k_1}{(k_2 - k_1)} \left[\left(\frac{k_1}{k_2}\right)^{\left(\frac{k_1}{k_2 - k_1}\right)} - \left(\frac{k_1}{k_2}\right)^{\left(\frac{k_2}{k_2 - k_1}\right)} \right]$$

This equation can be simplified as follows:

$$C_B^{max}/C_A^0 = \frac{k_1}{(k_2 - k_1)} \left[\left(\frac{k_1}{k_2}\right)^{\left(\frac{k_2}{k_2 - k_1}\right)} \right] \left[\left(\frac{k_1}{k_2}\right)^{\left(\frac{k_1 - k_2}{k_2 - k_1}\right)} - 1 \right]$$

or

$$C_B^{max}/C_A^0 = \frac{k_1}{(k_2 - k_1)} \left[\left(\frac{k_1}{k_2}\right)^{\left(\frac{k_2}{k_2 - k_1}\right)} \right] \left[\left(\frac{k_1}{k_2}\right)^{-1} - 1 \right] = \left(\frac{k_1}{k_2}\right)^{\left(\frac{k_2}{k_2 - k_1}\right)}$$

With $k_1 = 2k_2$,

$$C_B^{max} = 0.5 C_A^0$$

When dealing with multiple reactions that lead to various products, issues of selectivity and yield arise. The *instantaneous selectivity, s_i,* is a function of the local conditions and is defined as the production rate of species i divided by the production rates of all products of interest:

$$s_i = \frac{r_i}{\sum r_j}, \quad j = 1, \cdots, \text{all products of interest} \qquad (1.5.28)$$

where r_i is the rate of production of the species i. An *overall selectivity, S_i,* can be defined as:

$$S_i = \frac{\text{total amount of species } i}{\text{total amount of products of interest}} \qquad (1.5.29)$$

The *yield*, Y_i, is denoted as below:

$$Y_i = \frac{\text{total amount of product } i \text{ formed}}{\text{initial amount of reactant fed}} \tag{1.5.30}$$

where the initial amount of reactant fed is for the limiting component. For the network given by Equation (1.5.22):

$$S_B = \frac{\text{amount of } B \text{ formed}}{\text{amount of } B \text{ and } C \text{ formed}} = \frac{\text{amount of } B \text{ formed}}{\text{amount of } A \text{ reacted}}$$

or

$$S_B = \frac{C_B}{C_A^0 - C_A} \tag{1.5.31}$$

and

$$Y_B = \frac{\text{amount of } B \text{ formed}}{\text{initial amount of } A \text{ fed}} = \frac{C_B}{C_A^0} \tag{1.5.32}$$

The selectivity and yield should, of course, correctly account for the stoichiometry of the reaction in all cases.

EXAMPLE 1.5.7

Plot the percent selectivity and the yield of B [Equation (1.5.31) multiplied by 100 percent and Equation (1.5.32), respectively] as a function of time. Does the time required to reach $C_B^{\max}$ give the maximum percent selectivity to B and/or the maximum fractional yield of B? Let $k_1 = 2k_2$, $C_B^0 = C_C^0 = 0$.

■ Answer

From the plot shown below the answer is obvious. For practical purposes, what is important is the maximum yield.

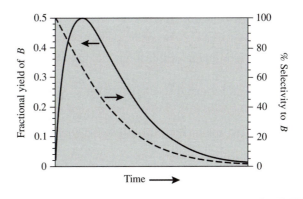

Consider the reaction network of two irreversible (one-way), first-order reactions in parallel:

$$A \begin{array}{c} \nearrow^{k_1} DP \\ \\ \searrow_{k_2} SP \end{array}$$
(1.5.33)

Again, like the series network shown in Equation (1.5.22), the parallel network of Equation (1.5.33) can represent a variety of important reactions. For example, dehydrogenation of alkanes can adhere to this reaction network where the desired product DP is the alkene and the undesired side-product SP is a hydrogenolysis (C—C bond-breaking reaction) product:

$$CH_3 - CH_3 \begin{array}{c} \nearrow CH_2 = CH_2 + H_2 \\ \\ \searrow CH_4 + \text{carbonaceous residue on catalyst} \end{array}$$

Using the Guldberg-Waage form of the reaction rates to describe the network in Equation (1.5.33) gives for constant volume:

$$\left. \begin{array}{c} \dfrac{dC_A}{dt} = -(k_1 + k_2)C_A \\[2mm] \dfrac{dC_{DP}}{dt} = k_1 C_A \\[2mm] \dfrac{dC_{SP}}{dt} = k_2 C_A \end{array} \right\}$$
(1.5.34)

with

$$C_A^0 + C_{DP}^0 + C_{SP}^0 = C^0 = C_A + C_{DP} + C_{SP}$$

Integration of the differential equation for C_A with $C_A = C_A^0$ at $t = 0$ gives:

$$C_A = C_A^0 \exp[-(k_1 + k_2)t]$$
(1.5.35)

Substitution of Equation (1.5.35) into the differential equation for C_{DP} yields:

$$\frac{dC_{DP}}{dt} = k_1 C_A^0 \exp[-(k_1 + k_2)t]$$

The solution to this differential equation with $C_{DP} = C_{DP}^0$ at $t = 0$ is:

$$C_{DP} = C_{DP}^0 + \frac{k_1 C_A^0}{(k_1 + k_2)} \left[1 - \exp[-(k_1 + k_2)t] \right]$$
(1.5.36)

Likewise, the equation for C_{SP} can be obtained and it is:

$$C_{SP} = C_{SP}^0 + \frac{k_2 C_A^0}{(k_1 + k_2)} \left[1 - \exp[-(k_1 + k_2)t]\right] \qquad (1.5.37)$$

The percent selectivity and yield of DP for this reaction network are:

$$s_{DP} \times 100\% = \frac{dC_{DP}}{-dC_A} \times 100\% = \frac{k_1 C_A}{k_1 C_A + k_2 C_A} \times 100\% = \frac{k_1}{k_1 + k_2} \times 100\% \qquad (1.5.38)$$

and

$$Y = \frac{C_{DP}}{C_A^0} \qquad (1.5.39)$$

EXAMPLE 1.5.8

The following reactions are observed when an olefin is epoxidized with dioxygen:

$$\text{alkene} + O_2 \Longrightarrow \text{epoxide}$$
$$\text{epoxide} + O_2 \Longrightarrow CO_2 + H_2O$$
$$\text{alkene} + O_2 \Longrightarrow CO_2 + H_2O$$

Derive the rate expression for this mixed-parallel series-reaction network and the expression for the percent selectivity to the epoxide.

■ Answer

The reaction network is assumed to be:

$$A + O \xrightarrow{k_1} EP + O \xrightarrow{k_2} CD$$
$$\downarrow{\scriptstyle k_3}$$
$$CD$$

where A: alkene, O: dioxygen, EP: epoxide, and CD: carbon dioxide. The rate expressions for this network are:

$$\frac{dC_A}{dt} = -k_1 C_A C_O - k_3 C_A C_O$$

$$\frac{dC_{EP}}{dt} = k_1 C_A C_O - k_2 C_{EP} C_O$$

$$\frac{dC_{CD}}{dt} = k_2 C_{EP} C_O + k_3 C_A C_O$$

The percent selectivity to EP is:

$$s_{EP} = \frac{r_{EP}}{r_{EP} + r_{CD}} \times 100\% = \frac{k_1 C_A C_O - k_2 C_{EP} C_O}{k_1 C_A C_O + k_3 C_A C_O} \times 100\%$$

EXAMPLE 1.5.9

In Example 1.5.6, the expression for the maximum concentration in a series reaction network was illustrated. Example 1.5.8 showed how to determine the selectivity in a mixed-parallel series-reaction network. Calculate the maximum epoxide selectivity attained from the reaction network illustrated in Example 1.5.8 assuming an excess of dioxygen.

■ **Answer**

If there is an excess of dioxygen then C_O can be held constant. Therefore,

$$s_{EP} = \frac{k_1 C_A C_O - k_2 C_{EP} C_O}{k_1 C_A C_O + k_3 C_A C_O} = \frac{k_1 C_A - k_2 C_{EP}}{(k_1 + k_3) C_A}$$

From this expression it is clear that the selectivity for any C_A will decline as $k_2 C_{EP}$ increases. Thus, the maximum selectivity will be:

$$s_{EP}^{max} = \frac{k_1 C_A}{(k_1 + k_3) C_A} = \frac{k_1}{k_1 + k_3}$$

and that this would occur at $t = 0$, as was illustrated in Example 1.5.7. Here, the maximum selectivity is not 100 percent at $t = 0$ but rather the fraction $k_1/(k_2 + k_3)$ due to the parallel portion of the network.

EXAMPLE 1.5.10

Find the maximum yield of the epoxide using the conditions listed for Example 1.5.9.

■ **Answer**

The maximum yield, C_{EP}^{max}/C_A^0 will occur at C_{EP}^{max}. If $\tilde{k}_1 = k_1 C_O$, $\tilde{k}_2 = k_2 C_O$, $\tilde{k}_3 = k_3 C_O$, $y = C_A/C_A^0$ and $x = C_{EP}/C_A^0$, the rate expressions for this network can be written as:

$$\frac{dy}{dt} = -\tilde{k}_1 y - \tilde{k}_3 y = -(\tilde{k}_1 + \tilde{k}_3) y$$

$$\frac{dx}{dt} = \tilde{k}_1 y - \tilde{k}_2 x$$

Note the analogy to Equation (1.5.23). Solving the differential equation for y and substituting this expression into the equation for x gives:

$$\frac{dx}{dt} + \tilde{k}_2 x = \tilde{k}_1 \exp[-(\tilde{k}_1 + \tilde{k}_3)t]$$

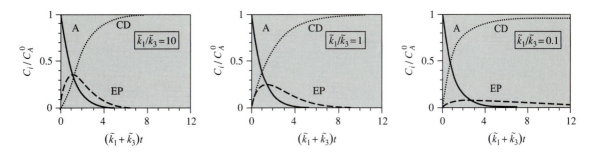

Figure 1.5.5 |
Normalized concentration of species i as a function of time for $\tilde{k}_1 = \tilde{k}_2$.

Solution of this differential equation by methods employed for the solution of Equation (1.5.23) gives:

$$x = \frac{\tilde{k}_1}{\tilde{k}_2 - (\tilde{k}_1 + \tilde{k}_3)} \left[\exp[-(\tilde{k}_1 + \tilde{k}_3)t] - \exp(-\tilde{k}_2 t) \right]$$

If $\tilde{k}_3 = 0$, then the expression is analogous to Equation (1.5.25). In Figure 1.5.5, the normalized concentration profiles for various ratios of $\tilde{k}_1/\tilde{k}_3$ are plotted. The maximum yields of epoxide are located at the x^{max} for each ratio of $\tilde{k}_1/\tilde{k}_3$. Note how increased reaction rate to deep oxidation of alkane decreases the yield to the epoxide.

Exercises for Chapter 1

1. Propylene can be produced from the dehydrogenation of propane over a catalyst. The reaction is:

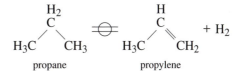

At atmospheric pressure, what is the fraction of propane converted to propylene at 400, 500, and 600°C if equilibrium is reached at each temperature? Assume ideal behavior.

Temperature (°C)	400	500	600
K_a	0.000521	0.0104	0.104

2. An alternative route to the production of propylene from propane would be through oxydehydrogenation:

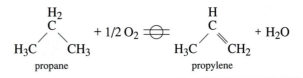

At atmospheric pressure, what is the fraction of propane converted to propylene at 400, 500, and 600°C if equilibrium is reached at each temperature? Compare the results to those from Exercise 1. What do you think is the major impediment to this route of olefin formation versus dehydrogenation? Assume ideal behavior.

Temperature (°C)	400	500	600
K_a	1.16×10^{26}	5.34×10^{23}	8.31×10^{21}

3. The following reaction network represents the isomerization of 1-butene to cis- and trans-2-butene:

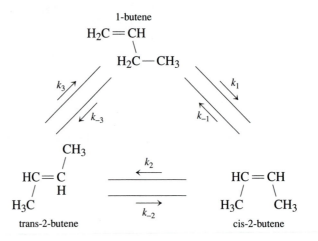

The equilibrium constants (K_a's) for steps 1 and 2 at 400 K are 4.30 and 2.15, respectively. Consider the fugacity coefficients to be unity.

(a) Calculate the equilibrium constant of reaction 3.

(b) Assuming pure 1-butene is initially present at atmospheric pressure, calculate the equilibrium conversion of 1-butene and the equilibrium composition of the butene mixture at 400 K. (Hint: only two of the three reactions are independent.)

4. Xylene can be produced from toluene as written schematically:

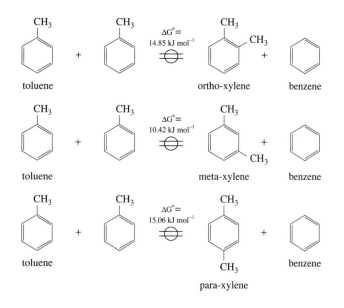

The values of $\Delta G°$ were determined at 700 K. What is the equilibrium composition (including all xylene isomers) at 700 K and 1.0 atm pressure? Propose a method to manufacture para-xylene without producing significant amounts of either ortho- or meta-xylene.

5. Vinyl chloride can be synthesized by reaction of acetylene with hydrochloric acid over a mercuric chloride catalyst at 500 K and 5.0 atm total pressure. An undesirable side reaction is the subsequent reaction of vinyl chloride with HCl. These reactions are illustrated below.

$$\underset{\text{acetylene}}{HC\equiv CH} + HCl \;\rightleftharpoons\; \underset{\text{vinyl chloride}}{H_2C=CHCl} \tag{1}$$

$$H_2C=CHCl + HCl \;\rightleftharpoons\; \underset{\text{1,2 dichloroethane}}{H_3C-CHCl_2} \tag{2}$$

The equilibrium constants at 500 K are 6.6×10^3 and 0.88 for reaction 1 and 2, respectively. Assume ideal behavior.

(a) Find the equilibrium composition at 5.0 atm and 500 K for the case when acetylene and HCl are present initially as an equimolar mixture. What is the equilibrium conversion of acetylene?

(b) Redo part (a) with a large excess of inert gas. Assume the inert gas constitutes 90 vol. % of the initial gas mixture.

6. Acetone is produced from 2-propanol in the presence of dioxygen and the photocatalyst TiO_2 when the reactor is irradiated with ultraviolet light. For a

reaction carried out at room temperature in 1.0 mol of liquid 2-propanol containing 0.125 g of catalyst, the following product concentrations were measured as a function of irradiation time (J. D. Lee, M.S. Thesis, Univ. of Virginia, 1993.) Calculate the first-order rate constant.

Reaction time (min)	20	40	60	80	100	120	140	160	180
Acetone produced $(g_{acetone}/g_{2\text{-propanol}}) \times 10^4$	1.9	3.9	5.0	6.2	8.2	10	11.5	13.2	14.0

7. The Diels-Alder reaction of 2,3-dimethyl-1,3-butadiene (DMB) and acrolein produces 3,4-dimethyl-Δ^3-tetrahydro-benzaldehyde.

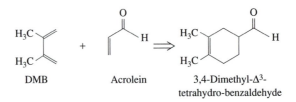

DMB   Acrolein   3,4-Dimethyl-Δ^3-
                               tetrahydro-benzaldehyde

This overall second-order reaction was performed in methanol solvent with equimolar amounts of DMB and acrolein (C. R. Clontz, Jr., M.S. Thesis, Univ. of Virginia, 1997.) Use the data shown below to evaluate the rate constant at each temperature.

Temperature (K)	Reaction time (hours)	[DMB] (mol L^{-1})
323	0	0.097
323	20	0.079
323	40	0.069
323	45	0.068
298	0	0.098
298	74	0.081
298	98	0.078
298	125	0.074
298	170	0.066
278	0	0.093
278	75	0.091
278	110	0.090
278	176	0.088
278	230	0.087

8. Some effluent streams, especially those from textile manufacturing facilities using dying processes, can be highly colored even though they are considered to be fairly nontoxic. Due to the stability of modern dyes, conventional

biological treatment methods are ineffective for decolorizing such streams. Davis et al. studied the photocatalytic decomposition of wastewater dyes as a possible option for decolorization [R. J. Davis, J. L. Gainer, G. O'Neal, and I.-W. Wu, *Water Environ. Res.* **66** (1994) 50]. The effluent from a municipal water treatment facility whose influent contained a high proportion of dyeing wastewater was mixed with TiO_2 photocatalyst (0.40 wt. %), sparged with air, and irradiated with UV light. The deep purple color of the original wastewater lightened with reaction time. Since the absolute concentration of dye was not known, the progress of the reaction was monitored colorimetrically by measuring the relative absorbance of the solution at various wavelengths. From the relative absorbance data collected at 438 nm (shown below), calculate the apparent order of the decolorization reaction and the rate constant. The relative absorbance is the absorbance at any time t divided by the value at $t = 0$.

Reaction time (min)	30	60	90	120	150	180	210
Relative absorbance	0.79	0.67	0.50	0.39	0.30	0.23	0.17

9. The following reaction is investigated in a constant density batch reactor:

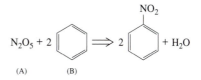

(A) (B)

The reaction rate is:

$$\frac{dC_A}{dt} = -kC_A C_B^2$$

The reactor is initially charged with C_A^0 and C_B^0 and $C_B^0/C_A^0 = 3.0$. Find the value of k if $C_A^0 = 0.01$ mol/L and after 10 min of reaction the conversion of N_2O_5 is 50 percent.

10. As discussed in Vignette 1.1.1, ammonia is synthesized from dinitrogen and dihydrogen in the presence of a metal catalyst. Fishel et al. used a constant volume reactor system that circulated the reactants over a heated ruthenium metal catalyst and then immediately condensed the product ammonia in a cryogenic trap [C. T. Fishel, R. J. Davis, and J. M. Garces, *J. Catal.* **163** (1996) 148]. A schematic diagram of the system is:

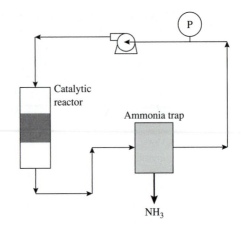

From the data presented in the following tables, determine the rates of ammonia synthesis (moles NH_3 produced per min per gcat) at 350°C over a supported ruthenium catalyst (0.20 g) and the orders of reaction with respect to dinitrogen and dihydrogen. Pressures are referenced to 298 K and the total volume of the system is 0.315 L. Assume that no ammonia is present in the gas phase.

$N_2:H_2:He = 3:1:0$

Pressure (torr)	766.2	731.9	711.9	686.2	661.5
Time (min)	0	18	30	42	54

$N_2:H_2:He = 1:1:2$

Pressure (torr)	753.4	737.5	726.6	709.5	700.3
Time (min)	0	15	30	45	60

$N_2:H_2:He = 1:3:0$

Pressure (torr)	707.1	700.2	693.2	683.5	675.5
Time (min)	0	15	30	45	55

11. In Example 1.5.6, the series reaction:

$$A \xrightarrow{k_1} B \xrightarrow{k_2} C$$

was analyzed to determine the time (t_{max}) and the concentration (C_B^{max}) associated with the maximum concentration of the intermediate B when $k_1 = 2k_2$. What are t_{max} and C_B^{max} when $k_1 = k_2$?

12. The Lotka-Volterra model is often used to characterize predator-prey interactions. For example, if R is the population of rabbits (which reproduce autocatalytically), G is the amount of grass available for rabbit food (assumed to be constant), L is the population of lynxes that feeds on the rabbits, and D represents dead lynxes, the following equations represent the dynamic behavior of the populations of rabbits and lynxes:

$$R + G \longrightarrow 2R \qquad (1)$$
$$L + R \longrightarrow 2L \qquad (2)$$
$$L \longrightarrow D \qquad (3)$$

Each step is irreversible since, for example, rabbits cannot turn back into grass.

(a) Write down the differential equations that describe how the populations of rabbits (R) and lynxes (L) change with time.

(b) Assuming G and all of the rate constants are unity, solve the equations for the evolution of the animal populations with time. Let the initial values of R and L be 20 and 1, respectively. Plot your results and discuss how the two populations are related.

13. Diethylamine (DEA) reacts with 1-bromobutane (BB) to form diethylbutylamine (DEBA) according to:

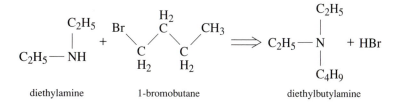

diethylamine   1-bromobutane   diethylbutylamine

From the data given below (provided by N. Leininger, Univ. of Virginia), find the effect of solvent on the second-order rate constant.

Time (min)	[DEBA] (mol L^{-1})
0	0.000
35	0.002
115	0.006
222	0.012
455	0.023

Solvent = 1,4-butanediol
T = 22°C
[DEA]^0 = [BB]^0 = 0.50 mol L^{-1}

Time (min)	[DEBA] (mol L^{-1})
0	0.000
31	0.004
58	0.007
108	0.013
190	0.026

Solvent = acetonitrile
T = 22°C
[DEA]^0 = 1.0 mol L^{-1}
[BB]^0 = 0.10 mol L^{-1}

14. A first-order homogeneous reaction of A going to $3B$ is carried out in a constant pressure batch reactor. It is found that starting with pure A the volume after 12 min is increased by 70 percent at a pressure of 1.8 atm. If the same reaction is to be carried out in a constant volume reactor and the initial pressure is 1.8 atm, calculate the time required to bring the pressure to 2.5 atm.

15. Consider the reversible, elementary, gas phase reaction shown below that occurs at 300 K in a constant volume reactor of 1.0 L.

$$A + B \underset{k_2}{\overset{k_1}{\rightleftharpoons}} C, \quad k_1 = 6.0 \text{ L mol}^{-1}\text{h}^{-1}, k_2 = 3.0 \text{ h}^{-1}$$

For an initial charge to the reactor of 1.0 mol of A, 2.0 mol of B, and no C, find the equilibrium conversion of A and the final pressure of the system. Plot the composition in the reactor as a function of time.

16. As an extension of Exercise 15, consider the reversible, elementary, gas phase reaction of A and B to form C occurring at 300 K in a variable volume (constant pressure) reactor with an *initial* volume of 1.0 L. For a reactant charge to the reactor of 1.0 mol of A, 2.0 mol of B, and no C, find the equilibrium conversion of A. Plot the composition in the reactor and the reactor volume as a function of time.

17. As an extension of Exercise 16, consider the effect of adding an inert gas, I, to the reacting mixture. For a reactant charge to the reactor of 1.0 mol of A, 2.0 mol of B, no C, and 3.0 mol of I, find the equilibrium conversion of A. Plot the composition in the reactor and the reactor volume as a function of time.

CHAPTER 2

Rate Constants of Elementary Reactions

2.1 | Elementary Reactions

Recall from the discussion of reaction networks in Chapter 1 that an elementary reaction must be written as it proceeds at the molecular level and represents an irreducible molecular event. An elementary reaction normally involves the breaking or making of a single chemical bond, although more rarely, two bonds are broken and two bonds are formed in what is denoted a four-center reaction. For example, the reaction:

$$OH + CH_3CH_3 \rightarrow CH_2CH_3 + H_2O$$

is a good candidate for possibly being an elementary reaction, while the reaction:

$$CH_3CH_3 + O_2 \rightarrow OH + CH_3CH_2O$$

is not. Whether or not a reaction is elementary must be determined by experimentation.

As stated in Chapter 1, an elementary reaction cannot be written arbitrarily and must be written the way it takes place. For example (see Table 1.4.3), the reaction:

$$2NO + O_2 \rightarrow 2NO_2 \qquad (2.1.1)$$

cannot be written as:

$$NO + \tfrac{1}{2}O_2 \rightarrow NO_2 \qquad (2.1.2)$$

since clearly there is no such entity as half a molecule of dioxygen. It is important to note the distinction between stoichiometric equations and elementary reactions (see Chapter 1) is that for the stoichiometric relation:

$$2NO + O_2 = 2NO_2 \qquad (2.1.3)$$

one can write (although not preferred):

$$NO + \tfrac{1}{2}O_2 = NO_2 \qquad\qquad (2.1.4)$$

The remainder of this chapter describes methods to determine the rate and temperature dependence of the rate of *elementary* reactions. This information is used to describe how reaction rates in general are appraised.

2.2 | Arrhenius Temperature Dependence of the Rate Constant

The rate constant normally depends on the absolute temperature, and the functional form of this relationship was first proposed by Arrhenius in 1889 (see Rule III in Chapter 1) to be:

$$k = \overline{A} \exp[-E/(R_g T)] \qquad\qquad (2.2.1)$$

where the activation energy, E, and the pre-exponential factor, $\overline{A}$, both do not depend on the absolute temperature. The Arrhenius form of the reaction rate constant is an empirical relationship. However, transition-state theory provides a justification for the Arrhenius formulation, as will be shown below. Note that the Arrhenius law (Equation 2.2.1) gives a linear relationship between $\ln k$ and T^{-1}.

EXAMPLE 2.2.1

The decomposition reaction:

$$2N_2O_5 \Rightarrow 2N_2O_4 + O_2$$

can proceed at temperatures below 100°C and the temperature dependence of the first-order rate constant has been measured. The data are:

T (K)	k (s^{-1})
288	1.04×10^{-5}
298	3.38×10^{-5}
313	2.47×10^{-4}
323	7.59×10^{-4}
338	4.87×10^{-3}

Suggest an experimental approach to obtain these rate constant data and calculate the activation energy and pre-exponential factor. (Adapted from C. G. Hill, *An Introduction to Chemical Engineering Kinetics & Reactor Design*, Wiley, New York, 1977.)

■ Answer
Note that the rate constants are for a first-order reaction. The material balance for a closed system at constant temperature is:

$$\frac{dn_{N_2O_5}}{dt} = -kn_{N_2O_5}$$

where $n_{N_2O_5}$ is the number of moles of N_2O_5. If the system is at constant volume (a closed vessel), then as the reaction proceeds the pressure will rise because there is a positive mole change with reaction. That is to say that the pressure will increase as N_2O_5 is reacted because the molar expansion factor is equal to 0.5. An expression for the total moles in the closed system can be written as:

$$n = n_0(1 + 0.5f_{N_2O_5})$$

where n is the total number of moles in the system. The material balance on the closed system can be formulated in terms of the fractional conversion and integrated (see Example 1.5.2) to give:

$$f_{N_2O_5} = 1 - \exp(-kt)$$

Since the closed system is at constant T and V ($PV = nR_gT$):

$$\frac{P}{P_0} = \frac{n}{n_0} = (1 + 0.5f_{N_2O_5})$$

and the pressure can therefore be written as:

$$P = P_0[1.5 - 0.5\exp(-kt)]$$

If the pressure rise in the closed system is monitored as a function of time, it is clear from the above expression how the rate constant can be obtained at each temperature.

In order to determine the pre-exponential factor and the activation energy, the $\ln k$ is plotted against T^{-1} as shown below:

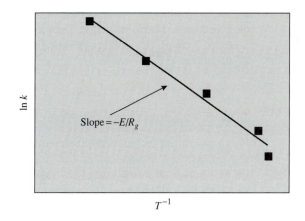

From a linear regression analysis of the data, the slope and intercept can be obtained, and they are 1.21×10^4 and 30.4, respectively. Thus,

$$\text{slope} = -E/R_g \quad : \quad E = 24 \text{ kcal/mol}$$
$$\text{intercept} = \ln A \quad : \quad A = 1.54 \times 10^{13} \text{ s}^{-1}$$

Consider the following elementary reaction:

$$A + B \underset{k_2}{\overset{k_1}{\rightleftharpoons}} S + W \tag{2.2.2}$$

At equilibrium,

$$k_1 C_A C_B = k_2 C_S C_W \tag{2.2.3}$$

and

$$K_C = \frac{k_1}{k_2} = \left(\frac{C_S C_W}{C_A C_B}\right)_{eq} \tag{2.2.4}$$

If the Arrhenius law is used for the reaction rate constants, Equation (2.2.4) can be written:

$$K_C = \frac{k_1}{k_2} = \left(\frac{\overline{A}_1}{\overline{A}_2}\right)\exp\left(\frac{E_2 - E_1}{R_g T}\right) = \left(\frac{C_S C_W}{C_A C_B}\right)_{eq} \tag{2.2.5}$$

It is easy to see from Equation (2.2.5) that if $(E_2 - E_1) > 0$ then the reaction is exothermic and likewise if $(E_2 - E_1) < 0$ it is endothermic (refer to Appendix A for temperature dependence of K_C). In a typical situation, the highest yields of products are desired. That is, the ratio $(C_S C_W/C_A C_B)_{eq}$ will be as large as possible. If the reaction is endothermic, Equation (2.2.5) suggests that in order to maximize the product yield, the reaction should be accomplished at the highest possible temperature. To do so, it is necessary to make $\exp[(E_2 - E_1)/(R_g T)]$ as large as possible by maximizing $(R_g T)$, since $(E_2 - E_1)$ is negative. Note that as the temperature increases, so do both the rates (forward and reverse). Thus, for endothermic reactions, the rate and yield must *both* increase with temperature. For exothermic reactions, there is always a trade-off between the equilibrium yield of products and the reaction rate. Therefore, a balance between rate and yield is used and the T chosen is dependent upon the situation.

2.3 | Transition-State Theory

For most elementary reactions, the rearrangement of atoms in going from reactants to products via a transition state (see, for example, Figure 1.1.1) proceeds through the movements of atomic nuclei that experience a potential energy field that is generated by the rapid motions of the electrons in the system. On this potential energy surface there will be a path of minimum energy expenditure for the reaction to proceed from reactants to products (reaction coordinate). The low energy positions of reactants and products on the potential energy surface will be separated by a higher energy region. The highest energy along the minimum energy pathway in going from reactants to products defines the transition state. As stated in Chapter 1, the transition state is not a reaction intermediate but rather a high energy configuration of a system in transit from one state to another.

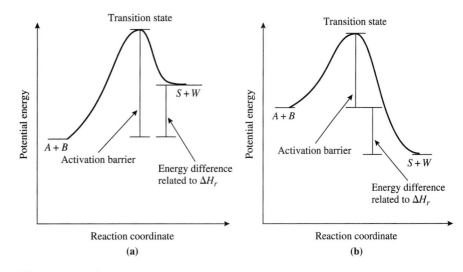

Figure 2.3.1 |
Potential energy profiles for the elementary reaction $A + B \rightarrow S + W$ for **(a)** an endothermic reaction and **(b)** an exothermic reaction.

The difference in energies of the reactants and products is related to the heat of reaction—a thermodynamic quantity. Figure 2.3.1 shows potential energy profiles for endothermic and exothermic elementary reactions.

Transition-state theory is used to calculate an elementary reaction that conforms to the energetic picture illustrated in Figure 2.3.1. How this is done is described next.

VIGNETTE 2.3.1

Since transition states exist for only picoseconds or less, the ability to observe them in real time requires femtosecond (10^{-15} s) time resolution. Femtochemistry, or chemistry on the femtosecond time scale, has been defined as the field of chemical dynamics concerned with the act of chemical transformation (i.e., the process of bond breaking and making). Femtochemistry requires ultrafast lasers to initiate and observe chemical reactions with femtosecond time resolution. Ahmed Zewail and his coworkers have carried out those types of experiments beginning in the late 1980s (*The Chemical Bond Structure and Dynamics*, ed. A. Zewail, Academic Press, San Diego, 1992). Figure 2.3.2 illustrates the femtochemistry experiment of "watching" excited ICN dissociate. The reaction is initiated by addition of light energy from a laser to bring ICN from the ground state to an excited state (upper to lower energy curves) where the I-CN is now completely in a repulsive state. As the I and CN depart from one another, the intermediates are observed. For example, at 100 femtoseconds, the distance between I and CN (I ⋯ CN) has increased 1 Å. These studies provide insights and help confirm theories of how reactions proceed at the molecular level.

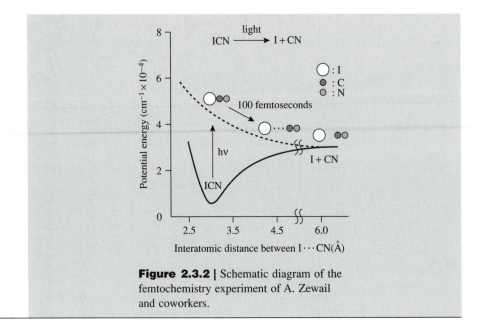

Figure 2.3.2 | Schematic diagram of the femtochemistry experiment of A. Zewail and coworkers.

Consider again the reaction given by Equation (2.2.2). At thermodynamic equilibrium, Equation (2.2.3) is satisfied. In the equilibrated system, there must also exist an equilibrium concentration of transition states: C_{TS}. The rates of the forward and the reverse reactions are equal, implying that there is an equivalent number of species traversing the activation barrier from either the reactant side or the product side. Thus,

$$k_1(C_A C_B)_{eq} = k_2(C_S C_W)_{eq} = \lambda C_{TS} \tag{2.3.1}$$

where λ is a frequency. Since C_{TS} is a thermodynamic quantity, it can be calculated from statistical thermodynamics. A fundamental assumption of transition-state theory is that if the system is perturbed from equilibrium by product removal, the rate of the forward reaction remains:

$$k_1 C_A C_B = \lambda C_{TS} \tag{2.3.2}$$

where C_{TS} is the concentration of transition-states in equilibrium with reactants A and B. The fictitious equilibrium:

$$A + B \rightleftharpoons TS \tag{2.3.3}$$

allows for the formulation of an equilibrium constant as shown below:

$$K^{\ddagger} = \frac{C_{TS}}{C_A C_B} \tag{2.3.4}$$

where

$$R_g T \ln K^{\ddagger} = -\Delta G_0^{\ddagger} = -\Delta H_0^{\ddagger} + T\Delta S_0^{\ddagger} \tag{2.3.5}$$

and $\Delta G_0^{\ddagger}$ is the change in the standard Gibbs function for the reaction given in Equation (2.3.3) and $\Delta H_0^{\ddagger}$ and $\Delta S_0^{\ddagger}$ are the corresponding changes in standard enthalpy and entropy, respectively. The subscript 0 is used to denote the standard state while the superscript $\ddagger$ denotes quantities pertaining to the transition state. By combining Equations (2.3.5) and (2.3.4), C_{TS} can be written as:

$$C_{TS} = \exp\left[\frac{\Delta S_0^{\ddagger}}{R_g}\right]\exp\left[\frac{-\Delta H_0^{\ddagger}}{R_g T}\right]C_A C_B \qquad (2.3.6)$$

Using this formulation for C_{TS} in Equation (2.3.2) allows the reaction rate to be of the form:

$$r = k_1 C_A C_B = \lambda \exp\left[\frac{\Delta S_0^{\ddagger}}{R_g}\right]\exp\left[\frac{-\Delta H_0^{\ddagger}}{R_g T}\right]C_A C_B \qquad (2.3.7)$$

A fundamental assumption of transition-state theory is that λ is a *universal frequency* and does not depend upon the nature of the reaction being considered. It can be proven that (see, for example, M. Boudart, *Kinetics of Chemical Processes,* Butterworth-Heinemann, 1991, pp. 41–45):

$$\lambda = \frac{\bar{k}T}{h} \qquad (2.3.8)$$

where h is Planck's constant and $\bar{k}$ is Boltzmann's constant. Using Equation (2.3.8) in Equation (2.3.7) gives:

$$r = \left(\frac{\bar{k}T}{h}\right)\exp\left[\frac{\Delta S_0^{\ddagger}}{R_g}\right]\exp\left[\frac{-\Delta H_0^{\ddagger}}{R_g T}\right]C_A C_B \qquad (2.3.9)$$

This is the general equation of transition-state theory in its thermodynamic form.

Equation (2.3.9) helps in the comprehension of how reactions proceed. In order for a reaction to occur, it is necessary to overcome not just an energy barrier but a free energy barrier. That is to say, a reaction involves not only energy but also requires reaching a favorable configuration associated with a change in entropy. These two effects can often compensate each other. For example, McKenzie et al. [*J. Catal.,* **138** (1992) 547] investigated the reaction:

$$\underset{\displaystyle CH_3CHCH_3}{\overset{\displaystyle OH}{|}} \Longrightarrow \underset{\displaystyle CH_3CCH_3}{\overset{\displaystyle O}{\|}} + H_2$$

over a series of solid catalysts called hydrotalcites, and obtained the following data:

Catalyst	$\bar{A}$ (s^{-1})	E (kJ(mol)$^{-1}$)
1	4.3×10^{12}	172
2	2.3×10^{11}	159
3	2.2×10^{10}	146
4	1.6×10^{9}	134

If Equation (2.3.9) is compared to Equation (2.2.1), then:

$$\overline{A} = \left(\frac{\overline{k}T}{h}\right)\exp\left[\frac{\Delta S_0^{\ddagger}}{R_g}\right] \qquad (2.3.10)$$

$$E = \Delta H_0^{\ddagger} \qquad (2.3.11)$$

The data of McKenzie et al. clearly show that as the energy barrier increases (higher E), the entropy of activation becomes more positive (larger $\Delta S_0^{\ddagger}$ implies more favorable configurational driving force for the reaction, since entropy will always attempt to be maximized). Thus, energy and entropy compensate in this example.

The Arrhenius form of the rate constant specifies that both $\overline{A}$ and E are independent of T. Note that when the formulation derived from transition-state theory is compared to the Arrhenius formulation [Equations (2.3.10) and (2.3.11)], both $\overline{A}$ and E do have some dependence on T. However, $\Delta H_0^{\ddagger}$ is very weakly dependent on T and the temperature dependence of:

$$\left(\frac{\overline{k}T}{h}\right)\exp\left[\frac{\Delta S_0^{\ddagger}}{R_g}\right] \qquad (2.3.12)$$

can normally be neglected as compared to the strong exponential dependence of:

$$\exp\left[\frac{-\Delta H_0^{\ddagger}}{R_g T}\right] \qquad (2.3.13)$$

Thus, the Arrhenius form is an excellent approximation to that obtained from transition-state theory, provided the temperature range does not become too large.

Previously, the concentration C_{TS} was expressed in simple terms assuming ideal conditions. However, for nonideal systems, $K^{\ddagger}$ must be written as:

$$K^{\ddagger} = \frac{a_{TS}}{a_A a_B} \qquad (2.3.14)$$

where a_i is the activity of species i. If activity coefficients, $\overline{\gamma}_i$, are used such that $a_i = \overline{\gamma}_i C_i$, then Equation (2.3.14) can be formulated as:

$$K^{\ddagger} = \left[\frac{\overline{\gamma}_{TS}}{\overline{\gamma}_A \overline{\gamma}_B}\right]\left[\frac{C_{TS}}{C_A C_B}\right] \qquad (2.3.15)$$

Following the substitutions and equation rearrangements used for ideal conditions, the nonideal system yields:

$$k = \left(\frac{\overline{k}T}{h}\right)\frac{\overline{\gamma}_A \overline{\gamma}_B}{\overline{\gamma}_{TS}} K^{\ddagger} \qquad (2.3.16)$$

At infinite dilution where the activity coefficients are one (ideal conditions), k_0 can be written as:

$$k_0 = \left(\frac{\overline{k}T}{h}\right)K^{\ddagger} \qquad (2.3.17)$$

Thus, the rate constant at nonideal conditions relative to the ideal system is:

$$\frac{k}{k_0} = \frac{\overline{\gamma}_A \overline{\gamma}_B}{\overline{\gamma}_{TS}}$$

(2.3.18)

EXAMPLE 2.3.1

Predict how the rate constant of the reaction:

$$A + B \rightarrow S$$

would vary as a function of ionic strength, $\overline{I}$, if A and B are both ions in aqueous solutions at 25°C. The Debye-Hückel theory predicts that:

$$-\log(\overline{\gamma}_i) = 0.5\overline{Z}_i^2 \sqrt{\overline{I}}$$

where $\overline{Z}_i$ is the charge of species i and

$$\overline{I} = \frac{1}{2} \sum_i \overline{Z}_i^2 c_i \qquad (i = A, B, S)$$

where c_i is the concentration of species i in units of molality.

■ Answer

Although the structure of the transition state is not given, it must have a charge of $\overline{Z}_A + \overline{Z}_B$. Using the Debye-Hückel equation in Equation (2.3.18) gives:

$$\log\left(\frac{k}{k_0}\right) = \log\left(\frac{\overline{\gamma}_A \overline{\gamma}_B}{\overline{\gamma}_{TS}}\right)$$

or

$$\log\left(\frac{k}{k_0}\right) = 0.5[\overline{Z}_A + \overline{Z}_B]^2 \sqrt{\overline{I}} - 0.5\overline{Z}_A^2 \sqrt{\overline{I}} - 0.5\overline{Z}_B^2 \sqrt{\overline{I}}$$

After simplification, the above equation reduces to:

$$\log\left(\frac{k}{k_0}\right) = [\overline{Z}_A \overline{Z}_B] \sqrt{\overline{I}}$$

This relationship has been experimentally verified. Note that if one of the reactants is uncharged, for example, B, then $\overline{Z}_B = 0$ and $k = k_0$. If the rate is mistakenly written:

$$\mathrm{r} = k_0 a_A a_B = k_0 \overline{\gamma}_A \overline{\gamma}_B c_A c_B$$

application of the Debye-Hückel equation gives:

$$\log\left(\frac{k}{k_0}\right) = \log\left(\frac{k_0 \overline{\gamma}_A \overline{\gamma}_B}{k_0}\right) = -0.5[\overline{Z}_A^2 + \overline{Z}_B^2] \sqrt{\overline{I}}$$

Note that this relationship gives the wrong slope of $\log\left(\frac{k}{k_0}\right)$ versus $\sqrt{\overline{I}}$, and if $\overline{Z}_B = 0$, it does not reduce to $k = k_0$.

It is expected that the transition state for a unimolecular reaction may have a structure similar to that of the reactant except for bond elongation prior to breakage. If this is the case, then $\Delta S_0^{\ddagger} \cong 0$ and

$$\overline{A} \cong \frac{\overline{k}T}{h} \cong 10^{13}\ \text{s}^{-1} \tag{2.3.19}$$

It has been experimentally verified that numerous unimolecular reactions have rate constants with pre-exponential factors on the order of $10^{13}\ \text{s}^{-1}$. However, the pre-exponential factor can be either larger or smaller than $10^{13}\ \text{s}^{-1}$ depending on the details of the transition state.

Although it is clear that transition-state theory provides a molecular perspective on the reaction and how to calculate the rate, it is difficult to apply since $\Delta S_0^{\ddagger}$, $\Delta H_0^{\ddagger}$, and $\overline{\gamma}_{TS}$ are usually not known *a priori*. Therefore, it is not surprising that the Arrhenius rate equation has been used to systemize the vast majority of experimental data.

Exercises for Chapter 2

1. For a series of similar reactions, there is often a trend in the equilibrium constants that can be predicted from the rate constants. For example, the reaction coordinate diagrams of two similar reactions are given below. If the difference in free energy of formation of the transition state is proportional to the difference in the free energy change upon reaction, that is, $\Delta G_1^{\ddagger} - \Delta G_2^{\ddagger} = \alpha(\Delta G_1 - \Delta G_2)$, derive the relationship between the rate constants k_1 and k_2 and the equilibrium constants (K_1 and K_2).

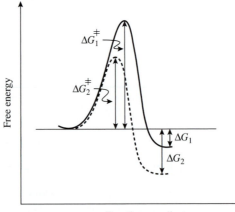

Schematic diagram of two similar reactions.

2. The decomposition of gaseous 2-propanol over a mixed oxide catalyst of magnesia and alumina produces both acetone and propene according to the following equations:

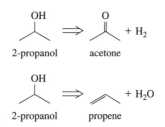

From the data presented below, calculate the activation energy for each reaction (A. L. McKenzie, M.S. Thesis, Univ. of Virginia, 1992). Assume the concentration of 2-propanol is constant for each experiment. Selectivity to acetone is defined with respect to the products acetone and propene.

Temperature (K)	573	583	594	603	612
Rate of acetone formation (mol gcat^{-1} s^{-1})	4.1×10^{-7}	7.0×10^{-7}	1.4×10^{-6}	2.2×10^{-6}	3.6×10^{-6}
Selectivity to acetone (%)	92	86	81	81	81

3. Use the data in Exercise 7 at the end of Chapter 1 to determine the activation energy and pre-exponential factor of the rate constant for the Diels-Alder reaction of 2,3-dimethyl-1,3-butadiene (DMB) and acrolein to produce 3,4-dimethyl-Δ^3-tetrahydro-benzaldehyde.

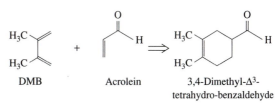

DMB   Acrolein   3,4-Dimethyl-Δ^3-tetrahydro-benzaldehyde

4. Explain how the pre-exponential factor of a unimolecular reaction can be greater than 10^{13} s^{-1}.

5. Discuss the strengths and weaknesses of transition-state theory.

6. Irradiation of water solutions with gamma rays can produce a very active intermediate known as a hydrated electron. This species can react with many different neutral and ionic species in solution. Devise an experiment to check the electrical charge of the hydrated electron. (Problem adapted from M. Boudart, *Kinetics of Chemical Processes,* Prentice Hall, Englewood Cliffs, NJ, 1968, p. 55.)

CHAPTER 3

Reactors for Measuring Reaction Rates

3.1 | Ideal Reactors

The confines in which chemical reactions occur are called reactors. A reactor can be a chemical reactor in the traditional sense or other entities, for example, a chemical vapor deposition apparatus for making computer chips, an organ of the human body, and the atmosphere of a large city. In this chapter, the discussion of reactors is limited to topics germane to the determination of reaction rates. Later in this text, strategies for attacking the problems of mathematically describing and predicting behavior of reactors in general are presented.

In practice, conditions in a reactor are usually quite different than the ideal requirements used in the definition of reaction rates. Normally, a reactor is not a closed system with uniform temperature, pressure, and composition. These ideal conditions can rarely if ever be met even in experimental reactors designed for the measurement of reaction rates. In fact, reaction rates cannot be measured directly in a closed system. In a closed system, the composition of the system varies with time and the rate is then inferred or calculated from these measurements.

There are several questions that can be put forth about the operation of reactors and they can be used to form the basis of classifying and defining ideal conditions that are desirable for the proper measurements of reaction rates.

The first question is whether the system exchanges *mass* with its surroundings. If it does not, then the system is called a *batch reactor*. If it does, then the system is classified as a *flow reactor*.

The second question involves the exchange of heat between the reactor and its surroundings. If there is no heat exchange, the reactor is then adiabatic. At the other extreme, if the reactor makes very good thermal contact with the surroundings it can be held at a constant temperature (in both time and position within the reactor) and is thus *isothermal*.

Table 3.1.1 | Limiting conditions of reactor operation.[1]

Event	Limiting conditions	
Exchange of mass	Batch	Flow
Exchange of heat	Isothermal	Adiabatic
Mechanical variables	Constant volume	Constant pressure
Residence time	Unique	Exponential distribution
Space-time behavior	Transient	Stationary

[1]From M. Boudart, *Kinetics of Chemical Processes,* Butterworth & Heinemann, 1991, p. 13.

The third question concerns the mechanical variables: pressure and volume. Is the reactor at constant pressure or constant volume? The fourth question is whether the time spent in the reactor by each volume element of fluid is the same. If it is not the same, there may exist a *distribution of residence times* and the opposite extreme of a unique residence time is an exponential distribution.

The fifth question focuses on a particular fixed volume element in the reactor and whether it changes as a function of time. If it does not, then the reactor is said to operate at a stationary state. If there are time variations, then the reactor is operating under transient conditions. A nontrivial example of the transient situation is designed on purpose to observe how a chemically reactive system at equilibrium relaxes back to the equilibrium state after a small perturbation. This type of relaxation experiment can often yield informative kinetic behavior.

The ten possibilities outlined above are collected in Table 3.1.1. Next, ideal reactors will be illustrated in the contexts of the limiting conditions of their operation.

3.2 | Batch and Semibatch Reactors

Consider the ideal batch reactor illustrated in Figure 3.2.1. If it is assumed that the contents of the reactor are perfectly mixed, a material balance on the reactor can be written for a species i as:

$$\underset{\text{accumulation}}{\frac{dn_i}{dt}} \quad = \quad \underset{\text{input}}{0} \quad - \quad \underset{\text{output}}{0} \quad + \quad \underset{\substack{\text{amount} \\ \text{produced} \\ \text{by reaction}}}{(v_i \text{r} V)}$$

or

$$\frac{dn_i}{dt} = v_i \text{r} V \quad \text{with } n_i = n_i^0 @ t = 0 \tag{3.2.1}$$

The material balance can also be written in terms of the fractional conversion and it is:

$$n_i^0 \frac{df_i}{dt} = -(v_i \text{r}) V^0 (1 + \varepsilon_i f_i) \quad \text{with } f_i = 0 @ t = 0 \tag{3.2.2}$$

where $|\varepsilon_i| > 0$ for nonconstant volume.

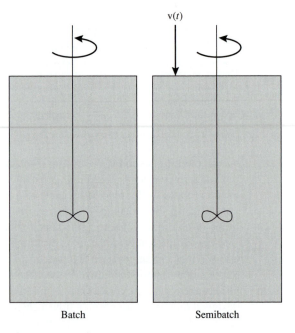

Batch   Semibatch

Figure 3.2.1 |
Ideal batch and semibatch reactors. $v(t)$ is a volumetric
flow rate that can vary with time.

EXAMPLE 3.2.1

An important class of carbon-carbon bond coupling reactions is the Diels-Alder reactions.
An example of a Diels-Alder reaction is shown below:

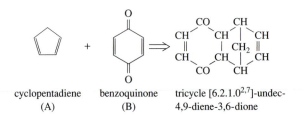

cyclopentadiene   benzoquinone   tricycle [6.2.1.0^{2,7}]-undec-
(A)   (B)   4,9-diene-3,6-dione

If this reaction is performed in a well-mixed isothermal batch reactor, determine the time necessary to achieve 95 percent conversion of the limiting reactant (from C. Hill, *An Introduction to Chemical Engineering Kinetics and Reactor Design*, Wiley, 1977, p. 259).

Data: $k = 9.92 \times 10^{-6}\ \text{m}^3/\text{mol/s}$

$C_A^0 = 100\ \text{mol/m}^3$

$C_B^0 = 80\ \text{mol/m}^3$

■ **Answer**

From the initial concentrations, benzoquinone is the limiting reactant. Additionally, since the reaction is conducted in a dilute liquid-phase, density changes can be neglected. The reaction rate is second-order from the units provided for the reaction rate constant. Thus,

$$(v_B r) = -k(C_B^0)^2(1 - f_B)(\overline{M} - f_B), \overline{M} = C_A^0/C_B^0$$

The material balance on the isothermal batch reactor is:

$$\frac{dn_B}{dt} = v_B r V$$

with $f_B = 0$ at $t = 0$. Integration of this first-order initial-value problem yields:

$$t = \int_0^{f_B} \frac{d\gamma}{kC_B^0(1 - \gamma)(\overline{M} - \gamma)}$$

where γ is the integration variable. The integration yields:

$$t = \frac{\ln\left[\left(\frac{\overline{M} - f_B}{1 - f_B}\right)\frac{1}{\overline{M}}\right]}{kC_B^0(\overline{M} - 1)}$$

Using the data provided, $t = 7.9 \times 10^3$ s or 2.2 h to reach 95 percent conversion of the benzoquinone. This example illustrates the general procedure used for solving isothermal problems. First, write down the reaction rate expression. Second, formulate the material balance. Third, substitute the reaction rate expression into the material balance and solve.

Consider the semibatch reactor schematically illustrated in Figure 3.2.1. This type of reactor is useful for accomplishing several classes of reactions. Fermentations are often conducted in semibatch reactors. For example, the concentration of glucose in a fermentation can be controlled by varying its concentration and flow rate into the reactor in order to have the appropriate times for: (1) the initial growth phase of the biological catalysts, and (2) the period of metabolite production. Additionally, many bioreactors are semibatch even if liquid-phase reactants are not fed to the reactor because oxygen must be continuously supplied to maintain the living catalyst systems (i.e., bacteria, yeast, etc.).

Alternatively, semibatch reactors of the type shown in Figure 3.2.1 are useful for reactions that have the stoichiometry:

$$A + B = \text{products}$$

where B is already in the reactor and A is slowly fed. This may be necessary to: (1) control heat release from exothermic reaction, for example, hydrogenations, (2) provide gas-phase reactants, for example, with halogenations, hydrogenations, or (3) alter reaction selectivities. In the network:

$$A + B \longrightarrow \text{desired product}$$

$$A \longrightarrow \text{undesired product}$$

maintaining a constant and high concentration of B would certainly aid in altering the selectivity to the desired product.

In addition to feeding of components into the reactor, if the sign on v(t) is negative, products are continuously removed, for example, reactive distillation. This is done for reactions: (1) that reveal product inhibition, that is, the product slows the reaction rate, (2) that have a low equilibrium constant (removal of product does not allow equilibrium to be reached), or (3) where the product alters the reaction network that is proceeding in the reactor. A common class of reactions where product removal is necessary is ester formation where water is removed,

$$2 \langle \bigcirc \rangle - \overset{\overset{\displaystyle O}{\|}}{C}OH + HOCH_2CH_2CH_2CH_2OH = \langle \bigcirc \rangle - \overset{\overset{\displaystyle O}{\|}}{C}OCH_2CH_2CH_2CH_2O\overset{\overset{\displaystyle O}{\|}}{C} - \langle \bigcirc \rangle + 2H_2O\uparrow$$

A very large-scale reaction that utilizes reactive distillation of desired liquid products is the hydroformylation of propene to give butyraldehyde:

$$CH_3-CH=CH_2 + CO + H_2 \Longrightarrow CH_3-CH_2CH_2\overset{\overset{\displaystyle O}{\|}}{C}-H \Longrightarrow \text{heavier products}$$

$$\searrow \quad CH_3-CH_2-\overset{\overset{\displaystyle O}{\|}}{\underset{\underset{\displaystyle CH_3}{|}}{C}}-H \Longrightarrow \text{heavier products}$$

A schematic illustration of the hydroformylation reactor is provided in Figure 3.2.2.

The material balance on a semibatch reactor can be written for a species i as:

$$\underset{\text{accumulation}}{\frac{dn_i}{dt}} = \underset{\text{input}}{v(t)C_i^0(t)} - \underset{\text{output}}{0} + \underset{\substack{\text{amount} \\ \text{produced} \\ \text{by reaction}}}{(v_i r)V}$$

or

$$\frac{dn_i}{dt} = (v_i r)V + v(t)C_i^0(t) \tag{3.2.2}$$

where $C_i^0(t)$ is the concentration of species i entering from the input stream of volumetric flow rate v(t).

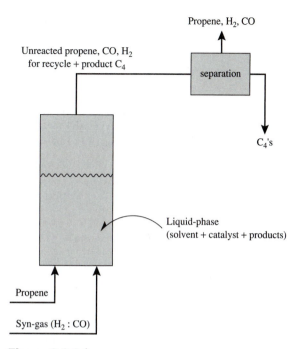

Figure 3.2.2 |
Schematic illustration of a propene hydroformylation reactor.

EXAMPLE 3.2.2 |

To a well-stirred tank containing 40 mol of triphenylmethylchloride in dry benzene (initial volume is 378 L) a stream of methanol in benzene at 0.054 mol/L is added at 3.78 L/min. A reaction proceeds as follows:

$$CH_3OH + (C_6H_5)_3CCl \Rightarrow (C_6H_5)_3COCH_3 + HCl$$
$$\quad (A) \qquad\qquad (B)$$

The reaction is essentially irreversible since pyridine is placed in the benzene to neutralize the formed HCl, and the reaction rate is:

$$r = 0.263 \, C_A^2 C_B \, (\text{mol/L/min})$$

Determine the concentration of the product ether as a function of time (problem adapted from N. H. Chen, *Process Reactor Design,* Allyn and Bacon, Inc., 1983, pp. 176–177).

■ Answer

The material balance equations for n_A and n_B are:

$$\frac{dn_A}{dt} = (0.054 \text{ mol/L})(3.78 \text{ L/min}) - 0.263 \, C_A^2 C_B V$$

$$\frac{dn_B}{dt} = -0.263 \, C_A^2 C_B V$$

The volume in the reactor is changing due to the input of methanol in benzene. Thus, the volume in the reactor at any time is:

$$V = 3.78(100 + t)$$

Therefore,

$$\frac{dn_A}{dt} = 0.204 - 0.263\, n_A^2 n_B / [3.78(100 + t)]^2$$

$$\frac{dn_B}{dt} = -0.263\, n_A^2 n_B / [3.78(100 + t)]^2$$

where

$$n_A^0 = 0 \; @ \; t = 0$$

$$n_B^0 = 40 \; @ \; t = 0$$

From Equation (1.2.6),

$$n_{\text{ether}} = n_B^0 - n_B$$

since no ether is initially present. The numerical solution to the rate equations gives $n_A(t)$ and $n_B(t)$ from which $n_{\text{ether}}(t)$ can be calculated. The results are plotted below.

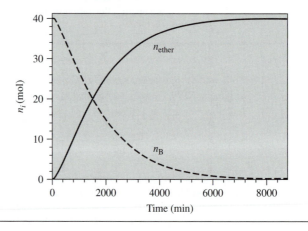

For other worked examples of semibatch reactors, see H. S. Fogler, *Elements of Chemical Reaction Engineering,* 3rd ed., Prentice-Hall, 1992, pp. 190–200, and N. H. Chen, *Process Reactor Design,* Allyn and Bacon, Inc., 1983, Chap. 6.

3.3 | Stirred-Flow Reactors

The ideal reactor for the direct measurement of reaction rates is an isothermal, constant pressure, flow reactor operating at steady-state with complete mixing such that the composition is uniform throughout the reactor. This ideal reactor is frequently called a *stirred-tank reactor,* a *continuous flow stirred-tank reactor (CSTR),* or a *mixed flow reactor (MFR).* In this type of reactor, the composition in the reactor is assumed to be that of the effluent stream and therefore all the reaction occurs at this constant composition (Figure 3.3.1).

Since the reactor is at steady-state, the difference in F_i^0 (input) and F_i (output) must be due to the reaction. (In this text, the superscript 0 on flow rates denotes the input to the reactor.) The material balance on a CSTR is written as:

$$\underset{\text{accumulation}}{0} = \underset{\substack{\text{input} \\ \text{of } i}}{F_i^0} - \underset{\substack{\text{output} \\ \text{of } i}}{F_i} + \underset{\substack{\text{amount} \\ \text{produced} \\ \text{by reaction}}}{(v_i r)V} \qquad (3.3.1)$$

(Note that V is the volume of the reacting system and V_R is the volume of the reactor; both are not necessarily equal.) Therefore, the rate can be measured directly as:

$$v_i r = \frac{F_i - F_i^0}{V} \qquad (3.3.2)$$

Equation (3.3.2) can be written for the limiting reactant to give:

$$v_l r = \frac{F_l - F_l^0}{V} \qquad (3.3.3)$$

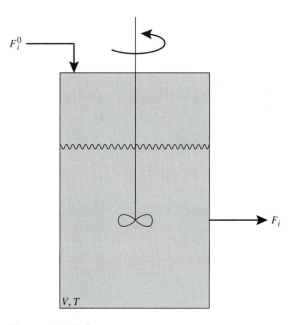

Figure 3.3.1 |
Stirred-flow reactor. The composition of the reacting volume, V, at temperature, T, is the same everywhere and at all times. F_i^0 is the molar flow rate of species i into the reactor while F_i is the molar flow rate of species i out of the reactor.

Recalling the definition of the fractional conversion:

$$f_l = \frac{n_l^0 - n_l}{n_l^0} = \frac{F_l^0 - F_l}{F_l^0} \tag{3.3.4}$$

Substitution of Equation (3.3.4) into Equation (3.3.3) yields:

$$(-\nu_l)\mathrm{r} = \left(\frac{F_l^0}{V}\right)f_l \tag{3.3.5}$$

If $\nu_l = -1$, then the reaction rate is equal to the number of moles of the limiting reactant fed to the reactor per unit time and per unit volume of the reacting fluid times the fractional conversion. For any product p not present in the feed stream, a material balance on p is easily obtained from Equation (3.3.1) with $F_i^0 = 0$ to give:

$$\nu_p\mathrm{r} = F_p/V \tag{3.3.6}$$

The quantity (F_p/V) is called the space-time yield.

The equations provided above describe the operation of stirred-flow reactors whether the reaction occurs at constant volume or not. In these types of reactors, the fluid is generally a liquid. If a large amount of solvent is used, that is, dilute solutions of reactants/products, then changes in volume can be neglected. However, if the solution is concentrated or pure reactants are used (sometimes the case for polymerization reactions), then the volume will change with the extent of reaction.

EXAMPLE 3.3.1

Write the material balance equation on comonomer A for the steady-state CSTR shown below with the two cases specified for the reaction of comonomer $A(CMA)$ and comonomer $B(CMB)$ to give a polymer (PM).

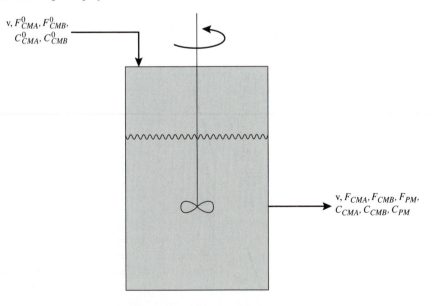

(I) $CMA + CMB \Rightarrow \text{polymer} - \text{I};\; r = k_I C_{CMA}^{\alpha_I} C_{CMB}^{\beta_I}$

(II) $2CMA + CMB \Rightarrow \text{polymer} - \text{II};\; r = k_{II} C_{CMA}^{\alpha_{II}} C_{CMB}^{\beta_{II}}$

■ Answer

The material balance equation is:

$$\text{input} = \text{output} - \text{removal by reaction} + \text{accumulation}$$

Since the reactor is at steady-state, there is no accumulation and:

$$F_{CMA}^0 = F_{CMA} - (\upsilon_{CMA} r)V$$

For case (I) the material balance is:

$$F_{CMA}^0 = F_{CMA} - (-k_I C_{CMA}^{\alpha_I} C_{CMB}^{\beta_I})V$$

while case (II) gives:

$$F_{CMA}^0 = F_{CMA} - (-2k_{II} C_{CMA}^{\alpha_{II}} C_{CMB}^{\beta_{II}})V$$

If changes in the volume due to reaction can be neglected, then the CSTR material balance can be written in terms of concentrations to give (v = volumetric flow rate; that is, volume per unit time):

$$0 = vC_l^0 - vC_l + (\upsilon_l r)V \tag{3.3.7}$$

$$(-\upsilon_l)r = \frac{C_l^0 - C_l}{(V/v)} \tag{3.3.8}$$

The ratio (V/v) is the volume of mixture in the reactor divided by the volume of mixture fed to the reactor per unit time and is called the *space time*, τ. The inverse of the space time is called the *space velocity*. In each case, the conditions for the volume of the feed must be specified: temperature, pressure (in the case of a gas), and state of aggregation (liquid or gas). Space velocity and space time should be used in preference to "contact time" or "holding time" since there is no unique residence time in the CSTR (see below). Why develop this terminology? Consider a batch reactor. The material balance on a batch reactor can be written [from Equation (3.2.1)]:

$$t = \int_{n_i^0}^{n_i^{\text{final}}} \frac{dn_i}{\upsilon_i r V} = C_i^0 \int_0^{f_i^{\text{final}}} \frac{df_i}{(-\upsilon_i)r(1+\varepsilon_i f_i)} \tag{3.3.9}$$

Equation (3.3.9) shows that the time required to reach a given fractional conversion does not depend upon the reactor volume or total amount of reagents. That is to say, for a given fractional conversion, as long as C_i^0 is the same, 1, 2, or 100 mol of i can be converted in the same time. With flow reactors, for a given C_i^0, the fractional conversion from different sized reactors is the same provided τ is the same. Table 3.3.1 compares the appropriate variables from flow and nonflow reactors.

Table 3.3.1 | Comparison of appropriate variables for flow and nonflow reactors.

Nonflow	Flow
t (time)	τ (time)
$V = V^0(1 + \varepsilon_i f_i)$ (volume)	$v = v^0(1 + \varepsilon_i f_i)$ (volume/time)
$n_i = n_i^0(1 - f_i)$ (mol)	$F_i = F_i^0(1 - f_i)$ (mol/time)

In order to show that there is not a unique residence time in a CSTR, consider the following experiment. A CSTR is at steady state and a tracer species (does not react) is flowing into and out of the reactor at a concentration C^0. At $t = 0$, the feed is changed to pure solvent at the same volumetric flow. The material balance for this situation is:

$$\frac{dn_i}{dt} = 0 - C_i v \tag{3.3.10}$$

$$(\text{accumulation}) = (\text{input}) - (\text{output})$$

or

$$-V\frac{dC_i}{dt} = C_i v \quad \text{with } C_i = C^0 \text{ at } t = 0.$$

Integration of this equation gives:

$$C = C^0 \exp[-t/\tau] \tag{3.3.11}$$

This exponential decay is typical of first-order processes as shown previously. Thus, there is an exponential distribution of residence times; some molecules will spend little time in the reactor while others will stay very long. The mean residence time is:

$$\langle t \rangle = \frac{\int_0^\infty tC(t)dt}{\int_0^\infty C(t)dt} \tag{3.3.12}$$

and can be calculated by substituting Equation (3.3.11) into Equation (3.3.12) to give:

$$\langle t \rangle = \frac{\int_0^\infty t \exp[-t/\tau]dt}{\int_0^\infty \exp[-t/\tau]dt} \tag{3.3.13}$$

Since

$$\int_0^\infty x \exp[-x]dx = 1$$

$$\langle t \rangle = \frac{\tau^2}{-\tau \exp[-t/\tau]\big|_0^\infty} = \tau \tag{3.3.14}$$

Thus, the mean residence time for a CSTR is the space time. The fact that $\langle t \rangle = \tau$ holds for reactors of any geometry and is discussed in more detail in Chapter 8.

EXAMPLE 3.3.2

The rate of the following reaction has been found to be first-order with respect to hydroxyl ions and ethyl acetate:

$$OH^- + CH_3\overset{\overset{\displaystyle O}{\displaystyle \|}}{C}OCH_2CH_3 \Longrightarrow CH_3\overset{\overset{\displaystyle O}{\displaystyle \|}}{C}O^- + CH_3CH_2OH$$

In a stirred-flow reactor of volume $V = 0.602$ L, the following data have been obtained at 298 K [Denbigh et al., *Disc. Faraday Soc.*, **2** (1977) 263]:

flow rate of barium hydroxide solution: 1.16 L/h

flow rate of ethyl acetate solution: 1.20 L/h

inlet concentration of OH^-:   0.00587 mol/L

inlet concentration of ethyl acetate: 0.0389 mol/L

outlet concentration of OH^-:   0.001094 mol/L

Calculate the rate constant. Changes in volume accompanying the reaction are negligible. (Problem taken from M. Boudart, *Kinetics of Chemical Processes,* Butterworth-Heinemann, Boston, 1991, pp. 23–24.)

■ **Answer**

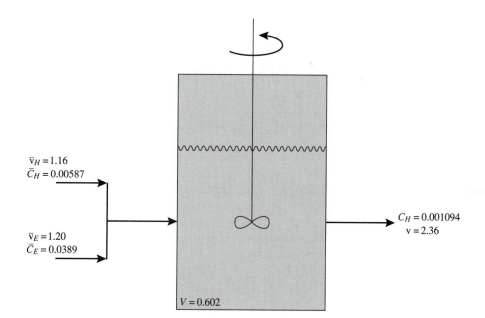

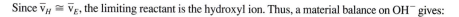

Since $\bar{v}_H \cong \bar{v}_E$, the limiting reactant is the hydroxyl ion. Thus, a material balance on OH^- gives:

$$0 = C_H^0 v - C_H v + (v_i r)V$$

where

$$v = \bar{v}_H + \bar{v}_E = 2.36 \text{ L/h}$$

$$C_H^0 = \bar{C}_H \bar{v}_H / v = 0.00289 \text{ mol/L}$$

$$(-v_H)r = k C_H C_E$$

Since the outlet value of C_H is known, the fractional conversion and C_E can be calculated as:

$$f_H = \frac{C_H^0 - C_H}{C_H^0} = \frac{0.00289 - 0.001094}{0.00289} = 0.62$$

and

$$C_E = C_E^0 - C_H^0 f_H = \frac{\bar{C}_E \bar{v}_E}{v} - C_H^0 f_H = 0.0180 \text{ mol/L}$$

Thus,

$$(-v_H)r = \frac{C_H^0 - C_H}{\tau} = \frac{0.00289 - 0.001094}{[0.602/2.36]} = 0.0070 \left(\frac{\text{mol}}{\text{L} \cdot \text{h}} \right)$$

and

$$k = \frac{(-v_H)r}{C_H C_E} = \frac{0.0070}{(0.001094)(0.0180)} = 360 \left(\frac{\text{L}}{\text{mol} \cdot \text{h}} \right)$$

3.4 | Ideal Tubular Reactors

Another type of ideal reactor is the tubular flow reactor operating isothermally at constant pressure and at steady state with a unique residence time. This type of reactor normally consists of a cylindrical pipe of constant cross-section with flow such that the fluid mixture completely fills the tube and the mixture moves as if it were a plug traveling down the length of the tube. Hence the name *plug flow reactor (PFR)*. In a PFR, the fluid properties are uniform over any cross-section normal to the direction of the flow; variations only exist along the length of the reactor. Additionally, it is assumed that *no* mixing occurs between adjacent fluid volume elements either radially (normal to flow) or axially (direction of flow). That is to say each volume element entering the reactor has the same residence time since it does not exchange mass with its neighbors. Thus, the CSTR and the PFR are the two *ideal limits of mixing* in that they are completely mixed and not mixed at all, respectively. *All* real flow reactors will lie somewhere between these two limits.

Since the fluid properties vary over the volume of the reactor, consider a material balance on a section of a steady-state isothermal PFR, *dL* (see Figure 3.4.1):

$$\underset{\text{(accumulation)}}{0} = \underset{\text{(input)}}{F_i} - \underset{\text{(output)}}{(F_i + dF_i)} + \underset{\substack{\text{(amount produced} \\ \text{by reaction)}}}{v_i r A_C \, dL} \qquad (3.4.1)$$

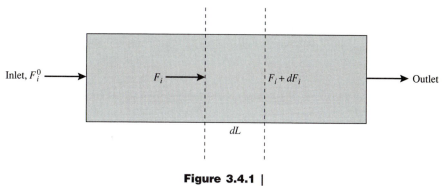

Figure 3.4.1 |
Tubular reactor.

where A_C is the cross-sectional area of the tube. Also, $A_C dL = dV_R$, so Equation (3.4.1) can be written as:

$$\frac{dF_i}{dV_R} = v_i \mathrm{r} \tag{3.4.2}$$

or

$$F_i^0 \frac{df_i}{dV_R} = -v_i \mathrm{r} \tag{3.4.3}$$

Integration of Equation (3.4.3) gives:

$$\frac{V_R}{F_i^0} = \int_{f_i^0}^{f_i^{\text{outlet}}} \frac{df_i}{(v_i \mathrm{r})} \quad \text{or} \quad \tau = \frac{V_R}{\mathrm{v}} = C_i^0 \int_{f_i^0}^{f_i^{\text{outlet}}} \frac{df_i}{(v_i \mathrm{r})} \tag{3.4.4}$$

If changes in volume due to reaction are negligible, then [$F_i = C_i \mathrm{v}$; moles of i/time = (moles of i/volume) (volume/time)]:

$$\frac{dC_i}{d\tau} = \frac{dC_i}{d(V_R/\mathrm{v})} = v_i \mathrm{r} \tag{3.4.5}$$

Note the analogy to batch reactors that have a unique residence time t and where

$$\frac{dC_i}{dt} = v_i \mathrm{r} \tag{3.4.6}$$

Clearly, the space-time, τ, in the ideal tubular reactor is the same as the residence time in the batch reactor *only* if volume changes are neglectable. This is easy to see from Equation (3.4.2) by substituting $C_i \mathrm{v}$ for F_i and recalling that for volume changes $\mathrm{v} = \mathrm{v}^0(1 + \varepsilon_i f_i)$:

$$\frac{d}{dV_R}(C_i \mathrm{v}) = \mathrm{v}\frac{dC_i}{dV_R} + C_i \frac{d\mathrm{v}}{dV_R} = v_i \mathrm{r} \tag{3.4.7}$$

Thus, if $dv/dV_R = 0$, then there is an analogy between τ and t in a PFR and batch reactor, respectively [Equations (3.4.5) and (3.4.6)], and if $dv/dV_R \neq 0$, then comparison of Equations (3.4.7) and (3.4.6) shows that there is none.

EXAMPLE 3.4.1

A PFR operating isothermally at 773 K is used to conduct the following reaction:

$$\underset{\substack{|\\ CH_3}}{CH_3\overset{\overset{O}{\|}}{C}O\overset{\ }{C}H\overset{\overset{O}{\|}}{C}OCH_3} \Longrightarrow CH_3\overset{\overset{O}{\|}}{C}OH + CH_2\!=\!CH\overset{\overset{O}{\|}}{C}OCH_3$$

methylacetoxypropionate   acetic acid   methyl acrylate

If a feed of pure methylacetoxypropionate enters at 5 atm and at a flow rate of 0.193 ft^3/s, what length of pipe with a cross-sectional area of 0.0388 ft^2 is necessary for the reaction to achieve 90 percent conversion (from C. G. Hill, *An Introduction to Chemical Engineering Kinetics & Reactor Design,* Wiley, 1977, pp. 266–267)?

Data: $k = 7.8 \times 10^9 \exp[-19{,}200/T]\ \text{s}^{-1}$

■ **Answer**

From Equation (3.4.2) and $F_A = F_A^0(1 - f_A) = C_A v$:

$$F_A^0 \frac{df_A}{dV_R} = (-v_A)\mathrm{r}$$

or

$$\tau = \frac{V_R}{v^0} = C_A^0 \int_0^{f_A = 0.9} \frac{df_A}{(-v_A)\mathrm{r}}$$

For this gas-phase reaction there is mole change with reaction and

$$\varepsilon_A = \frac{2-1}{|-1|} = 1$$

Therefore,

$$(-v_A)\mathrm{r} = kC_A = k\frac{n_A}{V} = k\frac{n_A^0(1 - f_A)}{V^0(1 + \varepsilon_A f_A)} = kC_A^0\left[\frac{1 - f_A}{1 + f_A}\right]$$

Combination of the material balance and reaction rate expressions yields:

$$\tau = \frac{1}{k}\int_0^{0.9} \frac{[1 + f_A]}{[1 - f_A]}df_A$$

Integration of this equation yields:

$$k\tau = -2\ln(1 - f_A) - f_A$$

and at 773 K, $k = 0.124 \text{ s}^{-1}$ to give at $f_A = 0.9$:

$$\tau = 29.9 \text{ s}$$

Now, if mole change with reaction is ignored (i.e., $\varepsilon_A = 0$), $\tau = 18.6$ s. Notice the large difference in the value of τ when mole change with reaction is properly accounted for in the calculations. Since the gas is expanding with increasing extent of reaction, its velocity through the tube increases. Therefore, τ must be likewise increased to allow for the specified conversion. The reactor volume and length of tube can be calculated as:

$$V_R = \tau v^0 = (29.9)(0.193) = 5.78 \text{ ft}^3$$

and

$$L = V_R/A_C = (5.78 \text{ ft}^3)/(0.0388 \text{ ft}^2) = 149 \text{ ft}$$

EXAMPLE 3.4.2

A first-order reaction occurs in an isothermal CSTR (MFR) and PFR of equal volume. If the space-time is the same in both reactors, which reactor gives the largest space-time yield and why?

■ Answer

Assume that any changes in volume due to reaction can be neglected.

Reactor	Material balance
MFR	$\tau_m = \dfrac{C_l^0 - C_l}{kC_l}$
PFR	$\tau_p = \displaystyle\int_{C_l^0}^{C_l} \dfrac{d\overline{C}_l}{k\overline{C}_l} = \dfrac{1}{k}\ln\left(\dfrac{C_l^0}{C_l}\right)$

If $\tau_m = \tau_p$ then

$$\frac{1}{k}\left[\frac{C_l^0 - C_l}{C_l}\right]_m = \frac{1}{k}\ln\left(\frac{C_l^0}{C_l}\right)_p$$

Rearranging this equation gives (C_l^p and C_l^m are C_l in the PFR and MFR, respectively):

$$\frac{C_l^0}{C_l^p} = \exp\left[\frac{C_l^0 - C_l^m}{C_l^m}\right]$$

If the approximation:

$$\exp[x] \cong 1 + x + \frac{x^2}{2!} + \cdots$$

is used then

$$\frac{C_l^0}{C_l^p} = 1 + \frac{C_l^0 - C_l^m}{C_l^m} + \tfrac{1}{2}\left(\frac{C_l^0 - C_l^m}{C_l^m}\right)^2 + \cdots$$

or

$$\frac{C_l^0}{C_l^p} = 1 + \frac{C_l^0}{C_l^m} - 1 + \tfrac{1}{2}\left(\frac{C_l^0 - C_l^m}{C_l^m}\right)^2 + \cdots$$

Thus,

$$C_l^p = C_l^m\left[\frac{1}{1 + \frac{1}{2c_l^0 C_l^m}(C_l^0 - C_l^m)^2 + \cdots}\right]$$

Since the term in the bracket will always be less than one, $C_l^p < C_l^m$, indicating that the fractional conversion and hence the space-time yield of the PFR is always higher than the CSTR. The reason for this is that for reaction-rate expressions that follow Rule I (r decreases with time or extent of reaction) the rate is maintained at a higher average value in the PFR than the CSTR. This is easy to rationalize since the concentration of the feed (highest value giving the largest rate) instantaneously drops to the outlet value (lowest value giving the lowest rate) in a CSTR [schematic (b) below] while in the PFR there is a steady progression of declining rate as the fluid element traverses the reactor [schematic (a) below]. In fact, the PFR and the CSTR are the ideal maximum and minimum space-time yield reactor configurations, respectively. This can be demonstrated by plotting the material balance equations for the PFR and CSTR as shown in the schematic (c) below. For the same conversion (i.e., the same outlet f_A), the CSTR always requires a larger reactor volume. In other words, the area in the graph is larger for the CSTR than the PFR.

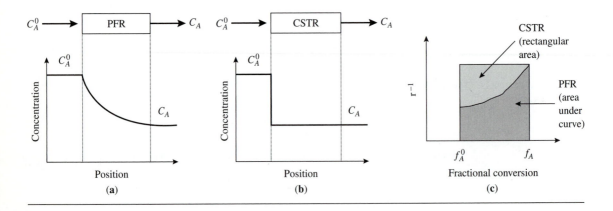

EXAMPLE 3.4.3

Show that a large number of CSTRs in series approaches the behavior of a PFR.

■ Answer

Consider the following series of CSTRs accomplishing a first-order reaction (reactors of equal size).

The material balance for reactor i is:

$$\tau^i = \tau = \frac{C^{i-1} - C^i}{kC^i} \quad \text{or} \quad \frac{C^{i-1}}{C^i} = 1 + k\tau$$

The ratio C^0/C^N can be written as:

$$\frac{C^0}{C^N} = \left(\frac{C^0}{C^1}\right)\left(\frac{C^1}{C^2}\right)\cdots\left(\frac{C^{i-1}}{C^i}\right)\cdots\left(\frac{C^{N-1}}{C^N}\right) = [1 + k\tau]^N$$

because $\tau^i = \tau$ (reactors of equal size). Also,

$$C^N \Big/ C^0 = 1 - f$$

giving:

$$C^0 \Big/ C^N = [1 + k\tau]^N = \frac{1}{1 - f}$$

If $\tau^T = N\tau$, then C^0/C^N can be written as:

$$C^0 \Big/ C^N = \left[1 + \frac{k\tau^T}{N}\right]^N = 1 + N\left(\frac{k\tau^T}{N}\right) + \frac{N(N-1)}{2!}\left(\frac{k\tau^T}{N}\right)^2 + \cdots \cong e^{k\tau^T}$$

Therefore, as N gets large the series better approximates the exponential and

$$C^0 \Big/ C^N = e^{k\tau^T}$$

or

$$\tau^T = \frac{1}{k} \ln\left(C^0 \Big/ C^N\right)$$

which is the material balance for a PFR. This result can be visualized graphically as follows. In Example 3.4.2, r^{-1} was plotted against f_A. Note that if the area under this curve is integrated as illustrated:

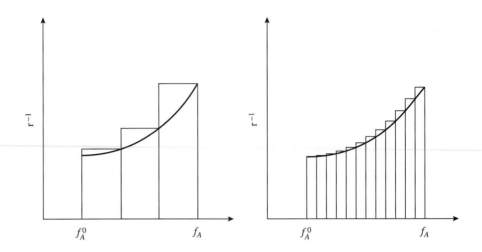

the larger number of rectangles that are used to approximate the area under the curve, the less error. Thus, if each rectangle (see Example 3.4.2) represents a CSTR, it is clear that the larger the number of CSTRs in series, the better their overall behavior simulates that of a PFR.

EXAMPLE 3.4.4

Calculate the outlet conversion for a series of CSTRs accomplishing a first-order reaction with $k\tau^T = 5$ and compare the results to that obtained from a PFR.

■ Answer

Using the equations developed in Example 3.4.3 gives with $k\tau^T = 5$ the following results:

N	$1 - f_A$
1	0.1667
5	0.0310
100	0.0076
1000	0.0068
PFR	0.0067

3.5 | Measurement of Reaction Rates

What are the types of problems that need to be addressed by measuring reaction rates? The answers to this question are very diverse. For example, in the testing of catalysts, a new catalyst may be evaluated for replacement of another catalyst in an existing process or for the development of a new process. Accurate, reliable laboratory reaction rate data are necessary for the design of an industrial reactor

Table 3.5.1 | Rate of reactions in ideal isothermal reactors.

Ideal reactor	General equation		Constant density equation	
Batch	$\dfrac{1}{V}\dfrac{dn_l}{dt} = v_l\,\mathrm{r}$	Equation (3.2.1)	$\dfrac{dC_l}{dt} = v_l\,\mathrm{r}$	
Stirred-flow	$\dfrac{F_l^0}{V}\,f_l = (-v_l)\mathrm{r}$	Equation (3.3.5)	$\dfrac{C_l^0 - C_l}{(V/\mathrm{v})} = (-v_l)\mathrm{r}$	Equation (3.3.8)
Tubular	$F_l^0\,\dfrac{df_l}{dV_R} = (-v_l)\mathrm{r}$	Equation (3.4.3)	$\dfrac{dC_l}{d(V_R/\mathrm{v})} = v_l\mathrm{r}$	Equation (3.4.5)

Nomenclature

C_l Molar concentration of limiting reactant, mol/volume

C_l^0 Initial value of C_l

f_l Fractional conversion of limiting reactant, dimensionless

n_l Number of moles of limiting reactant, mol

F_l^0 Initial value of the molar flow rate of limiting reactant, mol/time

r Reaction rate, mol/volume/time

v_l Stoichiometric coefficient of limiting reactant, dimensionless

V Volume of reacting system, volume

V_R Volume of reactor, volume

v Volumetric flow rate, volume/time

whether the process is new or old. Another example of why reaction rate data are needed is to make predictions about how large-scale systems behave (e.g., the appearance of ozone holes and the formation of smog). The key issue in all of these circumstances is the acquisition of high-quality reaction rate data. In order to do this, a laboratory-scale reactor must be used. Although deviations from ideal behavior still exist in laboratory reactors, deliberate efforts can be made to approximate ideal conditions as closely as possible. Table 3.5.1 summarizes the material balance equations for the ideal reactors described above. Examples of how these types of reactors are used to measure reaction rates are presented below.

When choosing a laboratory reactor for the measurement of reaction rate data, numerous issues must be resolved. The choice of the reactor is based on the characteristics of the reaction and for all practical matters by the availability of resources (i.e., reactors, analytical equipment, money, etc.). A good example of the issues involved in selecting a laboratory reactor and how they influence the ultimate choice is presented by Weekman [*AIChE J.*, **20** (1974) 833]. Methods for obtaining reaction rate data from laboratory reactors that approximate the ideal reactors listed in Table 3.5.1 are now discussed.

3.5.1 Batch Reactors

A batch reactor by its nature is a transient closed system. While a laboratory batch reactor can be a simple well-stirred flask in a constant temperature bath or a commercial laboratory-scale batch reactor, the direct measurement of reaction rates is not possible from these reactors. The observables are the concentrations of species from which the rate can be inferred. For example, in a typical batch experiment, the concentrations of reactants and products are measured as a function of time. From these data, initial reaction rates (rates at the zero conversion limit) can be obtained by calculating the initial slope (Figure 3.5.1b). Also, the complete data set can be numerically fit to a curve and the tangent to the curve calculated for any time (Figure 3.5.1a). The set of tangents can then be plotted versus the concentration at which the tangent was obtained (Figure 3.5.1c).

If, for example, the reaction rate function is first-order, then a plot of the tangents (dC/dt) versus concentration should be linear with a slope equal to the reaction rate constant and an intercept of zero. It is clear that the accuracy of the

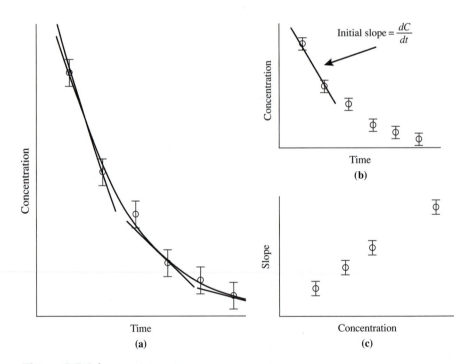

Figure 3.5.1 |
(**a**) Plot of concentration data versus time, a curve for a numerical fit of the data and lines at tangents to the curve at various times, (**b**) plot of the initial slope, (**c**) plot of the slopes of the tangent lines in (**a**) versus concentrations.

data (size of the error bars) is crucial to this method of determining good reaction rates. The accuracy will normally be fixed by the analytical technique used. Additionally, the greater the number of data points, the better the calculation of the rate. A typical way to measure concentrations is to sample the batch reactor and use chromatography for separation and determination of the amount of each component. In the best cases, this type of procedure has a time-scale of minutes. If the reaction is sufficiently slow, then this methodology can be used. Note, however, that only one datum point is obtained at each extent of reaction (i.e., at each time). If the reaction is fast relative to the time scale for sampling, then often it is not possible to follow the course of the reaction in a batch reactor.

EXAMPLE 3.5.1

P. Butler (Honors Thesis, Virginia Polytechnic Institute and State University, Blacksburg, VA, 1984) investigated the kinetics of the following reaction using rhodium catalysts:

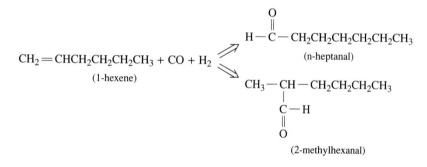

This homogeneous hydroformylation reaction was conducted in a batch reactor, and because of the nature of the catalyst, isomerization reactions of 1-hexene to 2- and 3-hexenes and hydrogenation reactions of hexenes to hexanes and aldehydes to alcohols were minimized. The following data were obtained at 323 K with an initial concentration of 1-hexene at 1 mol/L in toluene and $P_{CO} = P_{H_2} = P_{N_2}$ (inert) $= 0.33$ atm. Calculate the initial rates of formation of the linear, r_N, and branched, r_B, aldehydes from these data.

t(h)	n-heptanal (mol/L)	2-methylhexanal (mol/L)
0.17	0.0067	0.0000
0.67	0.0266	0.0058
1.08	0.0461	0.0109
1.90	0.1075	0.0184
2.58	0.1244	0.0279

■ **Answer**

Plot the concentration data as follows:

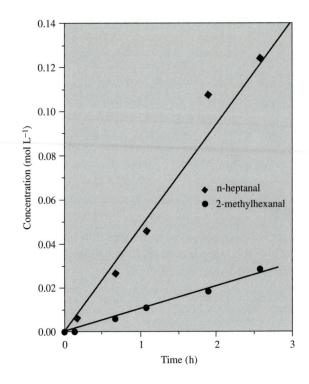

The slopes of the lines shown in the plot give $r_N = 0.0515$ mol/L/h and $r_B = 0.0109$ mol/L/h (from linear regression).

EXAMPLE 3.5.2

Butler obtained the initial rate data given in the following table in a manner analogous to that illustrated in Example 3.5.1. Show that a reaction rate expression that follows Rules III and IV can be used to describe these data.

P_{CO} (atm)	P_{H_2} (atm)	$C_{1 - \text{hexene}}$ (mol/L)	T(K)	r_N	r_B
0.50	0.50	1.00	323	0.0280	0.0074
0.33	0.33	1.00	323	0.0430	0.0115
0.66	0.33	1.00	323	0.0154	0.0040
0.33	0.33	1.00	313	0.0156	0.0040
0.33	0.33	1.00	303	0.0044	0.0016
0.33	0.33	0.45	323	0.0312	0.0069
0.33	0.33	1.00	323	0.0410	0.0100

■ Answer

An empirical expression of the rate takes the form:

$$r = k \exp[-E/(R_g T)]P_{CO}^{\alpha_1} P_{H_2}^{\alpha_2} C_{1-\text{hexene}}^{\alpha_3}$$

and was used to correlate the data to give (R_g in cal):

$$r_N = 2.0 \times 10^{13} \exp\left[-22{,}200/(R_g T)\right] P_{CO}^{-1.5} P_{H_2}^{0.45} C_{1-\text{hexene}}^{0.40}$$

$$r_B = 4.9 \times 10^{10} \exp\left[-19{,}200/(R_g T)\right] P_{CO}^{-1.5} P_{H_2}^{0.45} C_{1-\text{hexene}}^{0.64}$$

Ideally, much more data are required in order to obtain a higher degree of confidence in the reaction rate expressions. However, it is clear from Examples 3.5.1 and 3.5.2 how much experimental work is required to do so. Also, note that these rates are *initial* rates and cannot be used for integral conversions.

3.5.2 Flow Reactors

As pointed out previously, the use of flow reactors allows for the direct measurement of reaction rates. At steady state (unlike the batch reactor), the time scales of the analytical technique used and the reaction are decoupled. Additionally, since numerous samples can be acquired at the same conditions, the accuracy of the data dramatically increases.

Consider the following problem. In the petrochemical industry, many reactions are oxidations and hydrogenations that are very exothermic. Thus, to control the temperature in an industrial reactor the configuration is typically a bundle of tubes (between 1 and 2 inches in diameter and thousands in number) that are bathed in a heat exchange fluid. The high heat exchange surface area per reactor volume allows the large heat release to be effectively removed. Suppose that a new catalyst is to be prepared for ultimate use in a reactor of this type to conduct a gas-phase reaction. How are appropriate reaction rate data obtained for this situation?

Consider first the tubular reactor. From the material balance (Table 3.5.1), it is clear that in order to solve the mass balance the functional form of the rate expression must be provided because the reactor outlet is the *integral* result of reaction over the volume of the reactor. However, if only initial reaction rate data were required, then a tubular reactor could be used by noticing that if the differentials are replaced by deltas, then:

$$F_l^0 \frac{\Delta f_l}{\Delta V_R} = (-v_l)\mathrm{r}\bigg|_{C_l^0} \tag{3.5.1}$$

Thus, a small tubular reactor that gives differential conversion (i.e., typically below 5 percent) can yield a point value for the reaction rate. In this case, the reaction rate is evaluated at C_l^0. Actually, the rate could be better calculated with the arithmetic mean of the inlet and outlet concentrations:

$$C_l = \frac{C_l^0 + C_l^{\text{exit}}}{2}$$

However, since $C_l^0 \cong C_l^{\text{exit}}$ the inlet concentration is often used.

EXAMPLE 3.5.3

For the generic reaction $A \Rightarrow B$ the following three reaction rate expressions were proposed to correlate the initial rate data obtained. Describe how a differential tubular reactor could be used to discriminate among these models:

$$r_1 = \frac{k_1 C_A}{1 + k_2 C_B}$$

$$r_2 = \frac{k_3 C_A}{1 + k_4 C_B + k_5 C_A}$$

$$r_3 = \frac{k_6 C_A}{1 + k_7 C_A}$$

■ **Answer**

If initial rate data are obtained, and if there is no B in the feed stream, then the concentration of B at low conversion is small. Thus, at these conditions the rate expressions are:

$$r_1 = k_1 C_A$$

$$r_2 = \frac{k_3 C_A}{1 + k_5 C_A}$$

$$r_3 = \frac{k_6 C_A}{1 + k_7 C_A}$$

Clearly r_1 can be distinguished from r_2 and r_3 by varying C_A such that a plot of r versus C_A can be obtained. Now suppose that r_1 does not describe the data. In a second set of experiments, B can be added to the feed in varying amounts. If r_3 is the correct rate expression, then the measured rates will not change as C_B is varied. If there is a dependence of the observed rate on the concentration of feed B, then r_3 cannot describe the data.

Returning to the problem of obtaining reaction rate data from a new catalyst for a gas-phase reaction, if reaction rates are desired over the complete range of the extent of the reaction, the differential fixed bed is not an appropriate laboratory reactor for this purpose. However, the ideal stirred-flow reactor can accomplish this objective. By varying τ, r can be directly obtained (see Table 3.5.1) at any extent of reaction. The problem with a gas phase reaction is the mixing. If the reaction occurs in the liquid phase, thorough mixing can be achieved with high agitation in many cases. With gases the situation is more difficult. To overcome this, several reactor types have been developed and are commercially available on laboratory scale.

Referring to Figure 3.5.2, the Carberry reactor contains paddles in which the catalyst is mounted and the paddles are rapidly rotated via connection to a control shaft in order to obtain good mixing between the gas phase and the catalyst. A Berty reactor consists of a stationary bed of catalyst that is contacted via circulation of the gas phase by impeller blades. The quality of mixing in this type of configuration

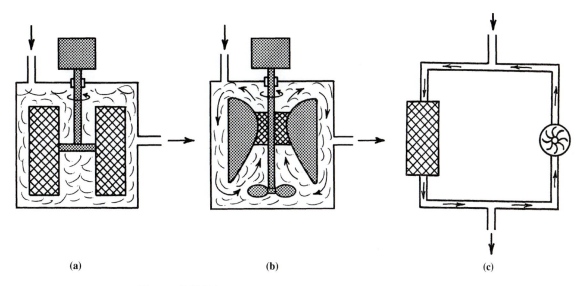

(a)   (b)   (c)

Figure 3.5.2 |
Stirred contained solids reactors. [Reproduced from V. W. Weekman, Jr., *AIChE J.,*
20 (1974) p. 835, with permission of the American Institute of Chemical Engineers.
Copyright © 1974 AIChE. All rights reserved.] **(a)** Carberry reactor, **(b)** Berty reactor
(internal recycle reactor), **(c)** external recycle reactor.

depends on the density of the gas. For low densities (i.e., low pressures), the mixing is poor. Thus, Berty-type internal recycle reactors are most frequently used for reactions well above atmospheric pressure. For low-pressure gas circulation, external recycle can be employed via the use of a pump. At high recycle, these reactors approximate the behavior of a CSTR. This statement is proven below. Thus, these types of laboratory reactors have become the workhorses of the petrochemical industry for measuring accurate reaction rate data.

Consider a generic recycle reactor schematically illustrated in Figure 3.5.3. First, denote R as the recycle ratio. The recycle ratio is defined as the volume of fluid returning to the reaction chamber entrance divided by the volume of fluid leaving the system. Simple material balances around the mixing point prior to the entrance of the reaction volume give:

$$v^i = v^0 + v^r \tag{3.5.2}$$

and

$$C_A^i \, v^i = C_A^i (v^0 + v^r) = C_A^0 v^0 + C_A^e \, v^r \tag{3.5.3}$$

If the density of the fluid is constant then $v^e = v^0$ and $v^r = Rv^0$. Using these relationships with Equation (3.5.3) gives:

$$C_A^i (v^0 + Rv^0) = C_A^0 v^0 + C_A^e Rv^0 \tag{3.5.4}$$

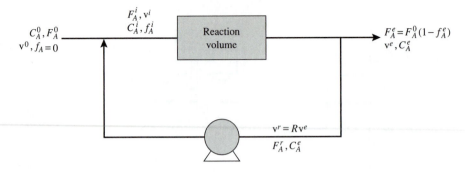

Figure 3.5.3 |
Schematic diagram of a general recycle reactor. Superscripts i, e, and r refer to inlet, exit, and recycle. R is the recycle ratio.

or

$$C_A^i = \frac{C_A^0}{1+R} + \frac{RC_A^e}{1+R} \qquad (3.5.5)$$

Notice that if $R \to \infty$, $C_A^i \to C_A^e$ or the result obtained from an ideal CSTR. Also, if $R \to 0$, $C_A^i \to C_A^0$, the inlet to the reaction volume. Thus, by fixing the value of R, the recycle reactor can behave like the two ideal limiting reactors (i.e., the CSTR and PFR, or anywhere between these limits). To see this further, a complete material balance on the reactor can be derived from Equation (3.4.2) and is:

$$\frac{dF_A}{dV_R} = v_A \mathrm{r} \qquad (3.5.6)$$

However, since

$$F_A = vC_A$$

$$C_A = C_A^0 \left(\frac{1 - f_A}{1 + \varepsilon_A f_A} \right)$$

$$v = v^e + v^r = (R + 1)v^e = (R + 1)v^0(1 + \varepsilon_A f_A)$$

then

$$F_A = v^0 C_A^0 (R + 1)(1 - f_A) = F_A^0 (R + 1)(1 - f_A)$$
$$-dF_A = F_A^0 (R + 1)df_A \qquad (3.5.7)$$

Substitution of Equation (3.5.7) into Equation (3.5.6) gives:

$$\frac{V_R}{F_A^0} = (R + 1) \int_{f_A^i}^{f_A^e} \frac{df_A}{(-v_A)\mathrm{r}} \qquad (3.5.8)$$

Now, f_A^i must be related to inlet and/or outlet variables for ease of evaluation from readily measurable parameters. To do so, notice that

$$C_A^i = \frac{F_A^i}{v^i} = \frac{F_A^0 + F_A^r}{v^0 + v^r} = \frac{F_A^0 + v^r C_A^e}{v^0 + v^r} = \frac{F_A^0 + Rv^e C_A^e}{v^0 + Rv^e}$$

In terms of f_A, C_A^i is then

$$C_A^i = \frac{F_A^0 + RF_A^0(1 - f_A^e)}{v^0 + Rv^0(1 + \varepsilon_A f_A^e)} = \frac{F_A^0}{v^0}\left[\frac{1 + R(1 - f_A^e)}{1 + R(1 + \varepsilon_A f_A^e)}\right]$$

Thus,

$$\frac{C_A^i}{C_A^0} = \frac{1 + R - Rf_A^e}{1 + R + R\varepsilon_A f_A^e} = \frac{1 - f_A^i}{1 + \varepsilon_A f_A^i}$$

Solving this equation for f_A^i gives:

$$f_A^i = \frac{R\varepsilon_A f_A^e + Rf_A^e}{1 + R + \varepsilon_A + \varepsilon_A R} = \frac{Rf_A^e}{1 + R} \tag{3.5.9}$$

Substitution of Equation (3.5.9) into Equation (3.5.8) yields:

$$\frac{V_R}{F_A^0} = (R + 1)\int_{\frac{Rf_A^e}{1+R}}^{f_A^e} \frac{df_A}{(-v_A)\mathrm{r}} \tag{3.5.10}$$

Clearly, if $R \to 0$, then Equation (3.5.10) reduces to the material balance for a PFR. However, it is not straightforward to recognize that Equation (3.5.10) reduces to the material balance for a CSTR as $R \to \infty$. To do so, notice that the bottom limit on the integral goes to f_A^e as $R \to \infty$. To obtain the value for the integral as $R \to \infty$, Leibnitz's Rule must be used, and it is:

$$\frac{d}{d\alpha}\int_{\varphi_1(\alpha)}^{\varphi_2(\alpha)} \overline{H}(x, \alpha)dx = \int_{\varphi_1(\alpha)}^{\varphi_2(\alpha)} \frac{\partial \overline{H}}{\partial \alpha}dx + \overline{H}(\varphi_2, \alpha)\frac{d\varphi_2}{d\alpha} - \overline{H}(\varphi_1, \alpha)\frac{d\varphi_1}{d\alpha}$$

Taking the limit of Equation (3.5.10) as $R \to \infty$ gives (L'Hopital's Rule):

$$\lim_{R\to\infty}\left(\frac{V_R}{F_A^0}\right) = \lim_{R\to\infty}\frac{\int_{\frac{Rf_A^e}{1+R}}^{f_A^e} \frac{df_A}{(-v_A)\mathrm{r}}}{\frac{1}{R + 1}}$$

To evaluate the numerator, use Leibnitz's Rule:

$$\frac{d}{dR}\int_{\frac{Rf_A^e}{1+R}}^{f_A^e} \frac{df_A}{(-v_A)\mathrm{r}} = \int_{\frac{Rf_A^e}{1+R}}^{f_A^e} \frac{\partial}{\partial R}\left[\frac{df_A}{(-v_A)\mathrm{r}}\right] + \frac{1}{(-v_A)\mathrm{r}}\bigg|_{f_A^e}\frac{df_A^e}{dR} - \frac{1}{(-v_A)\mathrm{r}}\bigg|_{\frac{Rf_A^e}{1+R}}\frac{d}{dR}\left(\frac{Rf_A^e}{1 + R}\right)$$

The first and second terms on the right-hand side of this equation are zero. Therefore,

$$\lim_{R \to \infty} \frac{V_R}{F_A^0} = \frac{-\frac{1}{(-v_A)r}\left|\frac{Rf_A^e}{1+R}\left[\frac{f_A^e}{(1+R)^2}\right]\right.}{-\frac{1}{(1+R)^2}} = \frac{f_A^e}{(-v_A)r\left|_{f_A^e}\right.}$$

Note that this equation *is* the material balance for a CSTR (see Table 3.5.1). Thus, when using any recycle reactor for the measurement of reaction rate data, the effect of stirring speed (that fixes recirculation rates) on extent of reaction must be investigated. If the outlet conditions do not vary with recirculation rates, then the recycle reactor can be evaluated as if it were a CSTR.

EXAMPLE 3.5.4

Al-Saleh et al. [*Chem. Eng. J.*, **37** (1988) 35] performed a kinetic study of ethylene oxidation over a silver supported on alumina catalyst in a Berty reactor. At temperatures between 513–553 K and a pressure of 21.5 atm, the observed reaction rates (calculated using the CSTR material balance) were independent of the impeller rotation speed in the range 350–1000 rpm (revolutions per minute). A summary of the data is:

Temperature (K)	v^0 [L(h)$^{-1}$]	f_{total}	Reaction rate $\times 10^4$ (mol gcat^{-1} h^{-1})	
			r_{EO}	r_{CO_2}
553	51.0	0.340	3.145	2.229
553	106.0	0.272	5.093	2.676
553	275.0	0.205	9.336	3.564
533	9.3	0.349	0.602	0.692
533	51.0	0.251	2.661	1.379
533	106.0	0.218	4.590	1.582
533	275.0	0.162	8.032	2.215
513	9.3	0.287	0.644	0.505
513	51.0	0.172	1.980	0.763
513	106.0	0.146	3.262	0.902
513	275.0	0.074	3.664	0.989

Derive the equations necessary to determine r_{EO} (rate of production of ethylene oxide) and r_{CO_2} (rate of production of CO$_2$). Assume for this example that the volumetric flow rate is unaffected by the reactions.

■ **Answer**

From Equation (3.3.6):

$$r_p = v_p r = F_p/V$$

Since the reaction rates are reported on a per mass of catalyst basis rather than per volume, V is replaced with W (the mass of catalyst in the reactor). The reaction network is:

$$CH_2{=}CH_2 + \tfrac{1}{2}O_2 \Longrightarrow \overset{O}{\overset{\displaystyle\triangle}{CH_2{-}CH_2}}$$

$$\overset{O}{\overset{\displaystyle\triangle}{CH_2{-}CH_2}} + \tfrac{5}{2}O_2 \Longrightarrow 2CO_2 + 2H_2O$$

$$CH_2{=}CH_2 + 3O_2 \Longrightarrow 2CO_2 + 2H_2O$$

For the molar flow rates,

$$F_{EO} = F_E^0 \left[\frac{C_{EO}^{\text{outlet}}}{C_E^0} \right]$$

$$F_{CO_2} = F_E^0 \left[\frac{C_{CO_2}^{\text{outlet}}}{C_E^0} \right]$$

where F_E^0 is the inlet molar flow rate of ethylene and C_E^0 is the inlet concentration of ethylene. Thus, the material balance equations are:

$$r_{EO} = \frac{F_E^0}{W} \left[\frac{C_{EO}^{\text{outlet}}}{C_E^0} \right]$$

$$r_{CO_2} = \frac{F_E^0}{W} \left[\frac{C_{CO_2}^{\text{outlet}}}{C_E^0} \right]$$

Notice that F_E^0, W, and C_E^0 are all fixed for a particular experiment. Thus, measurement of the concentrations in the outlet stream directly yield values for the observed rates.

VIGNETTE 3.5.1

Ethylene oxide (EO) is one of the largest volume chemicals produced from ethylene. The reason for its importance lies in its ability for ethoxylation of other molecules. Reactions of EO are based on the ring opening by a nucleophilic molecule such as water, alcohols, ammonia, amines, carboxylic acids, phenols, or mercaptans. For example, ring-opening reactions of EO with water and ammonia give ethylene glycol and ethanolamines, respectively:

$$\overset{O}{\overset{\displaystyle\triangle}{CH_2{-}CH_2}} + H_2O = HOCH_2CH_2OH$$

$$\overset{O}{\overset{\displaystyle\triangle}{CH_2{-}CH_2}} + NH_3 = NH_2CH_2CH_2OH$$

These products are manufactured at very large scale; for example, ethylene glycol production is in the millions of tons per year in the United States alone and is used as a heat transfer fluid such as antifreeze in the radiators of automobiles. With such large-scale use of EO, it is important that the selectivity to EO be as high as possible. Today, after years of catalyst optimization and advances in reactor technology, commercial Ag/Al$_2$O$_3$ catalysts give selectivities above 90 percent.

EXAMPLE 3.5.5

Using the data in Example 3.5.4 calculate the selectivity defined as the ratio of the moles of EO produced per mole of ethylene consumed times 100 percent, and plot the selectivity versus conversion.

Answer

The selectivity can be calculated as:

$$s_{EO} = \left[\frac{r_{EO}}{r_{EO} + r_{CO_2}}\right] \times 100\%$$

From the plot shown below, the selectivity declines as the conversion is increased because of combustion reactions that produce carbon dioxide.

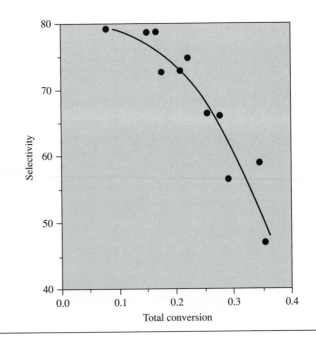

In addition to the laboratory-scale reactors described here, there are numerous more specialized reactors in use. However, as mentioned previously, the performance of these reactors must lie somewhere between the mixing limits of the PFR and the CSTR. Additionally, when using small laboratory reactors, it is often difficult to maintain ideal mixing conditions, and the state of mixing should always be verified (see Chapter 8 for more details) prior to use. A common problem is that flow rates sufficiently large to achieve PFR behavior cannot be obtained in a small laboratory system, and the flow is laminar rather than turbulent (necessary for PFR behavior). If such is the case, the velocity profile across the reactor diameter is parabolic rather than constant.

Exercises for Chapter 3

1. The space time necessary to achieve 70 percent conversion in a CSTR is 3 h. Determine the reactor volume required to process 4 ft^3 min^{-1}. What is the space velocity for this system?

2. The following parallel reactions take place in a CSTR:

$$A + B \xrightarrow{k_1} \text{Desired Product} \qquad k_1 = 2.0 \text{ L (mol min)}^{-1}$$

$$B \xrightarrow{k_2} \text{Undesired Product} \quad k_2 = 1.0 \text{ min}^{-1}$$

If a liquid stream of A (4 mol L^{-1}, 50 L min^{-1}) and a liquid stream of B (2 mol L^{-1}, 50 L min^{-1}) are co-fed to a 100 L reactor, what are the steady-state effluent concentrations of A and B?

3. The irreversible reaction $2A \rightarrow B$ takes place in the gas phase in a constant temperature plug flow reactor. Reactant A and diluent gas are fed in equimolar ratio, and the conversion of A is 85 percent. If the molar feed rate of A is doubled, what is the conversion of A assuming the feed rate of diluent is unchanged?

4. Consider the reversible first-order reaction of $A = B$ in a CSTR of volume $V = 2$ L with forward and reverse rate constants of $k_1 = 2.0$ min^{-1} and $k_{-1} = 1.0$ min^{-1}. At time $t = 0$, the concentrations of A and B in the tank are both zero. The incoming stream of A has a volumetric flow rate of 3 L min^{-1} at concentration $C_A^0 = 2$ mol L^{-1}. Find the concentrations of A and B as functions of time. You not need a computer to solve this problem.

5. Consider the liquid phase reaction: $A \Rightarrow$ products with rate r $= 0.20$ C_A^2 (mol L^{-1} min^{-1}) that takes place in a PFR of volume 30 L.

   (a) What is the concentration of A (C_A^e) exiting the PFR?

(b) What is C_A^e in a PFR with recycle shown below?

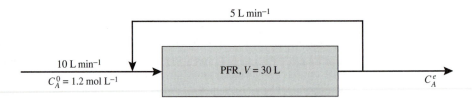

(c) Now, add a separator to the system. Find C_A^e, C_A^R, C_A^i and C_A^f. For the separator, assume $C_A^R = 5C_A^e$.

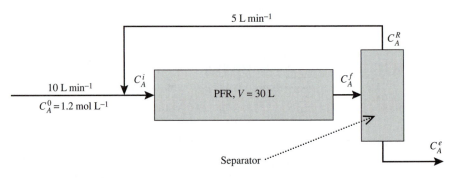

(Problem provided by Prof. J. L. Hudson, Univ. of Virginia.)

6. (Adapted from H. S. Fogler, *Elements of Chemical Reaction Engineering,* 3rd ed., Prentice Hall, Upper Saddle River, NJ, 1999.) Pure butanol is fed into a semibatch reactor containing pure ethyl acetate to produce butyl acetate and ethanol in the reversible reaction:

$$CH_3COOC_2H_5 + C_4H_9OH = CH_3COOC_4H_9 + C_2H_5OH$$

The reaction rate can be expressed in the Guldberg-Waage form. The reaction is carried out isothermally at 300 K. At this temperature, the equilibrium constant is 1.08 and the forward rate constant is 9.0×10^{-5} L (mol s)$^{-1}$. Initially, there are 200 L of ethyl acetate in the reactor and butanol is fed at a rate of 0.050 L s^{-1}. The feed and initial concentrations of butanol and ethyl acetate are 10.93 mol L^{-1} and 7.72 mol L^{-1}, respectively.

(a) Plot the equilibrium conversion of ethyl acetate as a function of time.

(b) Plot the conversion of ethylacetate, the rate of reaction, and the concentration of butanol as a function of time.

7. Dinitrogen pentoxide decomposes at 35°C with a first-order rate constant of 8×10^{-3} min^{-1} [F. Daniels and E. H. Johnston, *J. Am. Chem. Soc.,* **43** (1921) 53] according to:

$$2 \, N_2O_5 \Longrightarrow 4 \, NO_2 + O_2$$

However, the product NO_2 rapidly dimerizes to N_2O_4:

$$2NO_2 = N_2O_4$$

If the decomposition reaction is carried out in a constant volume at 35°C, plot the pressure rise in the reactor as a function of time, for an initial charge of pure N_2O_5 at 0.4 atm. Assume that the dimerization reaction equilibrates immediately. The equilibrium constant of the NO_2 dimerization reaction at 35°C is 3.68. Assume ideal behavior.

8. Ethanol can decompose in a parallel pathway:

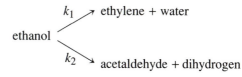

Assume that the reaction rates are both first-order in ethanol and that no products are initially present. After 100 s in a constant volume system, there is 30 percent of the ethanol remaining and the mixture contains 13.7 percent ethylene and 27.4 percent acetaldehyde. Calculate the rate constants k_1 and k_2.

9. Compound A is converted to B in a CSTR. The reaction rate is first-order with a reaction rate constant of 20 min.$^{-1}$ Compound A enters the reactor at a flow rate of 12 m^3/min (concentration of 2.0 kmol/m^3). The value of the product B is $1.50 per kmol and the cost of reactor operation is $2.50 per minute per cubic meter. It is not economical to separate unconverted A to recycle it back to the feed. Find the maximum profit.

10. A very simplified model for the dynamical processes occurring when an animal is given a drug can be represented by the single compartmental model shown below:

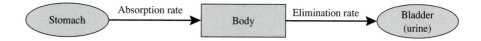

Upon administration of the drug, there is absorption into the body. Subsequently, the drug is converted to metabolites and/or is physically eliminated. As a result, the amount of drug in the body at any time is the net transient response to these input and output processes.

Find the maximum "body" alcohol level in grams and the time when it occurs for a 70 kg human who quickly consumes one can of beer. Assume the absorption and elimination rates are first-order.

Data: The mass of one can of beer is 400 g and contains 5 wt. % alcohol.

$$k_{absorption}/k_{elimination} = 5$$

$$k_{elimination} = 0.008 \text{ min}^{-1}$$

11. Titanium dioxide particles are used to brighten paints. They are produced by gas-phase oxidation of $TiCl_4$ vapor in a hydrocarbon flame. The dominant reaction is hydrolysis,

$$TiCl_4 + 2H_2O \Rightarrow TiO_2(s) + 4HCl$$

The reaction rate is first-order in $TiCl_4$ and zero-order in H_2O. The rate constant for the reaction is:

$$k = 8.0 \times 10^4 \exp\left[-\frac{88000 \text{ J/mol}}{R_g T}\right] s^{-1}$$

The reaction takes place at 1200 K in a constant pressure flow reactor at 1 atm pressure (1.01×10^5 Pa). The gas composition at the entrance to the reactor is:

CO_2	8%
H_2O	8%
O_2	5%
$TiCl_4$	3%
N_2	remainder

(a) What space time is required to achieve 99 percent conversion of the $TiCl_4$ to TiO_2?

(b) The reactor is 0.2 m diameter and 1.5 m long. Assuming that the reactor operates 80 percent of the time, how many kilograms of TiO_2 can be produced per year? (The molecular weight of TiO_2 is 80 g/mol.)

$$R_g = 8.3144 \text{ J/mol/K}$$

(Problem provided by Richard Flagan, Caltech.)

12. The autocatalytic reaction of A to form Q is one that accelerates with conversion. An example of this is shown below:

$$A + Q \xrightarrow{k_1} Q + Q$$

However, the rate decreases at high conversion due to the depletion of reactant A. The liquid feed to the reactor contains 1 mol L^{-1} of A and 0.1 mol L^{-1} of Q.

(a) To reach 50 percent conversion of A in the smallest reactor volume, would you use a PFR or a CSTR? Support your answer with appropriate calculations.

(b) To reach 95 percent conversion of A in the smallest reactor volume, would you use a PFR or a CSTR? Support your answer with appropriate calculations.

(c) What is the space time needed to convert 95 percent of A in a CSTR if $k_1 = 1$ L (mol s)$^{-1}$?

13. The irreversible, first-order, gas-phase reaction

$$A \Rightarrow 2B + C$$

takes place in a constant volume batch reactor that has a safety disk designed to rupture when the pressure exceeds 1000 psi. If the rate constant is $0.01\ s^{-1}$, how long will it take to rupture the safety disk if pure A is charged into the reactor at 500 psi?

14. If you have a CSTR and a PFR (both of the same volume) available to carry out an irreversible, first-order, liquid-phase reaction, how would you connect them in series (in what order) to maximize the conversion?

15. Find the minimum number of CSTRs connected in series to give an outlet conversion within 5 percent of that achieved in a PFR of equal total volume for:

   (a) first-order irreversible reaction of A to form B, $k\tau_{PFR} = 1$

   (b) second-order irreversible reaction of A to form B, $kC_A^0\tau_{PFR} = 1$.

16. Davis studied the hydrogenation of ethylene to ethane in a catalytic recycle reactor operated at atmospheric pressure (R. J. Davis, Ph.D. Thesis, Stanford University, 1989.) The recycle ratio was large enough so that the reactor approached CSTR behavior. Helium was used as a diluent to adjust the partial pressures of the gases. From the data presented, estimate the orders of the reaction rate with respect to ethylene and dihydrogen and the activation energy of the reaction.

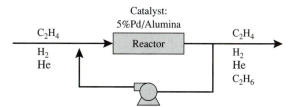

Hydrogenation of ethylene over 50 mg of Pd/alumina catalyst.

Reaction temperature (K)	Inlet flow C_2H_4 (mL min^{-1} STP)	Inlet flow H_2 (mL min^{-1} STP)	Inlet flow He (mL min^{-1} STP)	Conversion of C_2H_4 (%)
193	1.0	20	80	25.1
193	1.0	10	90	16.2
193	1.0	40	60	35.4
193	2.5	20	78.5	8.55
193	5.0	20	76	4.17
175	1.0	20	80	3.14

CHAPTER 4

The Steady-State Approximation: Catalysis

4.1 | Single Reactions

One-step reactions between stable molecules are rare since a stable molecule is by definition a quite unreactive entity. Rather, complicated rearrangements of chemical bonds are usually required to go from reactants to products. This implies that most reactions do not proceed in a single elementary step as illustrated below for NO formation from N_2 and O_2:

One-step (not observed)	Sequence of steps (observed)
$N_2 + O_2 \Longrightarrow 2NO$	$N_2 + O \rightleftharpoons NO + N$
$\begin{matrix} N \cdots O \\ ||| \quad || \Longrightarrow \\ N \cdots O \end{matrix} \begin{matrix} NO \\ + \\ NO \end{matrix}$	$N + O_2 \rightleftharpoons NO + O$
	$N_2 + O_2 == 2NO$

Normally, a sequence of elementary steps is necessary to proceed from reactants to products through the formation and destruction of reactive intermediates (see Section 1.1).

Reactive intermediates may be of numerous different chemical types (e.g., free radicals, free ions, solvated ions, complexes at solid surfaces, complexes in a homogeneous phase, complexes in enzymes). Although many reactive intermediates may be involved in a given reaction (see Scheme 1.1.1), the advancement of the reaction can still be described by a single parameter—the extent of reaction (see Section 1.2). If this is the case, the reaction is said to be single. Why an apparently complex reaction remains stoichiometrically simple or single, and how the kinetic treatment of such reactions can be enumerated are the two questions addressed in this chapter.

There are two types of sequences leading from reactants to products through reactive intermediates. The first type of sequence is one where a reactive intermediate is not reproduced in any other step of the sequence. This type of sequence is denoted as an *open* sequence. The second type of sequence is one in which a reactive intermediate is reproduced so that a cyclic reaction pattern repeats itself and a large number of product molecules can be made from only one reactive intermediate. This type of sequence is closed and is denoted a *catalytic* or *chain* reaction cycle. This type of sequence is the best definition of *catalysis*.

A few simple examples of sequences are listed in Table 4.1.1. The reactive intermediates are printed in boldface and the stoichiometrically simple or single reaction is in each case obtained by summation of the elementary steps of the sequence. While all reactions that are closed sequences may be said to be catalytic, there is a distinct difference between those where the reactive intermediates are provided by a separate entity called the *catalyst* that has a very long lifetime and those where the reactive intermediates are generated within the system and may survive only during a limited number of cycles. The first category encompasses truly catalytic reactions (catalytic reaction cycle) in the narrow sense of the word, while the second involves chain reactions (chain reaction cycle). Both types exhibit slightly different kinetic features. However, the two types are so closely related that it is conceptually straightforward to consider them together. In particular, both categories can be analyzed by means of the steady-state approximation that will be presented in the next section.

Chain and catalytic reaction cycles provide energetically favorable pathways for reactant molecules to proceed to product molecules. This point is illustrated below for both types of cycles. Consider the reaction between dihydrogen and dichlorine to produce HCl that can be brought about in the gas phase by irradiating the reactants with light. It is known that over 10^6 molecules of HCl can be formed per absorbed photon. The reaction proceeds as follows:

$$Cl_2 \xrightarrow{\text{light}} 2Cl \qquad \text{(initiation)}$$
$$Cl + H_2 \longrightarrow HCl + H \qquad \text{(propagation)}$$
$$H + Cl_2 \longrightarrow HCl + Cl \qquad \text{(propagation)}$$
$$2Cl \longrightarrow Cl_2 \qquad \text{(termination)}$$

Once chlorine atoms are produced (initiation), the propagation steps provide a closed cycle that can be repeated numerous times (e.g., 10^6) prior to the recombination of the chlorine atoms (termination).

The reason the chain reaction cycle dominates over a direct reaction between dihydrogen and dichlorine is easy to understand. The direct reaction between H_2 and Cl_2 has an activation energy of over 200 kJ/mol, while the activation energies of the two propagation steps are both less than 30 kJ/mol (see Figure 4.1.1). Thus, the difficult step is initiation and in this case is overcome by injection of photons.

Now consider what happens in a catalytic reaction cycle. For illustrative purposes, the decomposition of ozone is described. In the presence of oxygen atoms, ozone decomposes via the elementary reaction:

$$O + O_3 \rightarrow 2O_2$$

Table 4.1.1 | Sequences and reactive intermediates.[1]

Sequence	Type	Reactive intermediates
$O_3 \longrightarrow O_2 + O$ $O + O_3 \longrightarrow O_2 + O_2$ ———— $2O_3 \Longrightarrow 3O_2$	open	oxygen atoms in the gas phase
(benzhydryl chloride → benzhydryl cation $+ Cl^-$)	open	solvated (in liquid SO_2) benzhydryl ions
(benzhydryl cation $+ F^- \longrightarrow$ benzhydryl fluoride)		
(benzhydryl chloride $+ F^- \Longrightarrow$ benzhydryl fluoride $+ Cl^-$)		
$O + N_2 \longrightarrow NO + N$ $N + O_2 \longrightarrow NO + O$ ———— $N_2 + O_2 \Longrightarrow 2NO$	chain	oxygen and nitrogen atoms in the gas phase
$SO_3^- + O_2 \longrightarrow SO_5^-$ $SO_5^- + SO_3^{2-} \longrightarrow SO_5^{2-} + SO_3^-$ $SO_5^{2-} + SO_3^{2-} \longrightarrow 2SO_4^{2-}$ ———— $2SO_3^{2-} + O_2 \Longrightarrow 2SO_4^{2-}$	chain	free radical ions SO_3^- and SO_5^-
$* + H_2O \longrightarrow H_2 + O*$ $O* + CO \longrightarrow CO_2 + *$ ———— $H_2O + CO \Longrightarrow H_2 + CO_2$	catalytic	sites * on a catalyst surface and the surface complex $O*$
(cumyl radical $+ O_2 \longrightarrow$ cumylperoxy radical) (cumylperoxy radical + cumene → cumene hydroperoxide + cumyl radical) ———— (cumene $+ O_2 \Longrightarrow$ cumene hydroperoxide)	chain	cumyl and cumylperoxy free radicals in a solution of cumene

[1] Adapted from M. Boudart, *Kinetics of Chemical Processes,* Butterworth-Heinemann, 1991, pp. 61–62.

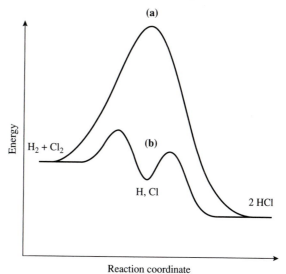

Figure 4.1.1 |
Energy versus reaction coordinate for $H_2 + Cl_2 \Rightarrow 2HCl$.
(a) direct reaction, **(b)** propagation reactions for photon assisted pathway.

The rate of the direct reaction can be written as:

$$r_d = k\,[O][O_3] \qquad (4.1.1)$$

where r_d is in units of molecule/cm³/s, $[O]$ and $[O_3]$ are the number densities (molecule/cm³) of O and O_3, respectively, and k is in units of cm³/s/molecule. In these units, k is known and is:

$$k = 1.9 \times 10^{-11}\exp[-2300/T]$$

where T is in Kelvin. Obviously, the decomposition of ozone at atmospheric conditions (temperatures in the low 200s in Kelvin) is quite slow.

The decomposition of ozone dramatically changes in the presence of chlorine atoms (catalyst):

$$\begin{array}{l} Cl + O_3 \xrightarrow{\;k_1\;} O_2 + ClO \\ \underline{ClO + O \xrightarrow{\;k_2\;} O_2 + Cl} \\ O + O_3 \Rightarrow 2O_2 \end{array}$$

where:

$$k_1 = 5 \times 10^{-11}\exp(-140/T) \qquad \text{cm}^3/\text{s/molecule}$$
$$k_2 = 1.1 \times 10^{-10}\exp(-220/T) \qquad \text{cm}^3/\text{s/molecule}$$

At steady state (using the steady-state approximation—developed in the next section), the rate of the catalyzed reaction r_c is:

$$r_c = \frac{k_1 k_2 [O][O_3][[Cl] + [ClO]]}{k_1[O_3] + k_2[O]} \tag{4.1.2}$$

However, since $[O] \ll [O_3]$ and $k_1 \cong k_2$:

$$r_c = k_2[O][[Cl] + [ClO]] \tag{4.1.3}$$

and

$$\frac{r_c}{r_d} = \frac{k_2[[Cl] + [ClO]]}{k[O_3]} \tag{4.1.4}$$

If

$$\frac{[Cl] + [ClO]}{[O_3]} \cong 10^{-3}$$

(a value typical of certain conditions in the atmosphere), then:

$$\frac{r_c}{r_d} = \frac{k_2}{k} \times 10^{-3} = 5.79 \times 10^{-3} \exp(2080/T) \tag{4.1.5}$$

At $T = 200$ K, $r_c/r_d = 190$. The enhancement of the rate is the result of the catalyst (Cl). As illustrated in the energy diagram shown in Figure 4.1.2, the presence of the catalyst lowers the activation barrier. The Cl catalyst first reacts with O to

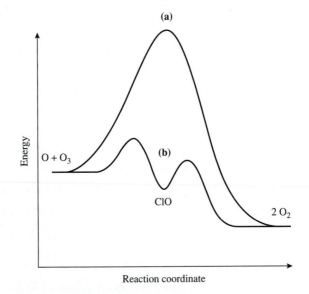

Figure 4.1.2 |
Energy versus reaction coordinate for ozone decomposition.
(a) direct reaction, **(b)** Cl catalyzed reaction.

give the reaction intermediate ClO, which then reacts with O_3 to give O_2 and regenerate Cl. Thus, the catalyst can perform many reaction cycles.

VIGNETTE 4.1.1

Ozone exists in a dynamic equilibrium in the stratosphere:

$$O_2 + light\ (<242\ nm) \rightarrow 2O \qquad (formation)$$
$$O + O_2 \rightarrow O_3$$

$$O_3 + light\ (<\sim300\ nm) \rightarrow O + O_2 \quad (destruction)$$
$$O + O \rightarrow O_2$$

It is an important species in the stratosphere (15–55 km) for the survival of life on the planet since it absorbs harmful ultraviolet radiation (UV-B, 280–320 nm). Every September over the past few years there has been a large loss of ozone (ozone hole) in the Antarctic stratosphere (losses near 100 percent at altitudes between 15–20 km). The losses account for about 3 percent of the entire global supply of stratospheric ozone in a period of 4 to 6 weeks. The primary cause for the Antarctic ozone loss is the increase of stratospheric chlorine. For example, the total organochlorine concentration has changed as follows: 1950—0.8 parts per billion by volume (ppbv), 1974—1.8 ppbv, 1990—4.0 ppbv, while the only natural organochlorine species CH_3Cl has remained 0.6 ppbv over the last 15 years. Where has the other chlorine come from? The chlorine arises from chlorofluorocarbons (CFCs). CFCs such as CFC-12 (CF_2Cl_2), CFC-11 ($CFCl_3$), and CFC-113 ($CF_2ClCFCl_2$) were used as refrigerants, blowing agents, and cleaning agents, respectively. More than \$28 billion/year of CFC-based products were manufactured. The total release (worldwide) to the atmosphere over the past two decades is estimated at CFC-12: 400 kt/year, CFC-11: 250 kt/year, and CFC-113: 300 kt/year. Because CFCs are very inert (property built into the molecule for numerous applications), they are not destroyed to any appreciable extent in the lower portions of the atmosphere and have lifetimes of many decades in the stratosphere. In the stratosphere, CFCs can be degraded to provide a source of chlorine atoms:

$$CF_2Cl_2 + light\ (<220\ nm) \rightarrow Cl + CF_2Cl$$

The reason the chlorine is not formed near the surface of the Earth is that O_2 and O_3 absorb UV radiation in the lower stratosphere. Once atomic chlorine is formed it can act as a catalyst for ozone decomposition as was shown in Section 4.1. In the presence of the chlorine catalyst, the ozone can be quickly consumed. Tremendous efforts are now underway worldwide to curb these effects.

4.2 | The Steady-State Approximation

Consider a closed system comprised of two, first-order, irreversible (one-way) elementary reactions with rate constants k_1 and k_2:

$$A \xrightarrow{\ k_1\ } B \xrightarrow{\ k_2\ } C$$

If C_A^0 denotes the concentration of A at time $t = 0$ and $C_B^0 = C_C^0 = 0$, the material balance equations for this system are:

$$\frac{dx}{dt} = -k_1 x \qquad \frac{dy}{dt} = k_1 x - k_2 y \qquad \frac{dw}{dt} = k_2 y \tag{4.2.1}$$

where $x = C_A/C_A^0$, $y = C_B/C_A^0$, and $w = C_C/C_A^0$. Integration of Equation (4.2.1) with $x = 1$, $y = 0$, $w = 0$ at $t = 0$ gives:

$$\left.\begin{aligned}
x &= \exp(-k_1 t) \\[6pt]
y &= \frac{k_1}{k_2 - k_1}\left[\exp(-k_1 t) - \exp(-k_2 t)\right] \\[6pt]
w &= 1 - \frac{k_2}{k_2 - k_1}\exp(-k_1 t) + \frac{k_1}{k_2 - k_1}\exp(-k_2 t)
\end{aligned}\right\} \tag{4.2.2}$$

EXAMPLE 4.2.1

Show how the expression for $y(t)$ in Equation (4.2.2) is obtained (see Section 1.5).

■ **Answer**

Placing the functional form of $x(t)$ into the equation for $\dfrac{dy}{dt}$ gives:

$$\frac{dy}{dt} + k_2 y = k_1 \exp(-k_1 t) \qquad y = 0 \text{ at } t = 0$$

This first-order initial-value problem can easily be solved by the use of the integration factor method. That is,

$$d(y e^{k_2 t}) = k_1 \exp[(k_2 - k_1)t]\,dt$$

or after integration:

$$y e^{k_2 t} = \frac{k_1 \exp[(k_2 - k_1)t]}{k_2 - k_1} + \gamma$$

Since $y = 0$ at $t = 0$:

$$\gamma = -\frac{k_1}{k_2 - k_1}$$

Substitution of the expression for γ into the equation for $y(t)$ gives:

$$y e^{k_2 t} = \frac{k_1}{k_2 - k_1}\left[\exp[(k_2 - k_1)t] - 1\right]$$

or

$$y = \frac{k_1}{k_2 - k_2}\left[\exp(-k_1 t) - \exp(-k_2 t)\right]$$

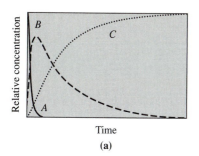
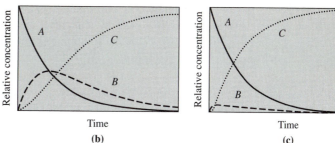

Figure 4.2.1 |

Two first-order reactions in series.

$$A \xrightarrow{k_1} B \xrightarrow{k_2} C$$

(a) $k_2 = 0.1\, k_1$, **(b)** $k_2 = k_1$, **(c)** $k_2 = 10\, k_1$.

It is obvious from the conservation of mass that:

$$x + y + w = 1 \tag{4.2.3}$$

or

$$\frac{dx}{dt} + \frac{dy}{dt} + \frac{dw}{dt} = 0 \tag{4.2.4}$$

The concentration of A decreases monotonically while that of B goes through a maximum (see, for example, Figure 4.2.1b). The maximum in C_B is reached at (Example 1.5.6):

$$t_{max} = \frac{1}{(k_2 - k_1)} \ln(k_2/k_1) \tag{4.2.5}$$

and is:

$$C_B^{max} = C_A^0 \left(\frac{k_1}{k_2}\right)^{\left[\frac{k_2}{k_2-k_1}\right]} \tag{4.2.6}$$

At t_{max}, the curve of C_c versus t shows an inflection point, that is, $(d^2w)/(dt^2) = 0$.

Suppose that B is not an intermediate but a reactive intermediate (see Section 1.1). Kinetically, this implies that $k_2 \gg k_1$. If such is the case, what happens to the solution of Equation (4.2.1), that is, Equation (4.2.2)? As $k_1/k_2 \to 0$, Equation (4.2.2) reduces to:

$$\left.\begin{array}{l} x = \exp(-k_1 t) \\[2mm] y = \dfrac{k_1}{k_2} \exp(-k_1 t) \\[2mm] w = 1 - \exp(-k_1 t) \end{array}\right\} \tag{4.2.7}$$

Additionally, $t_{max} \to 0$ as does y_{max} (see Figure 4.2.1 and compare as $k_1/k_2 \to 0$). Thus, the time required for C_B to reach its maximum concentration is also very small. Additionally, the inflection point in the curve of C_C versus time is translated back to the origin.

Equation (4.2.7) is the solution to Equation (4.2.8):

$$\left.\begin{array}{c} \dfrac{dx}{dt} = -k_1 x \\[2mm] 0 = k_1 x - k_2 y \\[2mm] \dfrac{dw}{dt} = k_2 y \end{array}\right\} \qquad (4.2.8)$$

Note that Equation (4.2.8) involves two differential and one algebraic equations. The algebraic equation specifies that:

$$\frac{dy}{dt} = 0 \qquad (4.2.9)$$

This is the analytical expression of the *steady-state approximation*: the time derivatives of the concentrations of reactive intermediates are equal to zero. Equation (4.2.9) must *not* be integrated since the result that y = constant is false [see Equation (4.2.7)]. What is important is that B varies with time implicitly through A and thus with the changes in A (a stable reactant). Another way to state the steady-state approximation is [Equation (4.2.4) with $dy/dt = 0$]:

$$\frac{dx}{dt} = -\frac{dw}{dt} \qquad (4.2.10)$$

Thus, in a sequence of steps proceeding through reactive intermediates, the rates of reaction of the steps in the sequence are equal. It follows from Equation (4.2.10) that the reaction, however complex, can be described by a single parameter, the extent of reaction (see Section 1.2):

$$\Phi(t) = \frac{n_i(t) - n_i^0}{v_i} \qquad (1.2.4)$$

so that:

$$\frac{d\Phi}{dt} = \frac{1}{v_1}\frac{dn_1}{dt} = \cdots = \frac{1}{v_i}\frac{dn_i}{dt} \qquad (4.2.11)$$

For the simple reaction $A \Rightarrow C$, Equation (4.2.11) simplifies to Equation (4.2.10). The steady-state approximation can be stated in three different ways:

1. The derivatives with respect to time of the concentrations of the reactive intermediates are equal to zero [Equation (4.2.9)].

2. The steady-state concentrations of the reaction intermediates are small since as $k_1/k_2 \ll 1$, $t_{max} \rightarrow 0$ and $C_B^{max} \rightarrow 0$.

3. The rates of all steps involving reactants, products, and intermediates are equal [Equation (4.2.10)].

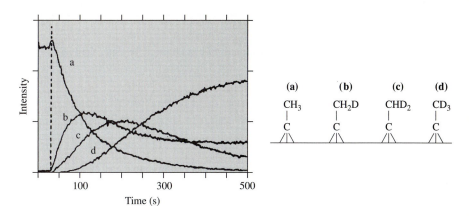

Figure 4.2.2 |
Display of the data of Creighton et al. (Left) reprinted from *Surface Science,* vol. 138, no. 1, J. R. Creighton, K. M. Ogle, and J. M. White, "Direct observation of hydrogen-deuterium exchange in ethylidyne adsorbed on Pt(111)," pp. L137–L141, copyright 1984, with permission from Elsevier Science. Schematic of species observed (right).

These conditions must be satisfied in order to correctly apply the steady-state approximation to a reaction sequence. Consider the H-D exchange with ethylidyne (CCH_3 from the chemisorption of ethylene) on a platinum surface. If the reaction proceeds in an excess of deuterium the backward reactions can be ignored. The concentrations of the adsorbed ethylidyne species have been monitored by a technique called secondary ion mass spectroscopy (SIMS). The concentrations of the various species are determined through mass spectroscopy since each of the species on the surface are different by one mass unit. Creighton et al. [*Surf. Sci.,* **138** (1984) L137] monitored the concentration of the reactive intermediates for the first 300 s, and the data are consistent with what are expected from three consecutive reactions. The results are shown in Figure 4.2.2.

Thus, the reaction sequence can be written as:

$$D_2 \rightleftharpoons 2D$$
$$Pt \equiv C - CH_3 + D \rightarrow Pt \equiv C - CH_2D + H$$
$$Pt \equiv C - CH_2D + D \rightarrow Pt \equiv C - CHD_2 + H$$
$$Pt \equiv C - CHD_2 + D \rightarrow Pt \equiv C - CD_3 + H$$
$$2H \rightleftharpoons H_2$$

If the rate of D_2-H_2 exchange were to be investigated for this reaction sequence,

$$-\frac{dC_{D_2}}{dt} = \frac{dC_{H_2}}{dt}$$

and the time derivative of all the surface ethylidyne species could be set equal to zero via the steady-state approximation.

EXAMPLE 4.2.2

Show how Equation (4.1.2) is obtained by using the steady-state approximation.

■ **Answer**

$$Cl + O_3 \xrightarrow{k_1} O_2 + ClO$$

$$ClO + O \xrightarrow{k_2} O_2 + Cl$$

$$\overline{O + O_3 \Rightarrow 2O_2}$$

The reaction rate expressions for this cycle are:

$$\frac{d[O_3]}{dt} = -k_1[Cl][O_3]$$

$$\frac{d[ClO]}{dt} = k_1[Cl][O_3] - k_2[ClO][O]$$

$$\frac{d[O_2]}{dt} = k_1[Cl][O_3] + k_2[ClO][O]$$

Using the steady-state approximation for the reactive intermediate [ClO] specifies that:

$$\frac{d[ClO]}{dt} = 0$$

and gives:

$$k_1[Cl][O_3] = k_2[ClO][O]$$

Also, notice that the total amount of the chlorine in any form must be constant. Therefore,

$$[Cl]_0 = [Cl] + [ClO]$$

By combining the mass balance on chlorine with the mass balance at steady-state for [ClO], the following expression is obtained:

$$[ClO] = \frac{k_1[Cl][O_3]}{k_2[O]} = \frac{k_1[O_3]}{k_2[O]}\{[Cl]_0 - [ClO]\}$$

After rearrangement:

$$[ClO] = \frac{k_1[O_3][Cl]_0}{k_2[O]\left\{1 + \frac{k_1[O_3]}{k_2[O]}\right\}} = \frac{k_1[Cl]_0[O_3]}{k_2[O] + k_1[O_3]}$$

Recall that:

$$r_c = \frac{1}{V}\frac{d\Phi}{dt} = \frac{1}{v_i}\frac{d[A_i]}{dt} = -\frac{d[O_3]}{dt} = \frac{1}{2}\frac{d[O_2]}{dt}$$

Thus, by use of the rate expressions for ozone and dioxygen:

$$2k_1[Cl][O_3] = k_1[Cl][O_3] + k_2[ClO][O]$$

or

$$k_1[Cl][O_3] = k_2[ClO][O]$$

The rate of ozone decomposition can be written as:

$$r_c = k_1[Cl][O_3]$$

or

$$r_c = k_2[ClO][O]$$

Substitution of the expression for [ClO] in this equation gives:

$$r_c = \frac{k_1 k_2 [Cl]_0 [O][O_3]}{k_2[O] + k_1[O_3]}$$

or Equation (4.1.2).

EXAMPLE 4.2.3

Explain why the initial rate of polymerization of styrene in the presence of $Zr(C_6H_5)_4$ in toluene at 303 K is linearly dependent on the concentration of $Zr(C_6H_5)_4$ [experimentally observed by D. G. H. Ballard, *Adv. Catal.* **23** (1988) 285].

■ Answer

Polymerization reactions proceed via initiation, propagation, and termination steps as illustrated in Section 4.1. A simplified network to describe the styrene polymerization is:

$$Zr(C_6H_5)_4 + styrene \xrightarrow{k_i} \underset{\underset{C_6H_5}{|}}{(C_6H_5)_3ZrCHCH_2C_6H_5} \tag{i}$$

$$\underset{\underset{C_6H_5}{|}}{(C_6H_5)_3Zr(polymer)_n} + styrene \xrightarrow{k_p} \underset{\underset{C_6H_5}{|}}{(C_6H_5)_3Zr(polymer)_{n+1}} \tag{ii}$$

$$\underset{\underset{C_6H_5}{|}}{(C_6H_5)_3Zr(polymer)_n} \xrightarrow{k_t} (C_6H_5)_3Zr + C_6H_5(polymer)_n \tag{iii}$$

where:

$$\underset{\underset{C_6H_5}{|}}{(C_6H_5)_3Zr(polymer)_{n=1}}$$

is the species shown on the right-hand side of Equation (i). Equations (i–iii) are the initiation, propagation, and termination reactions (β-hydrogen transfer terminates the growth of the polymer chains) respectively. The reaction rate equations can be written as:

$$r_i = k_i C_z C_s \qquad \text{(initiation)}$$

$$r_p = k_p C_s \left(\sum_{i=1}^{n} C_i \right) \qquad \text{(propagation)}$$

$$r_t = k_t \left(\sum_{i=1}^{n} C_i \right) \qquad \text{(termination)}$$

where C_z is the concentration of $Zr(C_6H_5)_4$, C_s is the styrene concentration, and C_i is the concentration of the Zr species containing $(\text{polymer})_{n=i}$. For simplicity, the rate constants for propagation and termination are assumed to be independent of polymer chain length (i.e., independent of the value of n).

If the steady-state approximation is invoked, then $r_i = r_t$ or

$$k_i C_z C_s = k_t \left(\sum_{i=1}^{n} C_i \right)$$

Solving for the sum of the reactive intermediates gives:

$$\sum_{i=1}^{n} C_i = \left(\frac{k_i}{k_t} \right) C_z C_s$$

Substitution of this expression into that for r_p yields:

$$r_p = \left(\frac{k_i k_p}{k_t} \right) C_z C_s^2$$

The rate of polymerization of the monomer is the combined rates of initiation and propagation. The long chain approximation is applicable when the rate of propagation is much faster than the rate of initiation. If the long chain approximation is used here, then the polymerization rate is equal to r_p. Note that the r_p is linearly dependent on the concentration of Zr(benzyl)$_4$, and that the polystyrene obtained (polystyrene is used to form styrofoam that can be made into cups, etc.) will have a distribution of molecular weights (chain lengths). That is, $C_6H_5(\text{polymer})_n$ in Equation (iii) has many values of n. The degree of polymerization is the average number of structural units per chain, and control of the molecular weight and its distribution is normally important to the industrial production of polymers.

The steady-state approximation applies only after a time t_r, the *relaxation time*. The relaxation time is the time required for the steady-state concentration of the reactive intermediates to be approached. Past the relaxation time, the steady-state approximation remains an approximation, but it is normally satisfactory. Below, a more quantitative description of the relaxation time is described.

Assume the actual concentration of species B in the sequence $A \xrightarrow{k_1} B \xrightarrow{k_2} C$, C_B, is different from its steady-state approximation C_B^* by an amount $\bar{\varepsilon}$:

$$C_B = C_B^*(1 + \bar{\varepsilon}) \qquad (4.2.12)$$

The expression:

$$\frac{dC_B}{dt} = k_1 C_A - k_2 C_B \qquad (4.2.13)$$

still applies as does the equation [from taking the time derivative of Equation (4.2.12)]:

$$\frac{dC_B}{dt} = C_B^* \frac{d\bar{\varepsilon}}{dt} + (1 + \bar{\varepsilon}) \frac{dC_B^*}{dt} \qquad (4.2.14)$$

According to the steady-state approximation [see Equation (4.2.8)]:

$$C_B^* = \frac{k_1}{k_2} C_A \qquad (4.2.15)$$

Since $dC_A/dt = -k_1 C_A$,

$$\frac{dC_B^*}{dt} = -\frac{k_1^2}{k_2} C_A \qquad (4.2.16)$$

Equating the right-hand sides of Equations (4.2.13) and (4.2.14) and substituting the values for C_B^* and dC_B^*/dt from Equations (4.2.15) and (4.2.16), respectively, gives:

$$\frac{d\bar{\varepsilon}}{dt} + (k_2 - k_1)\bar{\varepsilon} - k_1 = 0 \qquad (4.2.17)$$

Integration of this initial-value problem with $\bar{\varepsilon} = -1$ ($C_B = 0$) at $t = 0$ yields:

$$\bar{\varepsilon} = -\frac{1}{(\bar{K} - 1)} \left\{ \bar{K} - \exp[(\bar{K} - 1)k_2 t] \right\} \qquad (4.2.18)$$

where:

$$\bar{K} = k_1/k_2$$

Since B is a reactive intermediate, $\bar{K}$ must be smaller than one. With this qualification, and at "sufficiently large values" of time:

$$\bar{\varepsilon} = \bar{K} \qquad (4.2.19)$$

What is implied by "sufficiently large values" of time is easily seen from Equation (4.2.18) when $\bar{K} \ll 1$. For this case, Equation (4.2.18) reduces to:

$$\bar{\varepsilon} = -e^{-k_2 t} \qquad (4.2.20)$$

The relaxation time is the time required for a quantity to decay to a fraction $1/e$ of its original value. For the present case, $t_r = 1/k_2$. Intuitively, it would appear that the relaxation time (sometimes called the induction time) should be on the same order of magnitude as the turnover time, that is, the reciprocal of the turnover frequency (see Section 1.3). As mentioned previously, turnover frequencies around $1\ \mathrm{s}^{-1}$ are common. For this case, the induction time is short. However, if the turnover frequency is $10^{-3}\ \mathrm{s}^{-1}$, then the induction time could be very long. Thus, one should not assume *a priori* that the induction time is brief.

Now, let's consider an important class of catalysts—namely, enzymes. Enzymes are nature's catalysts and are made of proteins. The primary structure is the sequence of amino acids that are joined through peptide bonds (illustrated below) to create the protein polymer chain:

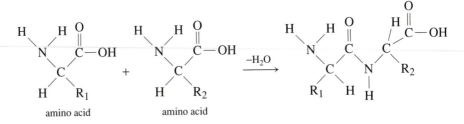

The primary structure gives rise to higher order levels of structure (secondary, tertiary, quaternary) and all enzymes have a three-dimensional "folded" structure of the polymer chain (or chains). This tertiary structure forms certain arrangements of amino acid groups that can behave as centers for catalytic reactions to occur (denoted as active sites). How an active site in an enzyme performs the chemical reaction is described in Vignette 4.2.1.

VIGNETTE 4.2.1

The elegance of enzyme reactivity is unparalleled in synthetic catalytic materials. Beginning from the time when Emil Fischer proposed his classic lock-and-key theory for enzyme specificity to the induced fit theory of Koshland, enzyme catalysis sets the standard to which all other catalytic transformations are compared.

The induced fit theory of Koshland is schematically represented in Figure 4.2.3. The critical issues of this theory are: (1) that there is a precise three-dimensional configuration of the amino acid functional groups that must interact with an appropriate reactant (substrate) in order for catalysis to occur, (2) that the binding (chemisorption) of the reactant produces an appreciable change in the three-dimensional conformation of the amino acids at the active site, and (3) that the changes in the protein structure produced by substrate binding bring about the proper alignment between protein functional groups and the substrate to allow catalytic reactions to occur. A nonsubstrate may still be able to bind but not react because of the misalignment of the appropriate interactions. Thus, as Koshland states, the process is not like a "lock-and-key" fit but rather a "hand-in-glove" fit, allowing for flexibility. Virtually all enzymes have been shown to have conformational changes upon binding. When a substrate binds to a particular area (domain) of an enzyme, thermal fluctuations can bring a second domain into contact with the bound substrate and the newly formed interactions stabilize what is called the "closed" or "bound" conformation. The protein-substrate interactions can be numerous and consist of combinations of interlocking salt bridges, hydrogen bonds, and van der Waals interactions that account for the stability and specificity of this state. These domain closures often exclude water from the active site giving rise to a plausible argument for why numerous enzymatic reactions can take place in aqueous media without utilizing water as a nucleophile. In addition to excluding water and providing the proper positioning of catalytic groups, the "closed" state

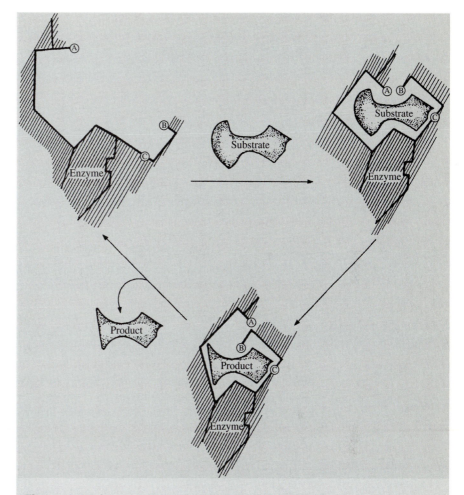

Figure 4.2.3 | Schematic model of the induced fit mechanism for enzyme catalysis. *A, B,* and *C* are particular functional groups in the enzyme binding site that must be properly aligned with the bound substrate for reaction to occur. The substrate binding induces a conformational change in the enzyme after which reaction of the bound complex takes place. Other conformational changes occur during the reaction of the substrate to produce the bound product. Desorption of the product returns the enzyme to its unbound conformation. [Adapted from D. E. Koshland, Jr., *Angew. Chem. Int. Ed. Engl.,* **33** (1994) 2375, with permission of WILEY-VCH Verlag GmbH and the author.]

confines substrates and prevents the escape of reaction intermediates. Domain closure must be fast because the energy barrier between the "open" and "closed" states must not be large. Also, these states are only slightly different in energy in that they are in dynamic equilibrium. An important feature of this concept is that the alignment of the catalytic and binding groups are optimized for the transition state and that the attainment of this state

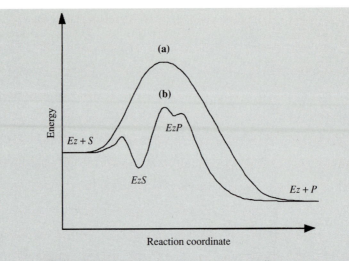

Figure 4.2.4 | Energy versus reaction coordinate.
(a) Direct reaction, **(b)** enzyme catalyzed reaction.

is energetically unfavorable unless it is supplied with the energy of substrate binding. Figure 4.2.4 is a schematic illustration of the energy diagram for a single-substrate, enzyme (Ez)-mediated reaction.

Proper substrate binding allows for the bound (closed) state (EzS) to be in dynamic equilibrium with free substrate. Upon domain closure, catalytic reaction can occur to transform the bound state (EzS) to an energetically less stable state than the open state of the protein (EzP) by altering the interactions between the protein and the bound molecule. Note that the upper limit on the rate of catalytic reaction should, therefore, be fixed by the rate of domain movements. Since the open state is more energetically favored, the product will desorb to return the enzyme to the open state.

As was advanced by Linus Pauling, enzymes accelerate reactions by lowering the activation barrier to the transition state (see Figure 4.2.4). How this is accomplished involves a complex sequence of events, as illustrated above. Upon initial reflection, it might be expected that evolution has selected enzymes to optimize for their bound (closed) state. However, Koshland states that the induced-fit model leads to the conclusion that evolution has selected the protein that optimized both the open and closed states. Although the aforementioned description of enzyme catalysis is simplified for illustrative purposes, the essential paradigms are captured. Thus, it is clear that an extremely complex set of events occur to allow enzyme catalysts to have high activity and selectivity.

VIGNETTE 4.2.2

Table sugar, sucrose, is just one of a family of natural sugars; for example, fructose (found in fruits), glucose (found in corn), and lactose (found in milk). Fructose has now become the sweetener of choice because of catalysis.

Glucose isomerase is an enzyme that converts d-glucose into d-fructose:

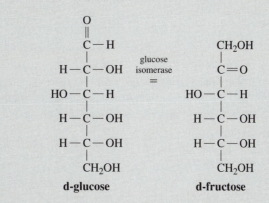

d-glucose   **d-fructose**

Although glucose, fructose, and sucrose are all sugars, they are not equally sweet. If sucrose would be ranked as 100 on an arbitrary sweetness scale, then fructose would be 173 and glucose only 74. Therefore, products like Coke and Pepsi now use fructose in their formulations. This fructose is produced from corn. Raw corn starch is "depolymerized" and treated to yield an aqueous solution of glucose that is then converted to fructose by the enzyme glucose isomerase. The enzyme is immobilized on a ceramic support and this heterogeneous catalyst can provide greater than 18 tons of fructose per ton of catalyst.

In order to describe the kinetics of an enzyme catalyzed reaction, consider the following sequence:

$$Ez + S \underset{k_{-1}}{\overset{k_1}{\rightleftharpoons}} EzS \underset{k_{-2}}{\overset{k_2}{\rightleftharpoons}} EzP \xrightarrow{k_3} Ez + P$$

where Ez is the enzyme, S is the substrate (reactant), EzS and EzP are enzyme bound complexes, and P is the product. An energy diagram for this sequence is shown in Figure 4.2.4. The rate equations used to describe this sequence are:

$$\left.\begin{aligned}
\frac{dC_S}{dt} &= -k_1 C_{Ez} C_S + k_{-1} C_{EzS} \\
\frac{dC_{EzS}}{dt} &= k_1 C_{Ez} C_S - k_{-1} C_{EzS} - k_2 C_{EzS} + k_{-2} C_{EzP} \\
\frac{dC_{EzP}}{dt} &= k_2 C_{EzS} - k_{-2} C_{EzP} - k_3 C_{EzP} \\
\frac{dC_P}{dt} &= k_3 C_{EzP}
\end{aligned}\right\} \qquad (4.2.21)$$

Using the steady-state approximation for the reactive intermediates C_{EzS} and C_{EzP} and the fact that:

$$C_{Ez}^0 = C_{Ez} + C_{EzS} + C_{EzP} \qquad (4.2.22)$$

where C_{Ez}^0 is the concentration of the enzyme in the absence of substrate gives:

$$r = \frac{k_2 k_3 C_{Ez}^0 C_S}{(k_2 + k_{-2} + k_3)\left[C_S + \dfrac{k_{-1}k_{-2} + k_{-1}k_3 + k_2 k_3}{k_1(k_2 + k_{-2} + k_3)}\right]} \tag{4.2.23}$$

If the product dissociates rapidly from the enzyme (i.e., k_3 is large compared to k_2 and k_{-2}), then a simplified sequence is obtained and is the one most commonly employed to describe the kinetics of enzyme catalyzed reactions. For this case,

$$Ez + S \underset{k_{-1}}{\overset{k_1}{\rightleftarrows}} EzS \xrightarrow{k_3} Es + P$$

with

$$\left.\begin{aligned}
\frac{dC_S}{dt} &= -k_1 C_S C_{Ez} + k_{-1} C_{EzS} \\[4pt]
\frac{dC_{EzS}}{dt} &= k_1 C_S C_{Ez} - k_{-1} C_{EzS} - k_3 C_{EzS} \\[4pt]
\frac{dC_P}{dt} &= k_3 C_{EzS}
\end{aligned}\right\} \tag{4.2.24}$$

where

$$C_{Ez}^0 = C_{Ez} + C_{EzS} \tag{4.2.25}$$

Using the steady-state approximation for C_{EzS} gives:

$$C_{EzS} = \frac{k_1 C_S C_{Ez}}{(k_{-1} + k_3)} = \frac{C_S C_{Ez}}{K_m} \tag{4.2.26}$$

with:

$$K_m = \frac{k_{-1} + k_3}{k_1} \tag{4.2.27}$$

K_m is called the *Michaelis constant* and is a measure of the binding affinity of the substrate for the enzyme. If $k_{-1} \gg k_3$ then $K_m = k_{-1}/k_1$ or the dissociation constant for the enzyme. The use of Equation (4.2.25) with Equation (4.2.26) yields an expression for C_{EzS} in terms of C_{Ez}^0 and C_S, that is, two measurable quantities:

$$C_{EzS} = \frac{C_{Ez}^0 C_S - C_{EzS} C_S}{K_m}$$

or

$$C_{EzS} = \frac{C^0_{Ez} C_S}{K_m + C_S} \qquad (4.2.28)$$

Substitution of Equation (4.2.28) into the expression for dC_P/dt gives:

$$\frac{dC_P}{dt} = \frac{k_3 C^0_{Ez} C_S}{K_m + C_S} \qquad (4.2.29)$$

If $r_{max} = k_3 C^0_{Ez}$, then Equation (4.2.29) can be written as:

$$\frac{dC_P}{dt} = -\frac{dC_S}{dt} = \frac{r_{max} C_S}{K_m + C_S} \qquad (4.2.30)$$

This form of the rate expression is called the Michaelis-Menton form and is used widely in describing enzyme catalyzed reactions. The following example illustrates the use of linear regression in order to obtain r_{max} and K_m from experimental kinetic data.

EXAMPLE 4.2.4

Para and Baratti [*Biocatalysis,* **2** (1988) 39] employed whole cells from *E. herbicola* immobilized in a polymer gel to catalyze the reaction of catechol to form *L*-dopa:

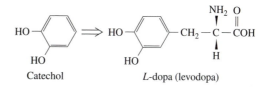

Catechol   *L*-dopa (levodopa)

Do the following data conform to the Michaelis-Menton kinetic model?

■ **Data**

The initial concentration of catechol was 0.0270 M and the data are:

Time (h)	Catechol conversion (%)
0.00	0.00
0.25	11.10
0.50	22.20
0.75	33.30
1.00	44.40
1.25	53.70
1.50	62.60
2.00	78.90
2.50	88.10
3.00	94.80
3.50	97.80
4.00	99.10
4.50	99.60
5.00	99.85

■ Answer

Notice that:

$$\left[-\frac{dC_S}{dt}\right]^{-1} = \frac{K_m}{r_{max}C_S} + \frac{1}{r_{max}}$$

Thus, if $\left[-\dfrac{dC_S}{dt}\right]^{-1}$ is plotted as a function of $1/C_S$, the data should conform to a straight line

with slope $= K_m/r_{max}$ and intercept $= 1/r_{max}$. This type of plot is called a Lineweaver-Burk plot.

First, plot the data for C_S (catechol) versus time from the following data [note that $C_S = C_S^0(1 - f_S)$]:

Time (h)	C_S (M)
0.00	0.027000
0.25	0.024003
0.50	0.021006
0.75	0.018009
1.00	0.015012
1.25	0.012501
1.50	0.010098
2.00	0.005697
2.50	0.003213
3.00	0.001404
3.50	0.000594
4.00	0.000243
4.50	0.000108
5.00	0.000041

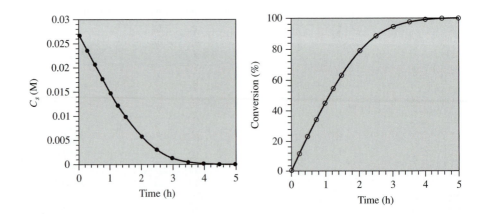

From the data of C_S versus time, dC_S/dt can be calculated and plotted as shown below. Additionally, the Lineweaver-Burk plot can be constructed and is illustrated next.

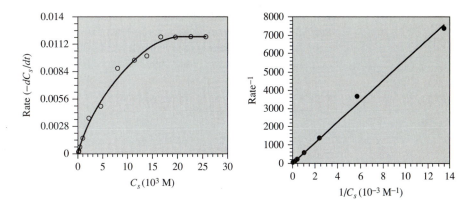

From the Lineweaver-Burk plot, the data do conform to the Michaelis-Menton rate law and

$$K_m = 6.80 \times 10^{-3} \frac{\text{kmol}}{\text{m}^3} \quad \text{and} \quad r_{max} = 1.22 \times 10^{-2} \frac{\text{kmol}}{\text{m}^3\text{-hr}}$$

The previous example illustrates the use of the Lineweaver-Burk plot. Notice that much of the data that determine K_m and r_{max} in the Lineweaver-Burk analysis originate from concentrations at high conversions. These data may be more difficult to determine because of analytical techniques commonly used (e.g., chromatography, UV absorbance) and thus contain larger errors than the data acquired at lower conversions. The *preferred* method of analyzing this type of data is to perform nonlinear regression on the untransformed data (see Appendix B for a brief overview of nonlinear regression). There are many computer programs and software packages available for performing nonlinear regression, and it is suggested that this method be used instead of the Lineweaver-Burk analysis. There are issues of concern when using nonlinear regression and they are illustrated in the next example.

EXAMPLE 4.2.5

Use the data given in Example 4.2.4 and perform a nonlinear regression analysis to obtain K_m and r_{max}.

■ **Answer**

A nonlinear least-squares fit of the Michaelis-Menton model (Equation 4.2.30) gives the following results:

Fitting method	Initial values		Result		
	r_{max} (kmol/m³-hr)	K_m (kmol/m³)	r_{max} (kmol/m³-hr)	K_m (kmol/m³)	Relative error[1]
Non-linear fitting	0	0	$(1.68 \pm 0.09) \times 10^{-2}$	$(8.51 \pm 1.19) \times 10^{-3}$	5.4%
Non-linear fitting	1	1	$(1.68 \pm 0.09) \times 10^{-2}$	$(8.51 \pm 1.19) \times 10^{-3}$	5.4%
Non-linear fitting	1.22×10^{-2}	6.80×10^{-3}	$(1.68 \pm 0.09) \times 10^{-2}$	$(8.51 \pm 1.19) \times 10^{-3}$	5.4%
Lineweaver-Burk	—	—	1.22×10^{-2}	6.80×10^{-3}	15.2%

[1]Relative error $= \sqrt{\sum \left(\text{rate} - \frac{r_{max}C_s}{K_m + C_s} \right)^2 \Big/ \sum (\text{rate})^2}$

Since the nonlinear least-squares method requires initial guesses to start the procedure, three different initial trials were performed: (1) (0,0), (2) (1,1), and (3) the values obtained from the Lineweaver-Burk plot in Example 4.2.4. All three initial trials give the same result (and thus the same relative error). Note the large differences in the values obtained from the nonlinear analysis versus those from the linear regression. If the solutions are plotted along with the experimental data as shown below, it is clear that the Lineweaver-Burk analysis does not provide a good fit to the data.

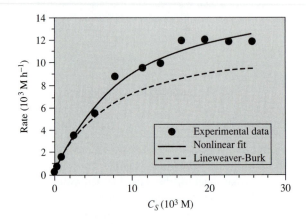

However, if the solutions are plotted as $(rate)^{-1}$ rather than (rate), the results are:

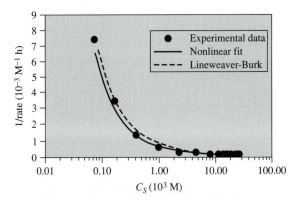

Thus, the Lineweaver-Burk method describes the behavior of $(rate)^{-1}$ but not (rate).

This example illustrates how the nonlinear least-squares method can be used and how initial guesses must be explored in order to provide some confidence in the solution obtained. It also demonstrates the problems associated with the Lineweaver-Burk method.

VIGNETTE 4.2.3

Parkinson's disease is an illness of the central nervous system where voluntary movements become slow and shaky. The cause of this disease is linked to the neurotransmitter

L-norepinephrine. In nerve cells *L*-dopa is converted to dopamine then to *L*-norepinephrine as follows using two enzyme catalysts:

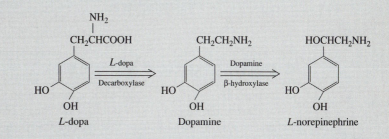

L-dopa   Dopamine   *L*-norepinephrine

At a synapse (junction) between nerves and at the junction between nerves and muscles, the nerve impulse stimulates the release of *L*-norepinephrine that then diffuses across the junction to bind at receptors. The binding event then triggers either another nerve impulse or a muscular event (contraction or relaxation). It has been found that patients with Parkinson's disease are lacking in dopamine. However, because dopamine cannot cross the blood-brain barrier, dopamine cannot be administered as a measure to relieve the symptoms of Parkinson's disease. Although dopamine is not able to penetrate the blood-brain barrier, *L*-dopa does. This breakthrough led to the need for *L*-dopa for the treatment of Parkinson's disease.

    L-dopa has one chiral center. Until the 1970s, only enzymes could produce chiral molecules. In the 1970s, workers at Monsanto developed the first nonbiological catalyst for the synthesis of chiral molecules. The Monsanto group showed that catalysts of the type:

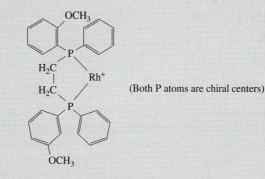

(Both P atoms are chiral centers)

could catalyze the following hydrogenation:

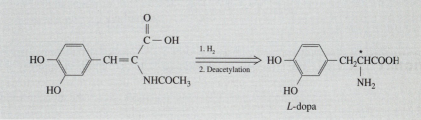

L-dopa

where the $\overset{*}{\text{C}}$ atom is a chiral center. Thus, synthetic catalysts could finally perform chiral reactions. This major breakthrough in catalytic science and technology provided the synthetic process for L-dopa which is now used worldwide to treat Parkinson's disease.

4.3 | Relaxation Methods

In the previous section the steady-state approximation was defined and illustrated. It was shown that this approximation is valid after a certain relaxation time that is a characteristic of the particular system under investigation. By perturbing the system and observing the recovery time, information concerning the kinetic parameters of the reaction sequence can be obtained. For example, with $A \xrightarrow{k_1} B \xrightarrow{k_2} C$, it was shown that the relaxation time when $k_1 \ll k_2$ was k_2^{-1}. Thus, relaxation methods can be very useful in determining the kinetic parameters of a particular sequence.

Consider the simple case of

$$ A \underset{k_{-1}}{\overset{k_1}{\rightleftharpoons}} B \tag{4.3.1} $$

It can be shown by methods illustrated in the previous section for $A \rightarrow B \rightarrow C$, that the relaxation time for the network in Equation (4.3.1) is:

$$ \frac{1}{t_r} = k_1 + k_{-1} \tag{4.3.2} $$

A perturbation in the concentration of either A or B from equilibrium would give rise to a relaxation that returned the system to equilibrium. Since

$$ K_a = k_1/k_{-1} \tag{4.3.3} $$

and K_a can be calculated from the Gibbs functions of A and B, experimental determination of t_r gives k_1 and k_{-1} via the use of Equations (4.3.2) and (4.3.3). Depending on the order of magnitude of t_r, the experimentalist must choose an analytical technique that has a time constant for analysis smaller than t_r. For very fast reactions this can be a problem.

A particularly useful method for determining relaxation times involves the use of flow reactors and labeled compounds. For example, say that the following reaction was proceeding over a solid catalyst:

$$ CO + 3H_2 \Rightarrow CH_4 + H_2O $$

At steady-state conditions, ^{12}CO can be replaced by ^{13}CO while maintaining all other process parameters (e.g., temperature, flow rate) constant. The outlet from the reactor can be continuously monitored by mass spectroscopy. The decay of the concentration of $^{12}CH_4$ and the increase in the concentration of $^{13}CH_4$ can provide

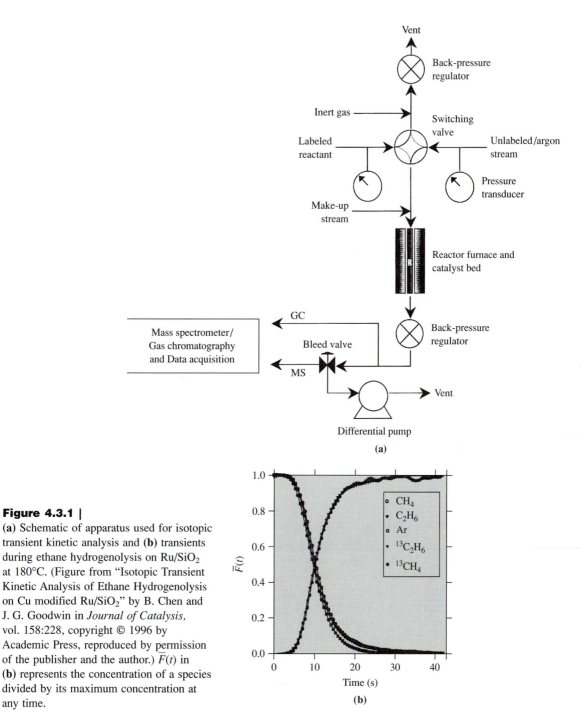

Figure 4.3.1 |
(a) Schematic of apparatus used for isotopic transient kinetic analysis and (b) transients during ethane hydrogenolysis on Ru/SiO$_2$ at 180°C. (Figure from "Isotopic Transient Kinetic Analysis of Ethane Hydrogenolysis on Cu modified Ru/SiO$_2$" by B. Chen and J. G. Goodwin in *Journal of Catalysis,* vol. 158:228, copyright © 1996 by Academic Press, reproduced by permission of the publisher and the author.) $\overline{F}(t)$ in (b) represents the concentration of a species divided by its maximum concentration at any time.

kinetic parameters for this system. This method is typically called "isotopic transient kinetic analysis."

Figure 4.3.1a shows a schematic of an apparatus to perform the steady-state, isotopic transient kinetic analysis for the hydrogenolysis of ethane over a Ru/SiO$_2$ catalysis:

$$CH_3CH_3 + H_2 \Rightarrow 2CH_4$$

A sampling of the type of data obtained from this experiment is given in Figure 4.3.1b. Kinetic constants can be calculated from these data using analyses like those presented above for the simple reversible, first-order system [Equation (4.3.1)].

Exercises for Chapter 4

1. N$_2$O$_5$ decomposes as follows:

$$2N_2O_5 \Rightarrow 4NO_2 + O_2$$

Experimentally, the rate of reaction was found to be:

$$r = \frac{d[O_2]}{dt} = k[N_2O_5]$$

Show that the following sequence can lead to a reaction rate expression that would be consistent with the experimental observations:

$$N_2O_5 \underset{k_{-1}}{\overset{k_1}{\rightleftarrows}} NO_2 + NO_3$$

$$NO_2 + NO_3 \xrightarrow{k_2} NO_2 + O_2 + NO$$

$$NO + NO_3 \xrightarrow{k_3} 2NO_2$$

2. The decomposition of acetaldehyde (CH$_3$CHO) to methane and carbon monoxide is an example of a free radical chain reaction. The overall reaction is believed to occur through the following sequence of steps:

$$CH_3CHO \xrightarrow{k_i} CH_3\bullet + CHO\bullet$$

$$CH_3CHO + CH_3\bullet \xrightarrow{k_2} CH_4 + CH_3CO\bullet$$

$$CH_3CO\bullet \xrightarrow{k_3} CO + CH_3\bullet$$

$$CH_3\bullet + CH_3\bullet \xrightarrow{k_t} C_2H_6$$

Free radical species are indicated by the • symbol. In this case, the free radical CHO• that is formed in the first reaction is kinetically insignificant. Derive a valid rate expression for the decomposition of acetaldehyde. State all of the assumptions that you use in your solution.

3. Chemical reactions that proceed through free radical intermediates can explode if there is a branching step in the reaction sequence. Consider the overall reaction of $A \Rightarrow B$ with initiator I and first-order termination of free radical R:

$$I \xrightarrow{k_1} R \qquad\qquad \text{initiation}$$

$$R + A \xrightarrow{k_2} B + R \qquad\qquad \text{propagation}$$

$$R + A \xrightarrow{k_3} B + R + R \qquad\qquad \text{branching}$$

$$R \xrightarrow{k_4} \text{side product} \qquad \text{termination}$$

Notice that two free radicals are created in the branching step for every one that is consumed.

(a) Find the concentration of A that leads to an explosion.

(b) Derive a rate expression for the overall reaction when it proceeds below the explosion limit.

4. Molecules present in the feed inhibit some reactions catalyzed by enzymes. In this problem, the kinetics of inhibition are investigated (from M. L. Shuler and F. Kargi, *Bioprocess Engineering, Basic Concepts,* Prentice Hall, Englewood Cliffs, NJ, 1992).

(a) Competitive inhibitors are often similar to the substrate and thus compete for the enzyme active site. Assuming that the binding of substrate S and inhibitor I are equilibrated, the following equations summarize the relevant reactions:

$$Ez + S \underset{K_m}{\rightleftharpoons} EzS \xrightarrow{k_2} Ez + P$$

$$Ez + I \underset{K_i}{\rightleftharpoons} EzI$$

Show how the rate of product formation can be expressed as:

$$r = \frac{r_{max} C_S}{K_m\left[1 + \dfrac{C_I}{K_i}\right] + C_S}$$

(b) Uncompetitive inhibitors do not bind to the free enzyme itself, but instead they react with the enzyme-substrate complex. Consider the reaction scheme for uncompetitive inhibition:

$$Ez + S \underset{K_m}{\rightleftharpoons} EzS \xrightarrow{k_2} Ez + P$$

$$EzS + I \underset{K_i}{\rightleftharpoons} EzSI$$

Show how the rate of product formation can be expressed as:

$$r = \frac{\dfrac{r_{max}}{\left(1 + \dfrac{C_I}{K_i}\right)} C_S}{\dfrac{K_m}{\left(1 + \dfrac{C_I}{K_i}\right)} + C_S}$$

(c) Noncompetitive inhibitors bind on sites other than the active site and reduce the enzyme affinity for the substrate. Noncompetitive enzyme inhibition can be described by the following reactions:

$$Ez + S \overset{K_m}{\rightleftarrows} EzS \overset{k_2}{\longrightarrow} Ez + P$$

$$Ez + I \overset{K_i}{\rightleftarrows} EzI$$

$$EzS + I \overset{K_i}{\rightleftarrows} EzSI$$

$$EzI + S \overset{K_m}{\rightleftarrows} EzSI$$

The Michaelis constant, K_m, is assumed to be unaffected by the presence of inhibitor I. Likewise, K_i is assumed to be unaffected by substrate S. Show that the rate of product formations is:

$$r = \frac{r_{max}}{\left(1 + \dfrac{C_I}{K_i}\right)\left(1 + \dfrac{K_m}{C_S}\right)}$$

(d) A high concentration of substrate can also inhibit some enzymatic reactions. The reaction scheme for substrate inhibition is given below:

$$Ez + S \overset{K_m}{\rightleftarrows} EzS \overset{k_2}{\longrightarrow} Ez + P$$

$$EzS + S \overset{K_{Si}}{\rightleftarrows} EzS_2$$

Show that the rate of product formation is:

$$r = \frac{r_{max} C_S}{K_m + C_S + \dfrac{C_S^2}{K_{Si}}}$$

5. Enzymes are more commonly involved in the reaction of two substrates to form products. In this problem, analyze the specific case of the "ping pong bi bi" mechanism [W. W. Cleland, *Biochim. Biophys. Acta,* **67** (1963) 104] for the irreversible enzymatic conversion of

$$A + B \Longrightarrow P + W$$

that takes place through the following simplified reaction sequence:

$$Ez + A \underset{K_{mA}}{\rightleftarrows} EzA \xrightarrow{k_2} Ez' + P$$

$$Ez' + B \underset{K_{mB}}{\rightleftarrows} Ez'B \xrightarrow{k_4} Ez + W$$

where Ez' is simply an enzyme that still contains a molecular fragment of A that was left behind after release of the product P. The free enzyme is regenerated after product W is released. Use the steady-state approximation to show that the rate of product formation can be written as:

$$r = \frac{r_{max}C_A C_B}{C_A C_B + K_\alpha C_A + K_\beta C_B}$$

where K_α, K_β, and r_{max} are collections of appropriate constants.
Hints: $C_{Ez}^0 = C_{Ez} + C_{EzA} + C_{Ez'} + C_{Ez'B}$ and $dC_P/dt = dC_W/dt$

6. Mensah et al. studied the esterification of propionic acid (P) and isoamyl alcohol (A) to isoamyl propionate and water in the presence of the lipase enzyme [P. Mensah, J. L. Gainer, and G. Carta, *Biotechnol. Bioeng.,* **60** (1998) 434.] The product ester has a pleasant fruity aroma and is used in perfumery and cosmetics. This enzyme-catalyzed reaction is shown below:

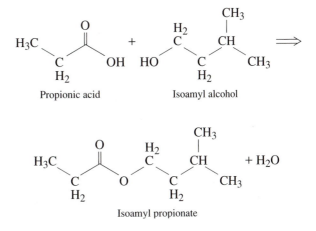

Propionic acid   Isoamyl alcohol

Isoamyl propionate

This reaction appears to proceed through a "ping pong bi bi" mechanism with substrate inhibition. The rate expression for the forward rate of reaction is given by:

$$r = \frac{r_{max}C_PC_A}{C_PC_A + K_1C_P\left(1 + \dfrac{C_P}{K_{Pi}}\right) + K_2C_A}$$

Use nonlinear regression with the following initial rate data to find values of r_{max}, K_1, K_2, and K_{Pi}. Make sure to use several different starting values of the parameters in your analysis. Show appropriate plots that compare the model to the experimental data.

Initial rate data for esterification of propionic acid and isoamyl alcohol in hexane with Lipozyme-IM (immobilized lipase) at 24°C.

C_P (mol L^{-1})	C_A (mol L^{-1})	Rate (mmol h^{-1} g^{-1})
0.15	0.10	1.19
0.15	0.20	1.74
0.15	0.41	1.92
0.15	0.60	1.97
0.15	0.82	2.06
0.15	1.04	2.09
0.33	0.10	0.90
0.33	0.11	1.00
0.33	0.20	1.29
0.33	0.41	1.63
0.33	0.60	1.88
0.33	0.81	1.94
0.33	1.01	1.97
0.60	0.13	0.80
0.60	0.13	0.79
0.60	0.20	1.03
0.60	0.42	1.45
0.60	0.62	1.61
0.60	0.83	1.74
0.60	1.04	1.89
0.72	0.14	0.73
0.72	0.20	0.90
0.72	0.41	1.27
0.72	0.61	1.51
0.72	0.82	1.56
0.72	0.85	1.69
0.72	1.06	1.75
0.93	0.21	0.70
0.93	0.42	1.16
0.93	0.65	1.37
0.93	0.93	1.51
0.93	1.13	1.70

From P. Mensah, J. L. Gainer and G. Carta, *Biotechnol. Bioeng.*, **60** (1998) 434.

7. Combustion systems are major sources of atmospheric pollutants. The oxidation of a hydrocarbon fuel proceeds rapidly (within a few milliseconds) and adiabatically to establish equilibrium among the H/C/O species (CO_2, H_2O, O_2, CO, H, OH, O, etc.) at temperatures that often exceed 2000 K. At such high temperatures, the highly endothermic oxidation of N_2:

$$N_2 + O_2 = 2NO \qquad \Delta H_r = 180.8 \ (kJ \ mol^{-1})$$

becomes important. The reaction mechanism by which this oxidation occurs begins with the attack on N_2 by O:

$$N_2 + O \underset{k_{-1}}{\overset{k_1}{\rightleftharpoons}} NO + N$$

with rate constants of:

$$k_1 = 1.8 \times 10^8 e^{-38400/T} \qquad (m^3 \ mol^{-1}s^{-1})$$
$$k_{-1} = 3.8 \times 10^7 e^{-425/T} \qquad (m^3 \ mol^{-1}s^{-1})$$

The oxygen radical is present in its equilibrium concentration with the major combustion products, that is,

$$1/2 \ O_2 \rightleftharpoons O \qquad K_O = 2170 \ e^{-30000/T} (atm^{1/2})$$

The very reactive N reacts with O_2:

$$N + O_2 \underset{k_{-2}}{\overset{k_2}{\rightleftharpoons}} NO + O$$

$$k_2 = 1.8 \times 10^4 \ Te^{-4680/T} \qquad (m^3 \ mol^{-1}s^{-1})$$
$$k_{-2} = 3.8 \times 10^3 \ Te^{-20800/T} \qquad (m^3 \ mol^{-1}s^{-1})$$

(a) Derive a rate expression for the production of NO in the combustion products of fuel lean (excess air) combustion of a hydrocarbon fuel. Express this rate in terms of the concentrations of NO and major species (O_2 and N_2).

(b) How much NO would be formed if the gas were maintained at the high temperature for a long time? How does that concentration relate to the equilibrium concentration?

(c) How would you estimate the time required to reach that asymptotic concentration?

(Problem provided by Richard Flagan, Caltech.)

8. The reaction of H_2 and Br_2 to form HBr occurs through the following sequence of elementary steps involving free radicals:

$$Br + H_2 \underset{k_{-1}}{\overset{k_1}{\rightleftharpoons}} HBr + H$$

$$H + Br_2 \overset{k_2}{\longrightarrow} HBr + Br$$

$$Br_2 \overset{K_3}{\rightleftharpoons} 2Br$$

Use the fact that bromine radicals are in equilibrium with Br_2 to derive a rate expression of the form:

$$r = \frac{\bar{a}_1[Br_2]^{1/2}[H_2]}{1 + \bar{a}_2\left(\dfrac{[HBr]}{[Br_2]}\right)}$$

9. As a continuation of Exercise 8, calculate the concentration of H radicals present during the HBr reaction at atmospheric pressure, 600 K and 50 percent conversion. The rate constants and equilibrium constant for the elementary steps at 600 K are given below.

$$k_1 = 1.79 \times 10^7 \text{ cm}^3 \text{ mol}^{-1} \text{ s}^{-1}$$

$$k_{-1} = 8.32 \times 10^{12} \text{ cm}^3 \text{ mol}^{-1} \text{ s}^{-1}$$

$$k_2 = 9.48 \times 10^{13} \text{ cm}^3 \text{ mol}^{-1} \text{ s}^{-1}$$

$$K_3 = 8.41 \times 10^{-17} \text{ mol cm}^{-3}$$

10. Consider the series reaction in which B is a reactive intermediate:

$$A \xrightarrow{k_1} B \xrightarrow{k_2} C$$

As discussed in the text, the steady-state approximation applies only after a relaxation time associated with the reactive intermediates. Plot the time dependence of $\bar{\varepsilon}$ (the deviation intermediate concentration from the steady-state value) for several values of k_1/k_2 (0.5, 0.1, 0.05) and $k_2 = 0.1 \text{ s}^{-1}$. What happens when: (a) $k_1 = k_2$ and (b) $k_1 > k_2$?

CHAPTER 5

Heterogeneous Catalysis

5.1 | Introduction

Catalysis is a term coined by Baron J. J. Berzelius in 1835 to describe the property of substances that facilitate chemical reactions without being consumed in them. A broad definition of catalysis also allows for materials that slow the rate of a reaction. Whereas catalysts can greatly affect the rate of a reaction, the equilibrium composition of reactants and products is still determined solely by thermodynamics. *Heterogeneous* catalysts are distinguished from *homogeneous* catalysts by the different phases present during reaction. Homogeneous catalysts are present in the same phase as reactants and products, usually liquid, while heterogeneous catalysts are present in a different phase, usually solid. The main advantage of using a heterogeneous catalyst is the relative ease of catalyst separation from the product stream that aids in the creation of continuous chemical processes. Additionally, heterogeneous catalysts are typically more tolerant of extreme operating conditions than their homogeneous analogues.

A heterogeneous catalytic reaction involves adsorption of reactants from a fluid phase onto a solid surface, surface reaction of adsorbed species, and desorption of products into the fluid phase. Clearly, the presence of a catalyst provides an alternative sequence of elementary steps to accomplish the desired chemical reaction from that in its absence. If the energy barriers of the catalytic path are much lower than the barrier(s) of the noncatalytic path, significant enhancements in the reaction rate can be realized by use of a catalyst. This concept has already been introduced in the previous chapter with regard to the Cl catalyzed decomposition of ozone (Figure 4.1.2) and enzyme-catalyzed conversion of substrate (Figure 4.2.4). A similar reaction profile can be constructed with a heterogeneous catalytic reaction.

For example, G. Ertl (*Catalysis: Science and Technology,* J. R. Anderson and M. Boudart, Eds., vol. 4, Springer-Verlag, Berlin, 1983, p. 245) proposed the thermochemical kinetic profile depicted in Figure 5.1.1 for the platinum-catalyzed oxidation of carbon monoxide according to the overall reaction $CO + \frac{1}{2} O_2 \Rightarrow CO_2$. The first step in the profile represents the adsorption of carbon monoxide and

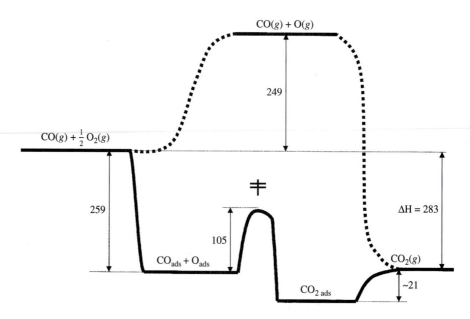

Figure 5.1.1 |

Schematic energy diagram for the oxidation of CO and a Pt catalyst. (From data presented by G. Ertl in *Catalysis: Science and Technology,* J. R. Anderson and M. Boudart, Eds., vol. 4, Springer-Verlag, Berlin, 1983, p. 245.) All energies are given in kJ mol^{-1}. For comparison, the heavy dashed lines show a noncatalytic route.

dioxygen onto the catalyst. In this case, adsorption of dioxygen involves dissociation into individual oxygen atoms on the Pt surface. The product is formed by addition of an adsorbed oxygen atom (O_{ads}) to an adsorbed carbon monoxide molecule (CO_{ads}). The final step in the catalytic reaction is desorption of adsorbed carbon dioxide (CO_{2ads}) into the gas phase. The Pt catalyst facilitates the reaction by providing a low energy path to dissociate dioxygen and form the product. The noncatalytic route depicted in Figure 5.1.1 is extremely slow at normal temperatures due to the stability of dioxygen molecules.

VIGNETTE 5.1.1

Perhaps one of the greatest triumphs of modern surface science is that rates of catalytic reactions on supported, transition metal particles can often be reproduced on very well-defined single crystal surfaces. The ability to reliably interrogate catalytic chemistry on single crystals allows for methodical exploration of the influence of atomic structure on catalytic activity [C. M. Friend, *Scientific American,* **268** (1993) 74]. Single crystals can be cut and processed to expose various low energy surface planes having specific atomic configurations. These various arrangements of surface atoms can be understood from the periodic nature of three-dimensional crystals. The three crystal structures of transition metals relevant to catalysis are called face-centered cubic, body-centered cubic, and hexagonal (see Figure 5.1.2).

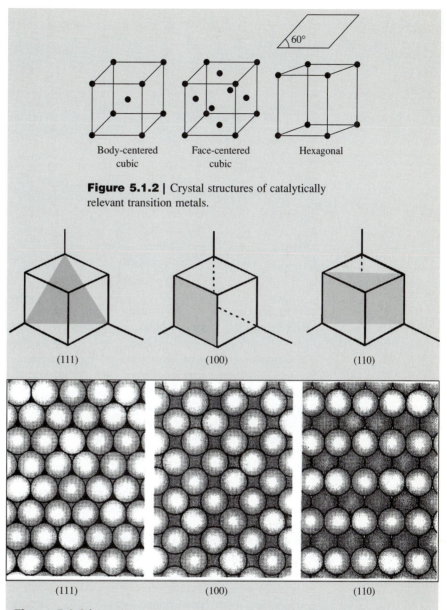

Figure 5.1.2 | Crystal structures of catalytically relevant transition metals.

Figure 5.1.3 | Atomic arrangements of the low-index surface planes of an FCC crystal. (Adapted from R. Masel, *Principles of Adsorption and Reaction on Solid Surfaces,* Wiley, New York, copyright © 1996, p. 38, by permission of John Wiley & Sons, Inc.)

Single crystal surfaces are associated with planes in the unit cells pictured in Figure 5.1.2 and are denoted by indices related to the unit cell parameters. Several examples of various low-index surface planes are shown in Figure 5.1.3 for the face-centered cubic structure.

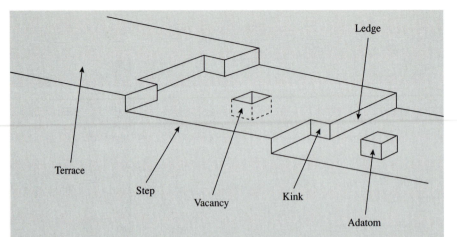

Figure 5.1.4 | Schematic representation of a single crystal surface.

On real, single crystal surfaces, atomic positions associated with ideal surface planes are perturbed by relaxation effects and the presence of atomic-scale defects. It is, therefore, more correct to think of single crystal surfaces being composed of terraces of stable planes separated by atomic-scale steps or ledges that may have kinks. In addition, various point defects like atomic vacancies or surface adatoms may be present. A schematic representation of single crystal surfaces is shown in Figure 5.1.4. Interestingly, the number density of surface atoms in the stable planes of transition metals is about 10^{15} cm^{-2}, regardless of crystal face. This value of surface atom density is a good starting point for estimating the total number of adsorption sites or active sites on metal surfaces.

Since the number of atoms on the surface of a bulk metal or metal oxide is extremely small compared to the number of atoms in the interior, bulk materials are often too costly to use in a catalytic process. One way to increase the effective surface area of a valuable catalytic material like a transition metal is to disperse it on a support. Figure 5.1.5 illustrates how Rh metal appears when it is supported as nanometer size crystallites on a silica carrier. High-resolution transmission electron microscopy reveals that metal crystallites, even as small as 10 nm, often expose the common low-index faces commonly associated with single crystals. However, the surface to volume ratio of the supported particles is many orders of magnitude higher than an equivalent amount of bulk metal. In fact, it is not uncommon to use catalysts with 1 nm sized metal particles where nearly every atom can be exposed to the reaction environment.

Estimation of the number of exposed metal atoms is rather straightforward in the case of a single crystal of metal since the geometric surface area can be measured and the number density of surface atoms can be found from the crystal structure.

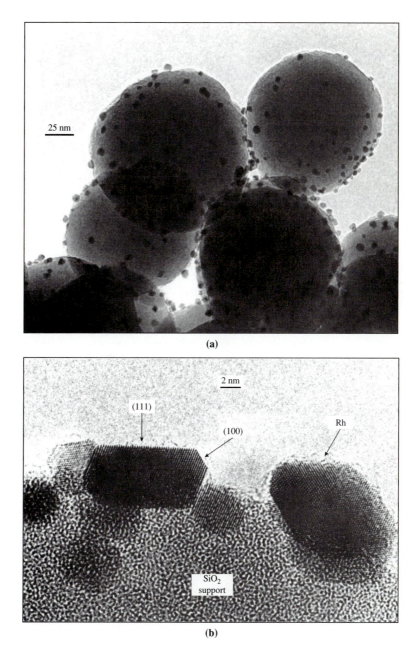

(a)

(b)

Figure 5.1.5 |

(a) Rhodium metal particles supported on silica carrier. (b) High-resolution electron micrograph shows how small supported Rh crystallites expose low-index faces. (Top photo courtesy of A. K. Datye. Bottom photo from "Modeling of heterogeneous catalysts using simple geometry supports" by A. K. Datye in *Topics in Catalysis,* vol. 13:131, copyright © 2000 by Kluwer Academic, reproduced by permission of the publisher and the author.)

For supported metal catalysts, no simple calculation is possible. A direct measurement of the metal crystallite size or a titration of surface metal atoms is required (see Example 1.3.1). Two common methods to estimate the size of supported crystallites are transmission electron microscopy and X-ray diffraction line broadening analysis. Transmission electron microscopy is excellent for imaging the crystallites, as illustrated in Figure 5.1.5. However, depending on the contrast difference with the support, very small crystallites may not be detected. X-ray diffraction is usually ineffective for estimating the size of very small particles, smaller than about 2 nm. Perhaps the most common method for measuring the number density of exposed metal atoms is selective *chemisorption* of a probe molecule like H_2, CO, or O_2.

Selective chemisorption uses a probe molecule that does not interact significantly with the support material but forms a strong chemical bond to the surface metal atoms of the supported crystallites. Chemisorption will be discussed in more detail in Section 5.2. Dihydrogen is perhaps the most common probe molecule to measure the fraction of exposed metal atoms. An example of H_2 chemisorption on Pt is shown below:

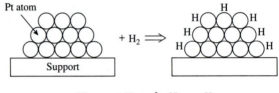

$$2Pt_{surface} + H_2 \Longrightarrow 2Pt_{surface}H$$

The exact stoichiometry of the Pt-H surface complex is still a matter of debate since it depends on the size of the metal particle. For many supported Pt catalysts, an assumption of 1 H atom adsorbing for every 1 Pt surface atom is often a good one. Results from chemisorption can be used to calculate the dispersion of Pt, or the fraction of exposed metal atoms, according to:

$$\text{Fraction exposed} = \text{Dispersion} = \frac{2 \times (H_2 \text{ molecules chemisorbed})}{\text{Total number of Pt atoms}}$$

If a shape of the metal particle is assumed, its size can be estimated from chemisorption data. For example, a good rule of thumb for spherical particles is to invert the dispersion to get the particle diameter in nanometers:

$$\text{Particle diameter (nm)} = \frac{1}{\text{Dispersion}}$$

Table 5.1.1 compares the average diameter of Pt particles supported on alumina determined by chemisorption of two different probe molecules, X-ray diffraction, and electron microscopy. The excellent agreement among the techniques used to

Table 5.1.1 | Determination of metal particle size on Pt/Al$_2$O$_3$ catalysts by chemisorption of H$_2$ and CO, X-ray diffraction, and transmission electron microscopy.

	Diameter of Pt particles (nm)			
% Pt	H$_2$	CO	X-ray diffraction	Electron microscopy
0.6[a]	1.2	1.3	1.3	1.6
2.0[a]	1.6	1.8	2.2	1.8
3.7[a]	2.7	2.9	2.7	2.4
3.7[b]	3.9	—	4.6	5.3

[a]Pretreatment temperature of 500°C.
[b]Pretreatment temperature of 800°C.
Source: Renouprez et al., *J. Catal.*, **34** (1974) 411.

characterize these model Pt catalysts shows the validity in using chemisorption to estimate particle size.

The reason for measuring the number of exposed metal atoms in a catalyst is that it allows reaction rates to be normalized to the amount of active component. As defined in Chapter 1, the rate expressed per active site is known as the *turnover frequency,* r$_t$. Since the turnover frequency is based on the number of active sites, it should not depend on how much metal is loaded into a reactor or how much metal is loaded onto a support. Indeed, the use of turnover frequency has made possible the comparison of rates measured on different catalysts in different laboratories throughout the world.

VIGNETTE 5.1.2

Dumesic and co-workers measured the rate of ethylene hydrogenation:

$$CH_2 = CH_2 + H_2 \Rightarrow CH_3 - CH_3$$

over a variety of Pt catalysts in a flow reactor with a feed of 25 torr of ethylene, 150 torr of dihydrogen, and 585 torr of helium [R. D. Cortright, S. A. Goddard, J. E. Rekoske, and J. A. Dumesic, *J. Catal.*, **127** (1991) 342]. Three of the catalysts consisted of Pt particles supported on Cab-O-Sil silica gel while the fourth was a bulk Pt wire. The fraction of Pt exposed on the supported catalysts, determined by chemisorption of dihydrogen, was 0.7 on the catalyst with 1.2 wt% Pt and unity on the lower loaded Pt catalysts. The number of Pt atoms on the surface of the Pt wire was estimated from the geometric surface area. Figure 5.1.6 shows the temperature dependence of the turnover frequency (TOF) on all of the catalysts at equivalent conditions. Over the entire temperature range, the turnover frequency was found to be independent of the dispersion of the metal. The TOF measured on a bulk Pt wire was virtually the same as that found on 1 nm supported Pt clusters. Reactions that reveal TOFs that are independent of dispersion are called structure insensitive.

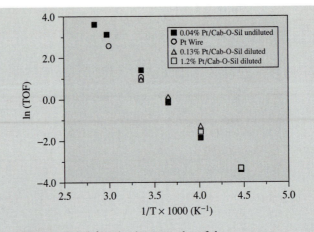

Figure 5.1.6 | Arrhenius-type plot of the turnover frequency for ethylene hydrogenation on bulk and supported Pt catalysts. (Figure from "Kinetic Study of Ethylene Hydrogenation" by R. D. Cortright, S. A. Goddard, J. E. Rekoske, and J. A. Dumesic, in *Journal of Catalysis*, Volume 127:342, copyright © 1991 by Academic Press, reproduced by permission of the publisher and the authors.)

5.2 | Kinetics of Elementary Steps: Adsorption, Desorption, and Surface Reaction

The necessary first step in a heterogeneous catalytic reaction involves activation of a reactant molecule by adsorption onto a catalyst surface. The activation step implies that a fairly strong chemical bond is formed with the catalyst surface. This mode of adsorption is called *chemisorption,* and it is characterized by an enthalpy change typically greater than 80 kJ mol^{-1} and sometimes greater than 400 kJ mol^{-1}. Since a chemical bond is formed with the catalyst surface, chemisorption is specific in nature, meaning only certain adsorbate-adsorbent combinations are possible. Chemisorption implies that only a single layer, or *monolayer,* of adsorbed molecules is possible since every adsorbed atom or molecule forms a strong bond with the surface. Once the available surface sites are occupied, no additional molecules can be chemisorbed.

Every molecule is capable of weakly interacting with any solid surface through van der Waals forces. The enthalpy change associated with this weak adsorption mode, called *physisorption,* is typically 40 kJ mol^{-1} or less, which is far lower than the enthalpy of chemical bond formation. Even though physisorbed molecules are not activated for catalysis, they may serve as precursors to chemisorbed molecules. More than one layer of molecules can physisorb on a surface since only van der Waals interactions are involved. The number of physisorbed molecules that occupy a monolayer on the surface of any material can be used to calculate its geometric surface area if the cross-sectional area of the adsorbate molecule is known.

VIGNETTE 5.2.1

BET Method for Evaluation of Surface Area

In 1919, the Fixed Nitrogen Laboratory was founded in Washington, D.C. to help the United States establish a synthetic ammonia industry. This was done to ensure a reliable supply of nitrate fertilizers and explosives. It was in this laboratory environment that Stephen Brunauer and Paul Emmet worked together to study the adsorption of dinitrogen on promoted iron, the catalyst commonly used in ammonia synthesis reactors today. It became clear to the researchers that some measure of the catalyst surface area was needed to place their studies on a more quantitative basis. After enlisting the help of Edward Teller, then a professor at George Washington University, Brunauer and Emmett were able to describe in very simple terms the phenomenon of multilayer adsorption of weakly adsorbed molecules, or physisorption. Figure 5.2.1 shows a typical physical adsorption isotherm of N_2 on a powdered catalyst surface.

There is no obvious transition in the isotherm that can be attributed to the formation of a complete monolayer of adsorbed N_2. Brunauer, Emmett, and Teller reasoned that molecules like N_2 do not adsorb on a surface in a layer by layer fashion, but instead begin to form multilayers before completion of a full monolayer. This mode of adsorption is schematically depicted in Figure 5.2.2. The following critical assumptions were made to enable solution of the problem:

1. The layers are densely packed.
2. The heat of adsorption for the first layer is greater than the second (and higher) layers.

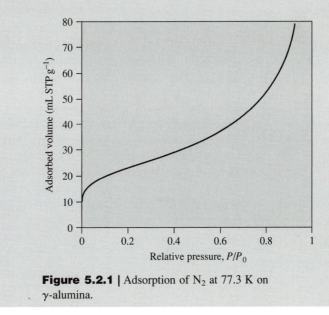

Figure 5.2.1 | Adsorption of N_2 at 77.3 K on γ-alumina.

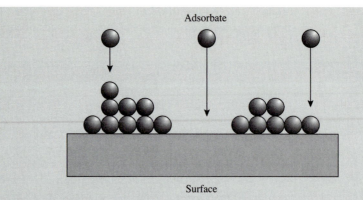

Adsorbate

Surface

Figure 5.2.2 | Illustration of multilayer adsorption.

3. The heat of adsorption is constant for all molecules in the first layer.
4. The heat of adsorption for the second (and higher) layers is the same as the heat of liquefaction.

A derivation of the multilayer isotherm is provided in the landmark paper by S. Brunauer, P. H. Emmett, and E. Teller [*J. Am. Chem. Soc.*, **60** (1938) 309] and will not be presented here. The commonly used linearized version of the isotherm developed by Brunauer, Emmett, and Teller bears their initials in the now famous "BET equation":

$$\frac{P}{V_{ads}(P_0 - P)} = \frac{1}{cV_m} + \left(\frac{c-1}{cV_m}\right)\frac{P}{P_0} \tag{5.2.1}$$

where P is the equilibrium pressure of gas with the surface, P_0 is the saturation vapor pressure, V_{ads} is the volume of gas (STP) adsorbed by the sample, V_m is the volume of gas (STP) corresponding to the formation of a monolayer, and c is a fitted constant. A plot of the left-hand side of Equation (5.2.1) versus P/P_0 yields a straight line having a slope equal to $(c-1)/(cV_m)$ and an intercept equal to $1/(cV_m)$. The volume of gas in a monolayer, V_m, can then be calculated from fitted values of both the slope and intercept. The best fits to experimental data are obtained by using a restricted pressure range (P/P_0) from 0.05 to 0.3. To arrive at a surface area, the number of molecules adsorbed in a monolayer is multiplied by the cross-sectional area of the adsorbing gas. For N_2, the molecular cross-sectional area is 0.162 nm^2. Even though the BET equation for multilayer adsorption was developed in the late 1930s and involved a few assumptions that are not strictly correct, the universal applicability of the method for determining surface area has allowed its use to continue even today. Indeed, nearly all surface area measurements on powdered catalysts are still analyzed by the BET method. Many commercial instruments perform the so-called one-point BET method. In doing so, the value of c is assumed to be so large that Equation (5.2.1) reduces to:

$$\frac{P}{V_{ads}(P_0 - P)} = \left(\frac{1}{V_m}\right)\frac{P}{P_0} \tag{5.2.2}$$

Notice that V_m can be obtained from one data point (V_{ads}, P/P_0) by the use of Equation (5.2.2). A problem with this method is that c may not be large enough to justify simplification of Equation (5.2.1) and is *a priori* usually unknown. Thus, the one-point method is normally useful for analysis of numerous samples of known composition for which c has been determined.

The potential energy diagram for the chemisorption of hydrogen atoms on nickel is schematically depicted in Figure 5.2.3. As molecular hydrogen approaches the surface, it is trapped in a shallow potential energy well associated with the physisorbed state having an enthalpy of physisorption ΔH_p. The deeper well found closer to the surface with enthalpy ΔH_c is associated with the hydrogen *atoms* chemisorbed on nickel. There can be an activation barrier to chemisorption, E, which must be overcome to reach a chemisorbed state from the physisorbed molecule. Since molecular hydrogen (dihydrogen) is dissociated to form chemisorbed hydrogen atoms, this phenomenon is known as *dissociative chemisorption*.

Because rates of heterogeneous catalytic reactions depend on the amounts of chemisorbed molecules, it is necessary to relate the fluid phase concentrations of

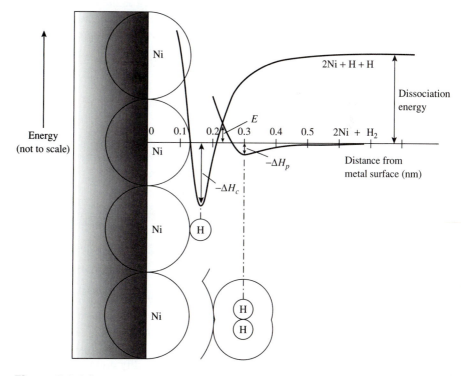

Figure 5.2.3 |
Potential energy diagram for the chemisorption of hydrogen on nickel.

reactants to their respective coverages on a solid surface. For simplicity, the following discussion assumes that the reactants are present in a single fluid phase (i.e., either liquid or gas), all surface sites have the same energetics for adsorption, and the adsorbed molecules do not interact with each other. The active site for chemisorption on the solid catalyst will be denoted by *, with a surface concentration [*]. The adsorption of species A can then be expressed as:

$$A + * \underset{k_{des}}{\overset{k_{ads}}{\rightleftarrows}} A*$$

where $A*$ corresponds to A chemisorbed on a surface site and k_{ads} and k_{des} refer to the rate constants for adsorption and desorption, respectively. Thus, the net rate of adsorption is given by:

$$r = k_{ads}[A][*] - k_{des}[A*] \tag{5.2.3}$$

Since the net rate of adsorption is zero at equilibrium, the equilibrium relationship is:

$$K_{ads} = \frac{k_{ads}}{k_{des}} = \frac{[A*]}{[A][*]} \tag{5.2.4}$$

The fractional coverage θ_A is defined as the fraction of total surface adsorption sites that are occupied by A. If $[*]_0$ represents the number density of all adsorption sites (vacant and occupied) on a catalyst surface, then:

$$[*]_0 = [*] + [A*] \tag{5.2.5}$$

and

$$\theta_A = \frac{[A*]}{[*]_0} \tag{5.2.6}$$

Combining Equations (5.2.4–5.2.6) gives expressions for $[*]$, $[A*]$, and θ_A in terms of the measurable quantities $[A]$ and K_{ads}:

$$[*] = \frac{[*]_0}{1 + K_{ads}[A]} \tag{5.2.7}$$

$$[A*] = \frac{K_{ads}[A][*]_0}{1 + K_{ads}[A]} \tag{5.2.8}$$

$$\theta_A = \frac{K_{ads}[A]}{1 + K_{ads}[A]} \tag{5.2.9}$$

Equation (5.2.9) is known as the *Langmuir Adsorption Isotherm,* and it is depicted in Figure 5.2.4. At low concentrations of A, the fractional surface coverage is proportional to $[A]$, with a proportionality constant of K_{ads}, whereas, at high concentrations of A, the surface coverage is unity and independent of $[A]$ (i.e., saturation of the available sites).

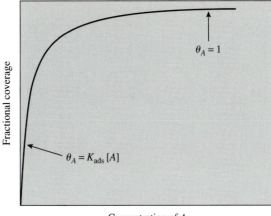

Figure 5.2.4 |
Langmuir Adsorption Isotherm: Fractional coverage θ_A
versus fluid phase concentration of A.

When more than one type of molecule can adsorb on the catalyst surface, the competition for unoccupied sites must be considered. For the adsorption of molecule A in the presence of adsorbing molecule B, the site balance becomes:

$$[*]_0 = [*] + [A*] + [B*]$$

and θ_A, θ_B can be expressed as:

$$\theta_A = \frac{K_{ads\,A}[A]}{1 + K_{ads\,A}[A] + K_{ads\,B}[B]} \tag{5.2.10}$$

$$\theta_B = \frac{K_{ads\,B}[B]}{1 + K_{ads\,A}[A] + K_{ads\,B}[B]} \tag{5.2.11}$$

where the equilibrium constants are now associated with adsorption/desorption of either A or B.

As mentioned above, some molecules like H_2 and O_2 can adsorb dissociatively to occupy two surface sites upon adsorption. The reverse reaction is known as associative desorption. In these cases, the rate of adsorption now depends on the number of *pairs* of available surface sites according to:

$$A_2 + ** \underset{k_{des}}{\overset{k_{ads}}{\rightleftarrows}} A**A$$

where ** refers to a pair of adjacent free sites and $A**A$ refers to a pair of adjacent occupied sites. M. Boudart and G. Djega-Mariadassou (*Kinetics of Heterogeneous Catalytic Reactions,* Princeton University Press, Princeton, 1984) present expressions for [**] and [$A**A$], based on the statistics of a partially occupied square

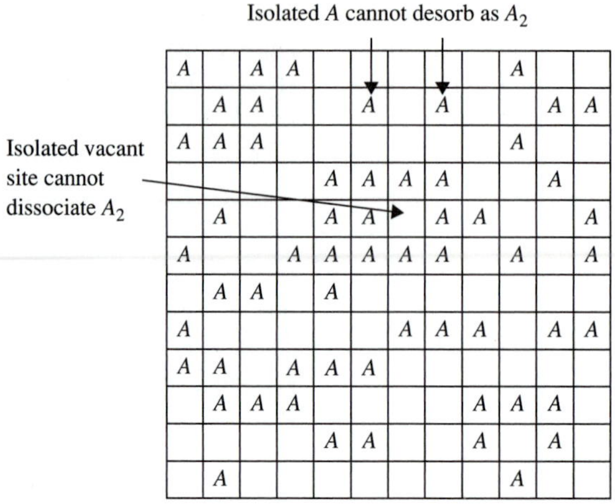

Isolated A cannot desorb as A_2

Isolated vacant
site cannot
dissociate A_2

Figure 5.2.5 |
Schematic depiction of a square lattice populated by A
atoms. Empty squares represent vacant surface sites.

lattice, shown in Figure 5.2.5. The surface concentrations of adjacent pairs of unoccupied and occupied sites are:

$$[**] = \frac{2[*]^2}{[*]_0} \tag{5.2.12}$$

$$[A**A] = \frac{2[A*]^2}{[*]_0} \tag{5.2.13}$$

Thus, the net rate of dissociative adsorption becomes:

$$r = k_{ads}[A_2][**] - k_{des}[A**A]$$

and upon substitution for $[**]$ and $[A**A]$ with Equations (5.2.12) and (5.2.13), respectively, gives:

$$r = \frac{2k_{ads}}{[*]_0}[A_2][*]^2 - \frac{2k_{des}}{[*]_0}[A*]^2 \tag{5.2.14}$$

The value of two before each rate constant in Equation (5.2.14) is a statistical factor
that contains the number of nearest neighbors to an adsorption site on a square lattice
and accounts for the fact that indistinguishable neighboring sites should not be double counted. For simplicity, this statistical factor will be lumped into the bimolecular
rate constant to yield the following rate expression for dissociative adsorption:

$$r = \frac{k_{ads}}{[*]_0}[A_2][*]^2 - \frac{k_{des}}{[*]_0}[A*]^2 \tag{5.2.15}$$

The total number of adsorption sites on the surface now appears explicitly in the rate expression for this elementary adsorption step. Since the site balance is the same as before (Equation 5.2.5), the equilibrium adsorption isotherm can be calculated in the manner described above:

$$K_{ads} = \frac{k_{ads}}{k_{des}} = \frac{[A*]^2}{[A_2][*]^2} \tag{5.2.16}$$

$$[*] = \frac{[*]_0}{1 + (K_{ads}[A_2])^{1/2}} \tag{5.2.17}$$

$$[A*] = \frac{(K_{ads}[A_2])^{1/2}[*]_0}{1 + (K_{ads}[A_2])^{1/2}} \tag{5.2.18}$$

$$\theta_A = \frac{(K_{ads}[A_2])^{1/2}}{1 + (K_{ads}[A_2])^{1/2}} \tag{5.2.19}$$

At low concentrations of A_2, the fractional surface coverage is proportional to $[A_2]^{1/2}$, which is quite different than the adsorption isotherm derived above for the case without dissociation. In terms of fractional surface coverage, the net rate of dissociative adsorption is expressed by:

$$r = k_{ads}[*]_0[A_2](1 - \theta_A)^2 - k_{des}[*]_0\theta_A^2 \tag{5.2.20}$$

As expected, the rate is proportional to the total number of adsorption sites $[*]_0$.

The next step in a catalytic cycle after adsorption of the reactant molecules is a surface reaction, the simplest of which is the unimolecular conversion of an adsorbed species into a product molecule. For example, the following two-step sequence represents the conversion of A into products through the irreversible surface reaction of A:

$$
\begin{array}{lll}
(1) & A + * \rightleftharpoons A* & \text{(reversible adsorption)} \\
(2) & A* \xrightarrow{k_2} \text{products} + * & \text{(surface reaction)} \\
\hline
& A \Longrightarrow \text{products} & \text{(overall reaction)}
\end{array}
$$

Since the surface reaction is irreversible (one-way), the mechanistic details involved in product formation and desorption are not needed to derive the rate equation. The rate of the overall reaction is:

$$r = k_2[A*] = k_2[*]_0\theta_A \tag{5.2.21}$$

If the adsorption of A is nearly equilibrated, the Langmuir adsorption isotherm can be used to find θ_A, and the final rate expression simplifies to:

$$r = \frac{k_2 K_{ads}[*]_0[A]}{1 + K_{ads}[A]} \tag{5.2.22}$$

EXAMPLE 5.2.1

Use the steady-state approximation to derive the rate expression given in Equation (5.2.22).

$$(1) \qquad A + * \underset{k_{des}}{\overset{k_{ads}}{\rightleftharpoons}} A* \qquad\qquad \text{(reversible adsorption)}$$

$$(2) \qquad \underline{A* \xrightarrow{k_2} \text{products} + *} \qquad \text{(surface reaction)}$$

$$\qquad A \Longrightarrow \text{products} \qquad\qquad \text{(overall reaction)}$$

■ **Answer**

The rate of the reaction is:

$$r = k_2[A*] \tag{5.2.23}$$

To solve the problem, $[A*]$ and $[*]$ must be evaluated by utilizing the site balance:

$$[*]_0 = [A*] + [*] \tag{5.2.24}$$

and the steady-state approximation:

$$\frac{d[A*]}{dt} = 0 = k_{ads}[A][*] - k_{des}[A*] - k_2[A*] \tag{5.2.25}$$

Solving Equations (5.2.24) and (5.2.25) simultaneously yields the following expression for the concentration of reactive intermediate:

$$[A*] = \frac{\left(\dfrac{k_{ads}}{k_{des}}\right)[A][*]_0}{1 + \left(\dfrac{k_{ads}}{k_{des}}\right)[A] + \dfrac{k_2}{k_{des}}} \tag{5.2.26}$$

The ratio k_{ads}/k_{des} is simply the adsorption equilibrium constant K_{ads}. Thus, the rate expression in Equation (5.2.23) can be rewritten as:

$$r = \frac{k_2 K_{ads}[A][*]_0}{1 + K_{ads}[A] + \dfrac{k_2}{k_{des}}} \tag{5.2.27}$$

by using Equation (5.2.26) together with K_{ads}, and is the same as Equation (5.2.22) except for the last term in the denominator. The ratio k_2/k_{des} represents the likelihood of $A*$ reacting to form product compared to simply desorbing from the surface. Equation (5.2.22) is based on the assumption that adsorption of A is nearly equilibrated, which means k_{des} is much greater than k_2. Thus, for this case, k_2/k_{des} can be ignored in the denominator of Equation (5.2.27).

The previous discussion involved a two-step sequence for which the adsorption of reactant is nearly equilibrated (quasi-equilibrated). The free energy change associated with the quasi-equilibrated adsorption step is negligible compared to that of the surface reaction step. The surface reaction is thus called the *rate-determining step* (RDS) since nearly all of the free energy change for the overall reaction is associated with that step. In general, if a rate determining step exists, all other steps in the catalytic sequence are assumed to be quasi-equilibrated. The two-step sequence discussed previously is written in the notation outlined in Chapter 1 as:

(1) $A + * \rightleftharpoons A*$ (quasi-equilibrated)

(2) $A* \xrightarrow{} \text{products} + *$ (rate-determining step)

$A \Longrightarrow \text{products}$ (overall reaction)

The intrinsic rate at which a catalytic cycle turns over on an active site is called the *turnover frequency,* r_t, of a catalytic reaction and is defined as in Chapter 1 [Equation (1.3.9)] as:

$$r_t = \frac{1}{\overline{S}} \frac{dn}{dt} \tag{5.2.28}$$

where $\overline{S}$ is the number of active sites in the experiment. As mentioned earlier, quantifying the number of active sites on a surface is problematic. With metals and some metal oxides, often the best one can do is to count the total number of exposed surface atoms per unit surface area as an approximation of $[*]_0$. Thus, the turnover frequency for the reaction rate expressed by Equation (5.2.22) is calculated by simply dividing the rate by $[*]_0$:

$$r_t = \frac{r}{[*]_0} = \frac{k_2 K_{ads}[A]}{1 + K_{ads}[A]} \tag{5.2.29}$$

Surface atoms on real catalysts reside in a variety of coordination environments depending on the exposed crystal plane (see Figure 5.1.3) and may exhibit different catalytic activities in a given reaction. Thus, a turnover frequency based on $[*]_0$ will be an average value of the catalytic activity. In fact, the calculated turnover frequency is a lower bound to the true activity because only a fraction of the total number of surface atoms may contribute to the reaction rate. Nevertheless, the concept of a turnover frequency on a uniform surface has proven to be very useful in relating reaction rates determined on metal single crystals, metal foils, and supported metal particles.

VIGNETTE 5.2.2

Ladas et al. measured the rate of CO oxidation by O_2 on a variety of Al_2O_3-supported Pd catalysts at 445 K, CO pressure equal to 1.2×10^{-4} Pa, and P_{O_2}/P_{CO} equal to 1.1 [S. Ladas, H. Poppa, and M. Boudart, *Surf. Sci.*, **102** (1981) 151]. The alumina was a flat single crystal onto which Pd particles were deposited by a metal evaporation method. The resulting Pd particle size was measured directly by transmission electron microscopy.

Table 5.2.1 | Turnover rates[a] of ammonia synthesis at 678 K, stoichiometric feed, 1 atm.

Catalyst	Particle size (nm)	r_t ($\times 10^3$ s^{-1})
1% Fe/MgO	1.5	1.0
5% Fe/MgO	4.0	9.0
40% Fe/MgO	30.0	35.0

[a]Rates measured at constant conversion (15% of equilibrium).
Source: J. A. Dumesic, H. Topsoe, and M. Boudart, *J. Catal.*, **37** (1975) 513.

Ladas et al. found that the turnover frequency, based on exposed Pd atoms determined by CO chemisorption, was 0.012 s^{-1} and independent of Pd particle size over the range of 1.5 to 8 nm. Thus, CO oxidation is classified as a *structure insensitive* reaction.

A very different result is found when ammonia synthesis is catalyzed by supported iron particles. The results in Table 5.2.1 indicate that the turnover frequency for ammonia synthesis varies by more than one order of magnitude as the Fe particle size changes from 1.5 to 30 nm. Ammonia synthesis on iron is therefore called *structure sensitive* since the turnover frequency varies significantly with particle size. Because the reaction rate does not scale proportionally with the number of surface metal atoms, the active site consists of a group, or *ensemble,* of surface atoms arranged in a particular configuration. Selective chemisorption of N_2 (instead of H_2 or CO) can be used to count these specific ensembles on Fe particles. Indeed, the dependence of the turnover frequency on Fe particle size vanishes when the active site density is based on N_2 chemisorption [H. Topsoe, N. Topsoe, H. Bohlbro, and J. A. Dumesic, *Proc. 7th Int. Cong. Catalysis*, T. Seiyama and K. Tanabe, Eds. (Kodansha: Tokyo) 1981, p. 247].

The presence of poisons or alloy additions can dramatically alter the observed rate of a structure sensitive reaction whereas the rates of structure insensitive reactions are much less affected. Sinfelt et al. studied ethane hydrogenolysis to form methane (a structure sensitive reaction) and cyclohexane dehydrogenation to benzene (a structure insensitive reaction) over a series of Ni-Cu alloys [J. H. Sinfelt, J. L. Carter, and D. J. C. Yates, *J. Catal.*, **24** (1972) 283]. Nickel is the active component for both reactions. The data provided in Figure 5.2.6 show that the rate of cyclohexane dehydrogenation was affected very little by alloying Cu into Ni over a wide range of alloy composition. In contrast, the rate of ethane hydrogenolysis decreased by many orders of magnitude with the small addition of copper.

These results can be easily rationalized if ethane hydrogenolysis requires an ensemble of multiple Ni surface atoms to form an active site. The concentration of active ensembles will decline sharply with composition as the active Ni atoms are diluted with inactive Cu atoms in the surface. Additional information on how bimetallic catalysts affect chemical reactions can be found in the excellent monograph by Sinfelt (J. H. Sinfelt, *Bimetallic Catalysts: Discoveries, Concepts, and Applications,* Wiley, New York, 1983).

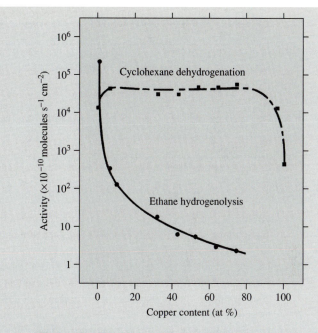

Figure 5.2.6 | Effect of alloy composition on the rates of ethane hydrogenolysis and cyclohexane dehydrogenation on Ni-Cu catalysts. (Figure from "Catalytic Hydrogenolysis and Dehydrogenation Over Copper-Nickel Alloys" by J. H. Sinfelt, J. L. Carter, and D. J. C. Yates in *Journal of Catalysis,* Volume 24:283, copyright © 1972 by Academic Press, reproduced by permission of the publisher.)

The rate of a structure sensitive reaction catalyzed by a metal single crystal is also a function of the exposed plane. As illustrated in Figure 5.1.3, the low-index planes of common crystal structures have different arrangements of surface atoms. Thus, caution must be exercised when the rates of structure sensitive reactions measured on single crystals are compared to those reported for supported metal particles.

As an extension to concepts discussed earlier, a rate expression for the reaction of A and B to form products can be developed by assuming an irreversible, rate-determining, bimolecular surface reaction:

(1) $A + * \rightleftharpoons A*$ (quasi-equilibrated adsorption of A)

(2) $B + * \rightleftharpoons B*$ (quasi-equilibrated adsorption of B)

(3) $A* + B* \xrightarrow{k_2} \text{products} + 2*$ (surface reaction, RDS)

        $A + B \Longrightarrow \text{products}$ (overall reaction)

Table 5.2.2 | Pre-exponential factors for selected unimolecular surface reactions.

Reaction	Surface	Pre-exponential factor (s^{-1})
$CO_{ads} \rightarrow CO_g$	Cu(001)	10^{14}
$CO_{ads} \rightarrow CO_g$	Ru(001)	$10^{19.5}$
$CO_{ads} \rightarrow CO_g$	Pt(111)	10^{14}
$Au_{ads} \rightarrow Au_g$	W(110)	10^{11}
$Cu_{ads} \rightarrow Cu_g$	W(110)	$10^{14.5}$
$Cl_{ads} \rightarrow Cl_g$	Ag(110)	10^{20}
$NO_{ads} \rightarrow NO_g$	Pt(111)	10^{16}
$CO_{ads} \rightarrow CO_g$	Ni(111)	10^{15}

Adapted from R. Masel, *Principles of Adsorption and Reaction on Solid Surfaces,* Wiley, New York, 1996, p. 607.

Before A and B can react, they must both adsorb on the catalyst surface. The next event is an elementary step that proceeds through a reaction of adsorbed intermediates and is often referred to as a *Langmuir-Hinshelwood* step. The rate expression for the bimolecular reaction depends on the number density of adsorbed A molecules that are adjacent to adsorbed B molecules on the catalyst surface. This case is similar to the one developed previously for the recombinative desorption of diatomic gases [reverse reaction step in Equation (5.2.20)] except that two different atomic species are present on the surface. A simplified rate expression for the bimolecular reaction is:

$$r = k_3 [A^*][B^*]/[^*]_0 = k_3[^*]_0 \, \theta_A \, \theta_B \tag{5.2.30}$$

where θ_A and θ_B each can be expressed in the form of the Langmuir isotherm for competitive adsorption of A and B that are presented in Equations (5.2.10) and (5.2.11) and k_3 is the rate constant for step 3. Thus, the overall rate of reaction of A and B can be expressed as:

$$r = \frac{k_3 K_{adsA} K_{adsB} [^*]_0 [A][B]}{(1 + K_{adsA}[A] + K_{adsB}[B])^2} \tag{5.2.31}$$

Notice that the denominator is squared for a bimolecular surface reaction. In general, the exponent on the denominator is equal to the number of sites participating in a rate-determining surface-catalyzed reaction. Since trimolecular surface events are uncommon, the exponent of the denominator rarely exceeds 2.

It is instructive to compare the values of pre-exponential factors for elementary step rate constants of simple surface reactions to those anticipated by transition state theory. Recall from Chapter 2 that the pre-exponential factor $\overline{A}$ is on the order of $\overline{k}T/h = 10^{13} \text{ s}^{-1}$ when the entropy change to form the transition state is negligible. Some pre-exponential factors for simple unimolecular desorption reactions are presented in Table 5.2.2. For the most part, the entries in the table are within a few orders of magnitude of 10^{13} s^{-1}. The very high values of the pre-exponential factor are likely attributed to large increases in the entropy upon formation of the transition state. Bimolecular surface reactions can be treated in the same way. However, one must explicitly account for the total number of surface

Table 5.2.3 | Pre-exponential factors for $2H_{ads} \rightarrow H_{2\,g}$ on transition metal surfaces.

Surface	Pre-exponential factor[a] (s^{-1})
Pd(111)	10^{12}
Ni(100)	$10^{13.5}$
Ru(001)	$10^{14.5}$
Pd(110)	$10^{13.5}$
Mo(110)	10^{13}
Pt(111)	10^{12}

[a]The values have been normalized by $[*]_0$.

Adapted from R. Masel, *Principles of Adsorption and Reaction on Solid Surfaces,* Wiley, New York, 1996, p. 607.

sites in the rate expression. As discussed above, the rate of associative desorption of H_2 from a square lattice can be written as:

$$r_{des} = \frac{2k_{des}[H*]^2}{[*]_0} \qquad (5.2.32)$$

where the pre-exponential factor of the rate constant is now multiplied by $2/[*]_0$ to properly account for the statistics of a reaction occurring on adjacent sites. For desorption of H_2 at 550 K from a W(100) surface, which is a square lattice with $[*]_0 = 5 \times 10^{14}$ cm^{-2}, the pre-exponential factor is anticipated to be 4.6×10^{-2} cm^2 s^{-2}. The reported experimental value of 4.2×10^{-2} cm^2 s^{-1} is very close to that predicted by transition state theory (M. Boudart and G. Djega-Mariadassou, *Kinetics of Heterogeneous Catalytic Reactions,* Princeton University Press, Princeton, 1984, p. 71). The measured pre-exponential factors for associative desorption of dihydrogen from other transition metal surfaces (normalized by the surface site density) are summarized in Table 5.2.3. Clearly, the values in Table 5.2.3 are consistent with transition state theory.

A fairly rare elementary reaction between *A* and *B*, often called a *Rideal-Eley* step, occurs by direct reaction of gaseous *B* with adsorbed *A* according to the following sequence:

$$A + * \rightleftharpoons A* \qquad \text{(reversible adsorption of } A)$$
$$\underline{A* + B \longrightarrow \text{products}} \qquad \text{(Rideal-Eley step)}$$
$$A + B \Longrightarrow \text{products} \qquad \text{(overall reaction)}$$

Theoretically, if reactions are able to proceed through either a Rideal-Eley step or a Langmuir-Hinshelwood step, the Langmuir-Hinshelwood route is much more preferred due to the extremely short time scale (picosecond) of a gas-surface collision. The kinetics of a Rideal-Eley step, however, can become important at extreme conditions. For example, the reactions involved during plasma processing of electronic materials are believed to occur through Rideal-Eley steps. Apparently, the conditions typical of semiconductor growth favor Rideal-Eley elementary steps whereas conditions normally encountered with catalytic reactions favor Langmuir-Hinshelwood steps. This point is thoroughly discussed by R. Masel (*Principles of Adsorption and Reaction on Solid Surfaces,* Wiley, New York, 1996, pp. 444–448).

EXAMPLE 5.2.2

A reaction that is catalyzed by a Bronsted acid site, or H^+, can often be accelerated by addition of a solid acid. Materials like ion-exchange resins, zeolites, and mixed metal oxides function as solid analogues of corrosive liquid acids (e.g., H_2SO_4 and HF) and can be used as acidic catalysts. For example, isobutylene (IB) reacts with itself to form dimers on cross-linked poly(styrene-sulfonic acid), a strongly acidic solid polymer catalyst:

$$CH_2=\underset{\underset{CH_3}{|}}{\overset{\overset{CH_3}{|}}{C}} + CH_2=\underset{\underset{CH_3}{|}}{\overset{\overset{CH_3}{|}}{C}} \Longrightarrow CH_3-\underset{\underset{CH_3}{|}}{\overset{\overset{CH_3}{|}}{C}}-CH_2-\underset{}{\overset{\overset{CH_3}{|}}{C}}=CH_2$$

The kinetics of IB dimerization are presented in Figure 5.2.7 for two different initial concentrations of IB in hexane solvent. The reaction appears to be first order at high

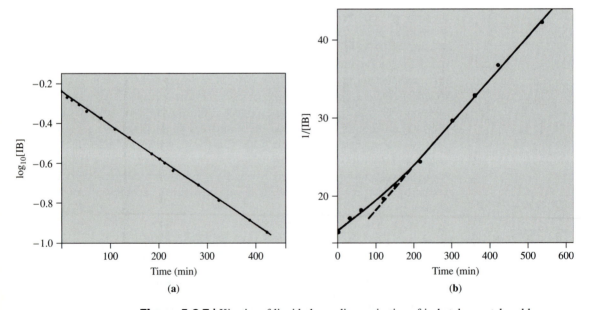

Figure 5.2.7 | Kinetics of liquid phase oligomerization of isobutylene catalyzed by poly(styrene-sulfonic acid) at 20°C in hexane solvent. **(a)** Corresponds to high initial concentration of isobutylene. **(b)** Corresponds to low initial concentration of isobutylene. [Figures from W. O. Haag, *Chem. Eng. Prog. Symp. Ser.,* **63** (1967) 140. Reproduced with permission of the American Institute of Chemical Engineers. Copyright © 1967 AIChE. All rights reserved.]

concentrations of IB since a logarithmic plot of [IB] with respect to reaction time is linear. At low concentrations, however, the reaction shifts to second order since a plot of $1/[IB]$ as a function of time becomes linear. A rate expression consistent with these results is:

$$r = \frac{\overline{\alpha}_1 [IB]^2}{1 + \overline{\alpha}_2 [IB]} \tag{5.2.33}$$

where $\overline{\alpha}_1$ and $\overline{\alpha}_2$ are constants. Show how a rate expression of this form can be derived for this reaction.

■ Answer

The following two-step catalytic cycle for dimerization of IB is proposed.

(1) $\quad IB + * \underset{}{\overset{}{\rightleftarrows}} IB*$ $\qquad$ (adsorption of IB on solid acid)

(2) $\quad \underline{IB* + IB \overset{k_2}{\longrightarrow} (IB)_2 + *}$ $\qquad$ (Rideal-Eley reaction)

$\qquad 2\ IB \Longrightarrow (IB)_2$ $\qquad$ (overall dimerization reaction)

In this case, * represents a surface acid site and IB* designates an adsorbed t-butyl cation. Notice that step 2 involves the combination of an adsorbed reactant with one present in the liquid phase. The rate equation takes the form:

$$r = k_2 [IB*][IB] \tag{5.2.34}$$

The concentration of adsorbed IB is found by combining the equilibrium relationship for step 1 and the overall site balance to give:

$$[IB*] = \frac{K_{ads}[*]_0[IB]}{1 + K_{ads}[IB]} \tag{5.2.35}$$

where $[*]_0$ represents the total number of acid sites on the catalyst surface, and K_{ads} is the adsorption equilibrium constant of IB. Substitution of Equation (5.2.35) into (5.2.34) yields a final form of the rate expression:

$$r = \frac{k_2 K_{ads}[*]_0[IB]^2}{1 + K_{ads}[IB]} \tag{5.2.36}$$

that has the same functional form as Equation (5.2.33) with $\overline{\alpha}_1 = k_2 K_{ads}[*]_0$ and $\overline{\alpha}_2 = K_{ads}$ and is consistent with the experimental data shown in Figure 5.2.7.

The rate of product desorption can also influence the kinetics of a surface-catalyzed reaction. Consider the following simple catalytic cycle:

$$A + * \rightleftarrows A*$$

$$A* \rightleftarrows B*$$

$$\underline{B* \rightleftarrows B + *}$$

$$A \Longrightarrow B$$

If desorption of B from the surface is rate-determining, then all elementary steps prior to desorption are assumed to be quasi-equilibrated:

$$(1) \qquad A + * \xrightleftharpoons{K_1} A*$$

$$(2) \qquad A* \xrightleftharpoons{K_2} B*$$

$$\underline{(3) \qquad B* \xrightarrow{k_3} B + *}$$

$$A \Longrightarrow B$$

The overall rate in the forward direction only is given by:

$$r = r_3 = k_3[B*] \tag{5.2.37}$$

The rate expression is simplified by eliminating surface concentrations of species through the use of appropriate equilibrium relationships. According to step (2):

$$[B*] = K_2[A*] \tag{5.2.38}$$

and thus:

$$r = k_3 K_2[A*] \tag{5.2.39}$$

To find $[A*]$, the equilibrium adsorption relationship is used:

$$K_1 = \frac{[A*]}{[A][*]} \qquad [A*] = K_1[A][*] \tag{5.2.40}$$

which gives:

$$r = k_3 K_2 K_1[A][*] \tag{5.2.41}$$

The concentration of vacant sites on the surface is derived from the total site balance:

$$[*]_0 = [*] + [A*] + [B*]$$
$$[*]_0 = [*] + [A*] + K_2[A*]$$
$$[*]_0 = [*] + K_1[A][*] + K_2 K_1[A][*]$$
$$[*] = \frac{[*]_0}{1 + (K_1 + K_2 K_1)[A]} \tag{5.2.42}$$

Substitution of Equation (5.2.42) into Equation (5.2.41) gives the final rate expression as:

$$r = \frac{k_3 K_2 K_1[A][*]_0}{1 + (K_1 + K_2 K_1)[A]} \tag{5.2.43}$$

Notice the rate expression for this case does not depend on the product concentration.

5.3 | Kinetics of Overall Reactions

Consider the entire sequence of elementary steps comprising a surface-catalyzed reaction: adsorption of reactant(s), surface reaction(s), and finally desorption of product(s). If the surface is considered uniform (i.e., all surface sites are identical kinetically and thermodynamically), and there are negligible interactions between adsorbed species, then derivation of overall reaction rate equations is rather straightforward.

For example, the reaction of dinitrogen and dihydrogen to form ammonia is postulated to proceed on some catalysts according to the following sequence of elementary steps:

Step		$\bar{\sigma}_i$
(1)	$N_2 + 2* \rightleftarrows 2N*$	1
(2)	$N* + H* \rightleftarrows NH* + *$	2
(3)	$NH* + H* \rightleftarrows NH_2* + *$	2
(4)	$NH_2* + H* \rightleftarrows NH_3 + 2*$	2
(5)	$H_2 + 2* \rightleftarrows 2H*$	3

$$N_2 + 3H_2 = 2NH_3$$

where $\bar{\sigma}_i$ is the stoichiometric number of elementary step i and defines the number of times an elementary step must occur in order to complete one catalytic cycle according to the overall reaction. In the sequence shown above, $\bar{\sigma}_2 = 2$ means that step 2 must occur twice for every time that a dinitrogen molecule dissociately adsorbs in step 1. The net rate of an overall reaction can now be written in terms of the rate of any one of the elementary steps, weighted by the appropriate stoichiometric number:

$$r = \frac{r_i - r_{-i}}{\bar{\sigma}_i} \tag{5.3.1}$$

The final form of a reaction rate equation from Equation (5.3.1) is derived by repeated application of the steady-state approximation to eliminate the concentrations of reactive intermediates.

In many cases, however, the sequence of kinetically relevant elementary steps can be reduced to two steps (M. Boudart and G. Djega-Mariadassou, *Kinetics of Heterogeneous Catalytic Reactions,* Princeton University Press, Princeton, 1984, p. 90). For example, the sequence given above for ammonia synthesis can be greatly simplified by assuming step 1 is rate-determining and all other steps are nearly equilibrated. The two relevant steps are now:

$$\bar{\sigma}_i$$

(1) $\qquad N_2 + 2* \underset{k_{-1}}{\overset{k_1}{\rightleftarrows}} 2N* \qquad 1 \qquad$ (rate-determining step)

(2) $\quad N* + 3/2\, H_2 \overset{K_2}{\rightleftarrows} NH_3 + * \quad 2 \qquad$ (quasi-equilibrated reaction)

It must be emphasized that step 2 is not an elementary step, but a sum of all of the quasi-equilibrated steps that must occur after dinitrogen adsorption. According to this abbreviated sequence, the only species on the surface of the catalyst of any kinetic relevance is N*. Even though the other species (H*, NH*, etc.) may also be present, according to the assumptions in this example only N* contributes to the site balance:

$$[*]_0 = [N*] + [*] \tag{5.3.2}$$

In a case such as this one where only one species is present in appreciable concentration on the surface, that species is often referred to as the *most abundant reaction intermediate,* or *mari*. The overall rate of reaction can be expressed as the rate of dissociative adsorption of N_2:

$$r = r_1 - r_{-1} = \frac{k_1[N_2][*]^2}{[*]_0} - \frac{k_{-1}[N*]^2}{[*]_0} \tag{5.3.3}$$

where [*] and [N*] are determined by the equilibrium relationship for step 2:

$$K_2 = \frac{[N*][H_2]^{3/2}}{[NH_3][*]} \tag{5.3.4}$$

and the site balance. The constant K_2 is written in such a way that it is large when [N*] is also large. Solving Equations (5.3.2) and (5.3.4) for [*] and [N*], respectively, yields:

$$[*] = \frac{[*]_0}{1 + K_2\left(\dfrac{[NH_3]}{[H_2]^{3/2}}\right)} \tag{5.3.5}$$

$$[N*] = \frac{K_2\left(\dfrac{[NH_3]}{[H_2]^{3/2}}\right)[*]_0}{1 + K_2\left(\dfrac{[NH_3]}{[H_2]^{3/2}}\right)} \tag{5.3.6}$$

Substitution of Equations (5.3.5) and (5.3.6) into (5.3.3) gives the rate equation for the reaction as:

$$r = \frac{k_1[*]_0[N_2] - k_{-1}K_2^2[*]_0\left(\dfrac{[NH_3]^2}{[H_2]^3}\right)}{\left[1 + K_2\left(\dfrac{[NH_3]}{[H_2]^{3/2}}\right)\right]^2} \tag{5.3.7}$$

At very low conversion (far from equilibrium), the reverse reaction can be neglected thus simplifying the rate expression to:

$$r = \frac{k_1[*]_0[N_2]}{\left[1 + K_2\left(\dfrac{[NH_3]}{[H_2]^{3/2}}\right)\right]^2} \tag{5.3.8}$$

EXAMPLE 5.3.1

Ruthenium has been investigated by many laboratories as a possible catalyst for ammonia synthesis. Recently, Becue et al. [T. Becue, R. J. Davis, and J. M. Garces, *J. Catal.,* **179** (1998) 129] reported that the forward rate (far from equilibrium) of ammonia synthesis at 20 bar total pressure and 623 K over base-promoted ruthenium metal is first order in dinitrogen and inverse first order in dihydrogen. The rate is very weakly inhibited by ammonia. Propose a plausible sequence of steps for the catalytic reaction and derive a rate equation consistent with experimental observation.

■ Answer

In the derivation of Equation (5.3.8), dissociative adsorption of dinitrogen was assumed to be the rate-determining step and this assumption resulted in a first-order dependence of the rate on dinitrogen. The same step is assumed to be rate-determining here. However, the rate expression in Equation (5.3.8) has an overall positive order in dihydrogen. Therefore, some of the assumptions used in the derivation of Equation (5.3.8) will have to be modified. The observed negative order in dihydrogen for ammonia synthesis on ruthenium suggests that H atoms occupy a significant fraction of the surface. If H* is assumed to be the most abundant reaction intermediate on the surface, an elementary step that accounts for adsorption of dihydrogen must be included explicitly. Consider the following steps:

$$(1) \qquad N_2 + 2* \xrightarrow{k_1} 2N* \qquad \text{(rate-determining step)}$$

$$(2) \qquad H_2 + 2* \underset{}{\overset{K_2}{\rightleftharpoons}} 2H* \qquad \text{(quasi-equilibrated)}$$

that are followed by many surface reactions to give an overall equilibrated reaction represented by:

$$(3) \qquad N* + 3H* \rightleftharpoons NH_3 + 4* \qquad \text{(quasi-equilibrated)}$$

As before, the forward rate of the reaction (far from equilibrium) can be expressed in terms of the rate-determining step:

$$r = r_1 = \frac{k_1[N_2][*]^2}{[*]_0} \tag{5.3.9}$$

To eliminate [*] from Equation (5.3.9), use the equilibrium relationship for step 2 combined with the site balance. Hence, the following equations:

$$K_2 = \frac{[H*]^2}{[*]^2[H_2]} \tag{5.3.10}$$

$$[*]_0 = [H*] + [*] \tag{5.3.11}$$

are solved simultaneously to give:

$$[*] = \frac{[*]_0}{1 + K_2^{1/2}[H_2]^{1/2}} \tag{5.3.12}$$

Substitution of Equation (5.3.12) into Equation (5.3.9) provides the final rate equation:

$$r = \frac{k_1[*]_0[N_2]}{(1 + K_2^{1/2}[H_2]^{1/2})^2} \tag{5.3.13}$$

If the surface is nearly covered with adsorbed H atoms, then:

$$K_2^{1/2}[H_2]^{1/2} \gg 1 \tag{5.3.14}$$

and the rate equation simplifies to:

$$r = \left(\frac{k_1[*]_0}{K_2}\right)\frac{[N_2]}{[H_2]} \tag{5.3.15}$$

This expression is consistent with the experimental observations. For this example, the reaction equilibrium represented by step 3 is never used to solve the problem since the most abundant reaction intermediate is assumed to be H* (accounted for in the equilibrated step 2.) Thus, a complex set of elementary steps is reduced to two kinetically significant reactions.

The concept that a multistep reaction can often be reduced to two kinetically significant steps is illustrated again by considering the dehydrogenation of methylcyclohexane to toluene on a Pt/Al_2O_3 reforming catalyst [J. H. Sinfelt, H. Hurwitz and R. A. Shulman, *J. Phys. Chem.*, **64** (1960)1559]:

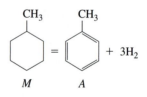

The observed rate expression for the reaction operated far from equilibrium can be written in the form:

$$r = \frac{\overline{\alpha}_1[M]}{1 + \overline{\alpha}_2[M]} \tag{5.3.16}$$

The reaction occurs through a complex sequence of elementary steps that includes adsorption of *M*, surface reactions of *M**, and partially dehydrogenated *M**, and finally desorption of both H_2 and toluene. It may be possible, however, to simplify this multistep sequence. Consider the following two step sequence that involves *M** as the *mari*:

(1) $M + * \overset{K_1}{\rightleftharpoons} M^*$ (quasi-equilibrated adsorption)

(2) $M^* \overset{k_2}{\not\rightarrow} \cdots$ (rate-determining step)

The site balance and the Langmuir adsorption isotherm can be used to derive the forward rate expression:

$$r = r_2 = k_2[M*] = \frac{k_2 K_1 [*]_0 [M]}{1 + K_1[M]} \tag{5.3.17}$$

that has the same form as the observed rate expression, Equation (5.3.16). A word of caution is warranted at this point. *The fact that a proposed sequence is consistent with observed kinetics does not prove that a reaction actually occurs by that pathway.* Indeed, an alternative sequence of steps can be proposed for the above reaction:

(1) $\quad M + * \xrightarrow{k_1} \cdots$ $\quad$ (irreversible adsorption)

$\quad$ ---------------------------- $\quad$ (surface reactions)

(2) $\quad A* \xrightarrow{k_2} * + A$ $\quad$ (irreversible desorption)

The application of the quasi-steady-state approximation and the site balance (assuming $A*$ is the *mari*) gives the following expression for the reaction rate:

$$r = r_2 = k_2[A*] = \frac{k_2(k_1/k_2)[*]_0[M]}{1 + (k_1/k_2)[M]} = \frac{k_1[*]_0[M]}{1 + (k_1/k_2)[M]} \tag{5.3.18}$$

The functional form of the rate equation in Equation (5.3.18) is identical to that of Equation (5.3.17), illustrating that two completely different sets of assumptions can give rate equations consistent with experimental observation. Clearly, more information is needed to discriminate between the two cases. Additional experiments have shown that benzene added to the methylcyclohexane feed inhibits the rate only slightly. In the first case, benzene is expected to compete with methylcyclohexane for available surface sites since M is equilibrated with the surface. In the second case, M is not equilibrated with the surface and the irreversibility of toluene desorption implies that the surface coverage of toluene is far above its equilibrium value. Benzene added to the feed will not effectively displace toluene from the surface since benzene will cover the surface only to the extent of its equilibrium amount. The additional information provided by the inclusion of benzene in the feed suggests that the second case is the preferred path.

It is possible to generalize the treatment of single-path reactions when a most abundant reaction intermediate (*mari*) can be assumed. According to M. Boudart and G. Djega-Mariadassou (*Kinetics of Heterogeneous Catalytic Reactions*, Princeton University Press, Princeton, 1984, p. 104) three rules can be formulated:

1. If in a sequence, the rate-determining step produces or destroys the *mari,* the sequence can be reduced to two steps, an adsorption equilibrium and the rate-determining step, with all other steps having no kinetic significance.

2. If all steps are practically irreversible and there exists a *mari*, only two steps need to be taken into account: the adsorption step and the reaction (or desorption) step

of the *mari*. All other steps have no kinetic significance. In fact, they may be reversible, in part or in whole.

3. All equilibrated steps following a rate-determining step that produces the *mari* may be summed up in an overall equilibrium reaction. Similarly, all equilibrated steps that precede a rate-determining step that consumes the *mari* may be represented by a single overall equilibrium reaction.

The derivation of a rate equation from two-step sequences can also be generalized. First, if the rate-determining step consumes the *mari,* the concentration of the latter is obtained from the equilibrium relationship that is available. Second, if the steps of the two-step sequence are practically irreversible, the steady-state approximation leads to the solution.

These simplifying assumptions must be adapted to some extent to explain the nature of some reactions on catalyst surfaces. The case of ammonia synthesis on supported ruthenium described in Example 5.3.1 presents a situation that is similar to rule 1, except the rate-determining step does not involve the *mari*. Nevertheless, the solution of the problem was possible. Example 5.3.2 involves a similar scenario. If a *mari* cannot be assumed, then a rate expression can be derived through repeated use of the steady-state approximation to eliminate the concentrations of reactive intermediates.

EXAMPLE 5.3.2

The oxidation of carbon monoxide on palladium single crystals at low pressures (between 10^{-8} and 10^{-6} torr) and temperatures ranging from about 450 to 550 K follows a rate law that is first order in O_2 and inverse first order in CO. An appropriate sequence of elementary steps is:

(1) $CO + * \rightleftharpoons CO^*$ (quasi-equilibrated adsorption)

(2) $O_2 + * \longrightarrow O^*O$ (irreversible adsorption)

(3) $O^*O + * \longrightarrow 2O^*$ (surface reaction)

(4) $CO^* + O^* \longrightarrow CO_2 + 2^*$ (surface reaction/desorption)

If CO^* is assumed to be the *mari,* derive the rate expression.

■ Answer

The rate of reaction is the rate of any single step in the sequence weighted by its appropriate stoichiometric number. Thus, for the reaction:

$$CO + \tfrac{1}{2}O_2 = CO_2$$

the rate can be written:

$$r = (r_1 - r_{-1}) = 2r_2 = 2r_3 = r_4 \tag{5.3.19}$$

The simplest solution involves the two-step sequence:

$$
(1) \qquad CO + * \underset{}{\overset{K_1}{\rightleftharpoons}} CO^* \qquad \text{(quasi-equilibrated adsorption)}
$$

$$
(2) \qquad O_2 + * \xrightarrow{k_2} O^*O \qquad \text{(irreversible adsorption)}
$$

where:

$$
r = 2r_2 = 2k_2[O_2][*] \tag{5.3.20}
$$

Application of the site balance, assuming CO* is the *mari* gives:

$$
[*]_0 = [*] + [CO^*] \tag{5.3.21}
$$

and the Langmuir isotherm for adsorption of CO can be written as:

$$
K_1 = \frac{[CO^*]}{[CO][*]} \tag{5.3.22}
$$

Substitution of Equations (5.3.21) and (5.3.22) into Equation (5.3.20) yields:

$$
r = \frac{2k_2[*]_0[O_2]}{1 + K_1[CO]} \tag{5.3.23}
$$

At high values of surface coverage, $K_1[CO] \gg 1$, the rate equation simplifies to:

$$
r = \frac{2k_2[*]_0[O_2]}{K_1[CO]} \tag{5.3.24}
$$

which is consistent with the observed rate law.

The rate constant for the reaction is composed of two terms, $2k_2$ and K_1^{-1}. Thus, the apparent activation energy contains contributions from the rate constant for O_2 adsorption and the equilibrium constant for CO adsorption according to:

$$
E_{\text{app}} = E_2 - \Delta H_{\text{adsCO}} \tag{5.3.25}
$$

Since adsorption of O_2 is essentially nonactivated, the apparent activation energy for CO oxidation is simply the negative of the enthalpy of CO adsorption on Pd. This result has been experimentally observed [M. Boudart, *J. Mol. Catal. A: Chem.,* **120** (1997) 271].

Example 5.3.2 demonstrates how the heat of adsorption of reactant molecules can profoundly affect the kinetics of a surface catalyzed chemical reaction. The experimentally determined, apparent rate constant $(2k_2/K_1)$ shows typical Arrhenius-type behavior since it increases exponentially with temperature. The apparent activation energy of the reaction is simply $E_{\text{app}} = E_2 - \Delta H_{\text{adsCO}} \cong -\Delta H_{\text{adsCO}}$ (see Example 5.3.2), which is a positive number. A situation can also arise in which a *negative* overall activation energy is observed, that is, the observed reaction rate decreases with increasing temperature. This seemingly odd phenomenon can be understood in terms of the multistep mechanism of surface catalyzed reactions. Consider the rate of conversion of A occurring through a rate-determining surface reaction as described earlier:

$$r = \frac{k_2 K_{ads}[*]_0[A]}{1 + K_{ads}[A]} \tag{5.3.26}$$

Experimental conditions can arise so that $1 \gg K_{ads}[A]$, and the reaction rate expression reduces to:

$$r = k_2 K_{ads}[*]_0[A] \tag{5.3.27}$$

with an apparent rate constant $k_2 K_{ads}$. The apparent activation energy is now:

$$E_{app} = E_2 + \Delta H_{ads} \tag{5.3.28}$$

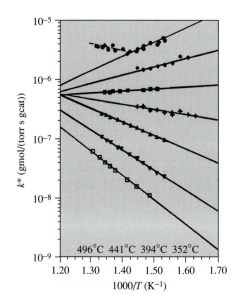

Figure 5.3.1 |
Temperature dependence of the cracking of n-alkanes. [Reprinted from J. Wei, "Adsorption and Cracking of N-Alkanes over ZSM-5: Negative Activation Energy of Reaction," *Chem. Eng. Sci.,* **51** (1996) 2995, with permission from Elsevier Science.] Open square—C_8, inverted triangle—C_{10}, triangle—C_{12}, plus—C_{14}, filled square—C_{16}, diamond—C_{18}, circle—C_{20}.

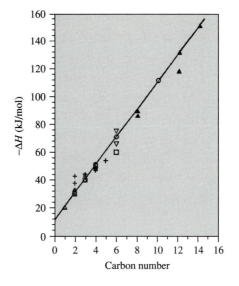

Figure 5.3.2 |
Heats of adsorption of n-alkanes on ZSM-5.
[Reprinted from J. Wei, "Adsorption and
Cracking of N-Alkanes over ZSM-5:
Negative Activation Energy of Reaction,"
Chem. Eng. Sci., **51** (1996) 2995, with
permission from Elsevier Science.]

Since the enthalpy of adsorption is almost always negative, the apparent activation energy can be either positive or negative, depending on the magnitudes of E_2 and ΔH_{ads}.

The cracking of n-alkanes over H-ZSM-5, an acidic zeolite, provides a clear illustration of how the apparent activation energy can be profoundly affected by the enthalpy of adsorption. An Arrhenius plot of the pseudo-first-order rate constant, $k^* = k_2 K_{ads}$, for n-alkanes having between 8 and 20 carbon atoms is shown in Figure 5.3.1. The reaction of smaller alkanes (C_8—C_{14}) has a positive apparent activation energy that declines with chain length, whereas the reaction of larger alkanes (C_{18}, C_{20}) has a negative apparent activation energy that becomes more negative with chain length. Interestingly, the reaction of C_{16} is almost invariant to temperature. The linear relationship in Figure 5.3.2 demonstrates that the adsorption enthalpy is proportional to the number of carbons in the alkanes. Thus, if the activation energy of the surface reaction step, E_2, is not sensitive to the chain length, then Equation (5.3.28) predicts that the apparent activation energy will decrease with increasing length of the alkane. The temperature invariance of the pseudo-first-order rate constant for cracking of C_{16} apparently results from the cancellation of E_2 by ΔH_{ads}. It is also worth mentioning that the magnitude of k^*, at a constant temperature, will be profoundly affected by carbon chain length through the equilibrium adsorption constant.

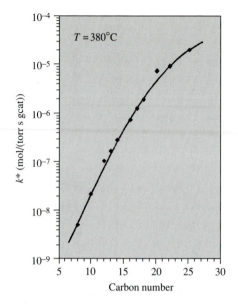

Figure 5.3.3 |
The apparent first-order rate constant of
n-alkane cracking over ZSM-5 at 380°C.
[Reprinted from J. Wei, "Adsorption and
Cracking of N-Alkanes over ZSM-5:
Negative Activation Energy of Reaction,"
Chem. Eng. Sci., **51** (1996) 2995, with
permission from Elsevier Science.]

Indeed, Figure 5.3.3 illustrates the exponential dependence of $k*$ on the size of
the alkane molecule. The high enthalpies of adsorption for long chain alkanes means
that their surface coverages will far exceed those associated with short chain molecules, which translates into much higher reaction rates.

VIGNETTE 5.3.1

Zeolites are highly porous, crystalline, aluminosilicates that are widely used as industrial
catalysts. In 1756, the Swedish minerologist A. F. Cronstedt coined the term zeolite, which
is derived from the Greek words "zeo," to boil, and "lithos," stone, after he observed that
the mineral stilbite gave off steam when heated. Today, it is known that zeolites are constructed from TO_4 tetrahedral (T = tetrahedral atom, e.g., Si, Al) with each apical oxygen
atom shared with an adjacent tetrahedron.

    To understand the function of zeolites in a catalytic reaction, it is necessary to
first describe the crystal chemistry of their framework. From the valency of silicon it

follows that silicon atoms generally prefer bonds with four neighboring atoms in tetrahedral geometry. If a SiO_4 entity could be isolated, its formal charge would be -4 since silicon is $+4$ and each oxygen anion is -2. However, a defect-free, pure SiO_2 framework will not contain any charge since an oxygen atom bridges two silicon atoms and shares electron density with each. If aluminum is tetrahedrally coordinated to four oxygen atoms in a framework, the net formal charge is -1 since aluminum carries a $+3$ valency. When tetrahedra containing silicon and aluminum are connected to form an aluminosilicate framework, there is a negative charge associated with each aluminum atom and it is balanced by a positive ion to give electrical neutrality. Typical cations are alkali metals (e.g., Na^+, K^+), alkaline earth metals (e.g., Ca^{2+}, Ba^{2+}), and the proton H^+. Figure 5.3.4 illustrates an example of a common cage structure known as sodalite and how it is constructed of silicon, aluminum, and oxygen atoms. Figure 5.3.4 also shows how several different zeolites are comprised of these sodalite cages. Each line in the figure represents a bridging oxygen atom while the intersection locates a silicon or aluminum atom.

Many zeolite structures exist. There are natural zeolites, synthetic analogues of natural zeolites, and synthetic zeolites with no natural counterparts. In all, there are more than 100 structures. Figure 5.3.5 illustrates some typical framework projections containing various rings (pores) of different sizes. Notice that the sizes range from approximately 4 to 12 Å and that the topologies may contain channels and/or cages. What

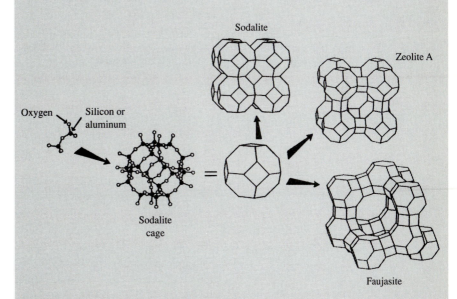

Figure 5.3.4 | Schematic of zeolite frameworks. The synthetic faujasites are NaX and NaY (difference between NaX and NaY is the ratio of Si to Al: NaX ~ 1.1, NaY ~ 2.4). These constructions do not represent how the materials are synthesized but rather their structural features alone. [Reprinted with permission from M. E. Davis, *Ind. Eng. Chem. Res.,* **30** (1991) 1676. Copyright 1991 American Chemical Society.]

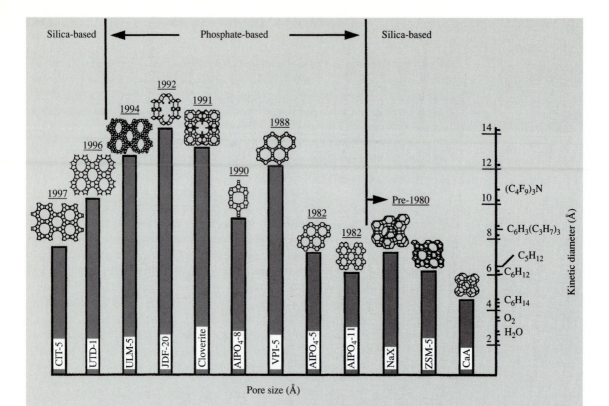

Figure 5.3.5 | Correlation between pore size of molecular sieves and the diameter of various molecules. [Reprinted from M. E. Davis, "Zeolite-Based Catalysts for Chemicals Synthesis," *Microporous and Mesoporous Mater.*, **21** (1998) 179, with permission of Elsevier Science.]

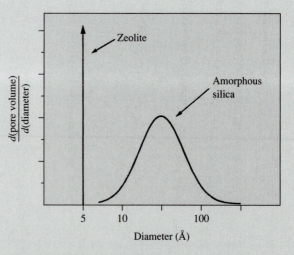

Figure 5.3.6 | Pore size distribution of zeolite compared to amorphous silica. [Reprinted with permission from M. E. Davis, *Ind. Eng. Chem. Res.*, **30** (1991) 1677. Copyright 1991 American Chemical Society.]

makes zeolites unique is that their pores are uniform in size (see Figure 5.3.6) and that they are in the same size range as small molecules (see Figure 5.3.5). Thus, zeolites are *molecular sieves* since they can discriminate molecules on the basis of size. Molecules smaller than the aperture size are admitted to the crystal interior (adsorbed) while those larger are not.

If the charge balancing cation in a zeolite is H^+, then the material is a solid acid that can reveal shape selective properties due to the confinement of the acidic proton within the zeolite pore architecture. An example of shape selective acid catalysis is provided in Figure 5.3.7. In this case, normal butanol and isobutanol were dehydrated over CaX and CaA zeolites that contained protons in the pore structure. Both the primary and secondary alcohols were dehydrated on the X zeolite whereas only the primary one reacted on the A zeolite. Since the secondary alcohol is too large to diffuse through the pores of CaA, it cannot reach the active sites within the CaA crystals.

The rate of hexane cracking on the zeolite ZSM-5 has been evaluated as a function of proton content (achieved by varying the Al content in the framework) over

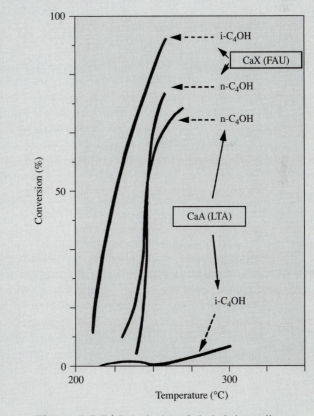

Figure 5.3.7 | Dehydration of alcohols by zeolites. [Reprinted with permission from M. E. Davis, *Ind. Eng. Chem. Res.,* **30** (1991) 1677. Copyright 1991 American Chemical Society.]

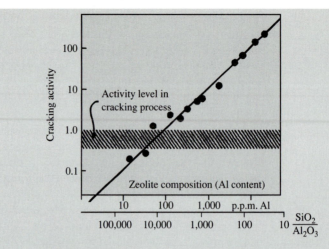

Figure 5.3.8 | Variation in the hexane cracking activity with aluminum content in ZSM-5. [Reprinted with permission from W. O. Haag, R. M. Lago, and P. B. Weisz, *Nature,* **309** (1984) 589.]

4 orders of magnitude of proton loading (see Figure 5.3.8). Since the rate was found to be strictly proportional to the proton content, these catalytic sites were clearly identical and noninteracting. W. O. Haag concluded that the possibility of synthesizing zeolite catalysts with a well-defined, predetermined number of active sites of uniform activity was certainly without parallel in hetereogeneous catalysis (W. O. Haag in "Zeolites and Related Microporous Materials: State of the Art 1994," J. Weitkamp et al., Eds., *Studies in Surface Science and Catalysis,* Vol. 84B, Elsevier Science B.V., 1994, p. 1375).

The discussion to this point has emphasized kinetics of catalytic reactions on a uniform surface where only one type of active site participates in the reaction. Bifunctional catalysts operate by utilizing two different types of catalytic sites on the same solid. For example, hydrocarbon reforming reactions that are used to upgrade motor fuels are catalyzed by platinum particles supported on acidified alumina. Extensive research revealed that the metallic function of Pt/Al_2O_3 catalyzes hydrogenation/dehydrogenation of hydrocarbons, whereas the acidic function of the support facilitates skeletal isomerization of alkenes. The isomerization of n-pentane (N) to isopentane (I) is used to illustrate the kinetic sequence associated with a bifunctional Pt/Al_2O_3 catalyst:

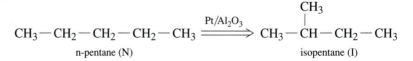

The sequence involves dehydrogenation of n-pentane on Pt particles to form intermediate n-pentene ($N^=$), which then migrates to the acidic alumina and reacts to i-pentene ($I^=$) in a rate-determining step. The i-pentene subsequently migrates back to the Pt particles where it is hydrogenated to the product i-pentane. The following sequence describes these processes (where * represents an acid site on alumina):

$$\text{(1)} \qquad N \underset{}{\overset{K_1}{\rightleftharpoons}} N^= + H_2 \qquad \text{(Pt-catalyzed)}$$

$$\text{(2)} \qquad N^= + * \underset{}{\overset{K_2}{\rightleftharpoons}} N^=* $$

$$\text{(3)} \qquad N^=* \overset{k_3}{\longrightarrow} I^=* $$

$$\text{(4)} \qquad I^=* \rightleftharpoons I^= + * $$

$$\text{(5)} \qquad I^= + H_2 \rightleftharpoons I \qquad \text{(Pt-catalyzed)}$$

$$\overline{\qquad N \Longrightarrow I \qquad}$$

Interestingly, the rate is inhibited by dihydrogen even though it does not appear in the stoichiometric equation. The rate is simply:

$$r = r_3 = k_3[N^=*] \tag{5.3.29}$$

The equilibrium relationships for steps 1 and 2 give:

$$[N^=*] = K_2[N^=][*] \tag{5.3.30}$$

$$[N^=] = K_1\frac{[N]}{[H_2]} \tag{5.3.31}$$

The site balance on alumina (assuming $N^=*$ is the *mari*, far from equilibrium):

$$[*]_0 = [*] + [N^=*] \tag{5.3.32}$$

and its use along with Equations (5.3.30) and (5.5.31) allow the rate expression for the forward reaction to be written as:

$$r = \frac{K_1K_2k_3[*]_0[N]}{[H_2] + K_1K_2[N]} \tag{5.3.33}$$

Thus, the inhibitory effect of H_2 in pentane isomerization arises from the equilibrated dehydrogenation reaction in step 1 that occurs on the Pt particles.

5.4 | Evaluation of Kinetic Parameters

Rate data can be used to postulate a kinetic sequence for a particular catalytic reaction. The general approach is to first propose a sequence of elementary steps consistent with the stoichiometric reaction. A rate expression is derived using the steady-state approximation together with any other assumptions, like a rate-determining step, a most abundant reaction intermediate, etc. and compared to the rate data. If the functional dependence of the data is similar to the proposed rate expression, then the sequence of elementary steps is considered plausible. Otherwise, the proposed sequence is discarded and an alternative one is proposed.

Consider the isomerization of molecule A far from equilibrium:

$$A \Rightarrow B$$

that is postulated to occur through the following sequence of elementary steps:

(1) $A + * \rightleftharpoons A*$

(2) $A* \rightleftharpoons B*$

(3) $B* \rightleftharpoons B + *$

Case 1.   If the rate of adsorption is rate determining, then the forward rate of reaction can be simplified to two steps:

(1) $A + * \xrightarrow{k_1} A*$

(2a) $A* \underset{}{\overset{K_{2a}}{\rightleftharpoons}} B + *$

where step 2a represents the overall equilibrium associated with surface reaction and desorption of product. A rate expression consistent with these assumptions is (derived according to methods described earlier):

$$r = r_1 = k_1[A][*] \tag{5.4.1}$$

$$K_{2a} = \frac{[A*]}{[B][*]} \tag{5.4.2}$$

$$[*]_0 = [*] + [A*] \tag{5.4.3}$$

The equilibrium constant is written such that K_{2a} is large when $[A*]$ is large. Combining Equations (5.4.1–5.4.3) results in the following expression for the forward rate:

$$r = \frac{k_1[*]_0[A]}{1 + K_{2a}[B]} \tag{5.4.4}$$

Case 2.   If the surface reaction is rate-determining, the following sequence for the forward rate is appropriate:

(1) $A + * \underset{}{\overset{K_1}{\rightleftharpoons}} A*$

(2) $A* \xrightarrow{k_2} B*$

(3) $B* \underset{}{\overset{K_3}{\rightleftharpoons}} B + *$

This particular sequence assumes both A^* and B^* are present on the surface in kinetically significant amounts. The rate expression for this case is:

$$r = r_2 = k_2[A^*] \tag{5.4.5}$$

$$K_1 = \frac{[A^*]}{[A][*]} \tag{5.4.6}$$

$$K_3 = \frac{[B^*]}{[B][*]} \tag{5.4.7}$$

$$[*]_0 = [*] + [A^*] + [B^*] \tag{5.4.8}$$

$$r = \frac{k_2[*]_0[A]}{1 + K_1[A] + K_3[B]} \tag{5.4.9}$$

Case 3. In this last case, the desorption of product is assumed to be rate determining. Similar to Case 1, two elementary steps are combined into an overall equilibrated reaction:

$$(1a) \qquad A + * \xrightleftharpoons{K_{1a}} B^*$$

$$(3) \qquad B^* \xrightarrow{k_2} B + *$$

The expression for the forward rate is derived accordingly:

$$r = r_3 = k_3[B^*] \tag{5.4.10}$$

$$K_{1a} = \frac{[B^*]}{[A][*]} \tag{5.4.11}$$

$$[*]_0 = [*] + [B^*] \tag{5.4.12}$$

$$r = \frac{k_3 K_{1a}[*]_0[A]}{1 + K_{1a}[A]} \tag{5.4.13}$$

A common method used to distinguish among the three cases involves the measurement of the initial rate as a function of reactant concentration. Since B is not present at early reaction times, the initial rate will vary proportionally with the concentration of A when adsorption of A is the rate-determining step (Figure 5.4.1).

Similarly, the initial rate behavior is plotted in Figure 5.4.2 for Cases 2 and 3, in which the rate-determining step is either surface reaction or desorption of product. Since the functional form of the rate expression is the same in Cases 2 and 3 when B is not present, additional experiments are required to distinguish between the two cases. Adding a large excess of product B to the feed allows for the difference between kinetic mechanisms to be observed. As shown in Figure 5.4.3, the presence of product in the feed does not inhibit the initial rate when desorption of B is the rate-determining step. If surface reaction of adsorbed A were the rate-determining

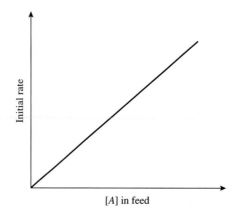
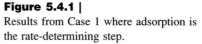

Figure 5.4.1 |
Results from Case 1 where adsorption is
the rate-determining step.

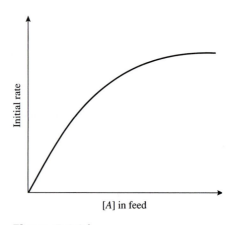
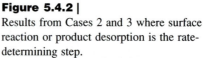

Figure 5.4.2 |
Results from Cases 2 and 3 where surface
reaction or product desorption is the rate-
determining step.

step, then extra B in the feed effectively competes for surface sites, displaces A from
the catalyst, and lowers the overall rate.

Once a rate expression is found to be consistent with experimental observations, then
rate constants and equilibrium constants are obtained from quantitative rate data. One
way to arrive at numerical values for the different constants in a Langmuir-Hinshelwood
rate expression is to first invert the rate expression. For Case 2, the rate expression:

$$r = \frac{k_2[*]_0[A]}{1 + K_1[A] + K_3[B]} \tag{5.4.14}$$

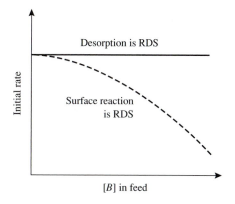

Figure 5.4.3 |
The influence of extra B on the initial rate
(at constant $[A]$) for cases with different
rate-determining steps (RDS).

becomes:

$$\frac{1}{r} = \frac{1}{k_2[*]_0[A]} + \frac{K_1[A]}{k_2[*]_0[A]} + \frac{K_2[B]}{k_2[*]_0[A]} \qquad (5.4.15)$$

Multiplying by $[A]$ gives an equation in which the groupings of constants can be calculated by a linear least squares analysis:

$$\frac{[A]}{r} = \frac{1}{k_2[*]_0} + \frac{K_1}{k_2[*]_0}[A] + \frac{K_2}{k_2[*]_0}[B] \qquad (5.4.16)$$

The above expression is of the form:

$$y = \overline{\alpha}_1 + \overline{\alpha}_2[A] + \overline{\alpha}_3[B]$$

with

$$y = \frac{[A]}{r}, \quad \overline{\alpha}_1 = \frac{1}{k_2[*]_0}, \quad \overline{\alpha}_2 = \frac{K_1}{k_2[*]_0}, \quad \overline{\alpha}_3 = \frac{K_2}{k_2[*]_0}$$

If an independent measure of $[*]_0$ is available from chemisorption, the constants k_2, K_1, and K_2 can be obtained from linear regression. It should be noted that many kineticists no longer use the linearized form of the rate equation to obtain rate constants. Inverting the rate expression places greater statistical emphasis on the lowest measured rates in a data set. Since the lowest rates are usually the least precise, a nonlinear least squares analysis of the entire data set using the normal rate expression is *preferred*.

EXAMPLE 5.4.1

The reaction $CO + Cl_2 \Rightarrow COCl_2$ has been studied over an activated carbon catalyst. A surface reaction appears to be the rate-determining step. Fashion a rate model consistent with the following data:

Rate ($\times 10^3 \frac{mol}{gcat\text{-}h}$)	P_{CO} (atm)	P_{Cl_2} (atm)	P_{COCl_2} (atm)
4.41	0.406	0.352	0.226
4.4	0.396	0.363	0.231
2.41	0.310	0.320	0.356
2.45	0.287	0.333	0.376
1.57	0.253	0.218	0.522
3.9	0.610	0.113	0.231
2.0	0.179	0.608	0.206

■ Answer

The strategy is to propose a reasonable sequence of steps, derive a rate expression, and then evaluate the kinetic parameters from a regression analysis of the data. As a first attempt at solution, assume both Cl_2 and CO adsorb (nondissociatively) on the catalyst and react to form adsorbed product in a Langmuir-Hinshelwood step. This will be called Case 1. Another possible sequence involves adsorption of Cl_2 (nondissociatively) followed by reaction with CO to form an adsorbed product in a Rideal-Eley step. This scenario will be called Case 2.

Case 1

$$Cl_2 + * \overset{K_1}{\rightleftharpoons} Cl_2*$$

$$CO + * \overset{K_2}{\rightleftharpoons} CO*$$

$$CO* + Cl_2* \overset{k_2}{\longrightarrow} COCl_2* + *$$

$$COCl_2* \overset{K_4}{\rightleftharpoons} COCl_2 + *$$

The rate expression derived from the equilibrium relations for steps 1, 2, and 4, assuming all three adsorbed species are present in significant quantities, is:

$$r = \frac{k_3 K_2 K_1 [*]_0 [CO][Cl_2]}{(1 + K_1[Cl_2] + K_2[CO] + K_4[COCl_2])^2}$$

The data fit well the above expression. However, some of the constants (not shown) have negative values and are thus unrealistic. Therefore, Case 1 is discarded.

Case 2

$$Cl_2 + * \overset{K_1}{\rightleftarrows} Cl_2*$$

$$CO + Cl_2* \overset{k_2}{\longrightarrow} COCl_2*$$

$$COCl_2* \overset{K_1}{\rightleftarrows} COCl_2 + *$$

Assuming only Cl_2* and $COCl_2*$ are present on the surface, the following rate expression is derived:

$$r = \frac{k_2 K_1 [*]_0 [CO][Cl_2]}{1 + K_1[Cl_2] + K_3[COCl_2]}$$

Fitting the data to the above equation results in the following rate model:

$$r = \frac{1.642[CO][Cl_2]}{1 + 124.4[Cl_2] + 58.1[COCl_2]} \quad \left(\frac{mol}{gcat\text{-}h}\right)$$

where concentrations are actually partial pressures expressed in atm. Even though all the kinetic parameters are positive and fit the data set reasonably well, this solution is not guaranteed to represent the actual kinetic sequence. Reaction kinetics can be consistent with a mechanism but they cannot prove it. Numerous other models need to be constructed and tested against one another (as illustrated previously in this section) in order to gain confidence in the kinetic model.

Rate constants and equilibrium constants should be checked for thermodynamic consistency if at all possible. For example, the heat of adsorption ΔH_{ads} derived from the temperature dependence of K_{ads} should be negative since adsorption reactions are almost always exothermic. Likewise, the entropy change ΔS_{ads} for nondissociative adsorption must be negative since every gas phase molecule loses translational entropy upon adsorption. In fact, $|\Delta S_{ads}| < S_g$ (where S_g is the gas phase entropy) must also be satisfied because a molecule cannot lose more entropy than it originally possessed in the gas phase. A proposed kinetic sequence that produces adsorption rate constants and/or equilibrium constants that do not satisfy these basic principles should be either discarded or considered very suspiciously.

Exercises for Chapter 5

1. (a) Calculate the BET surface area per gram of solid for Sample 1 using the full BET equation and the one-point BET equation. Are the values the same? What is the BET constant?

    (b) Calculate the BET surface area per gram of solid for Sample 2 using the full BET equation and the one-point BET equation. Are the values the same? What is the BET constant and how does it compare to the value obtained in (a)?

Dinitrogen adsorption data

	Volume adsorbed (cm^3/g)	
P/P_0	Sample 1	Sample 2
0.02	23.0	0.15
0.03	25.0	0.23
0.04	26.5	0.32
0.05	27.7	0.38
0.10	31.7	0.56
0.15	34.2	0.65
0.20	36.1	0.73
0.25	37.6	0.81
0.30	39.1	0.89

2. A 0.5 wt. % Pt on silica catalyst gave the data listed below for the sorption of H_2. Upon completion of Run 1, the system was evacuated and then Run 2 was performed. Find the dispersion and average particle size of the Pt particles. *Hint*: Run 1 measures the total sorption of hydrogen (reversible + irreversible) while Run 2 gives only the reversible hydrogen uptake. Calculate the dispersion based on the chemisorbed (irreversible) hydrogen.

Run 2		Run 1	
Pressure (torr)	H/Pt	Pressure (torr)	H/Pt
10.2	1.09	10.7	1.71
12.9	1.30	14.2	2.01
16.1	1.60	18.2	2.33
20.4	1.93	23.2	2.73
25.2	2.30	28.9	3.17
30.9	2.75	35.9	3.71
37.9	3.30	44.4	4.38
46.5	3.96	55.2	5.22
57.2	4.79	66.2	6.05
68.2	5.64	77.2	6.90
79.2	6.49	88.3	7.72
90.2	7.32	99.3	8.57
101	8.18	110	9.42
112	9.03	121	10.3
123	9.87	132	11.1
134	10.7	143	12.0
145	11.6	154	12.8
156	12.4		

3. A. Peloso et al. [*Can. J. Chem. Eng.*, **57** (1979) 159] investigated the kinetics of the following reaction over a $CuO/Cr_2O_3/SiO_2$ catalyst at temperatures of 225–285°C:

$$CH_3CH_2OH = CH_3CHO + H_2$$

$$\text{(Et)} \qquad\qquad \text{(Ad)} \qquad \text{(DH)}$$

A possible rate expression to describe the data is:

$$r = \frac{k\left[P_{Et} - (P_{Ad}P_{DH})/K_e\right]}{\left[1 + K_{Et}P_{Et} + K_{Ad}P_{Ad}\right]^2}$$

Write a reaction sequence that would give this rate expression.

4. J. Franckaerts and G. F. Froment [*Chem. Eng. Sci.*, **19** (1964) 807] investigated the reaction listed in Exercise 3 over the same temperature range using a $CuO/CoO/Cr_2O_3$ catalyst. The rate expression obtained in this work was of the form:

$$r = \frac{k\left[P_{Et} - (P_{Ad}P_{DH})/K_e\right]}{\left[1 + K_{Et}P_{Et} + K_{Ad}P_{Ad} + K_{DH}P_{DH}\right]^2}$$

Write a reaction sequence that gives a rate expression of this form. What is different from the sequence used in Exercise 3?

5. The following oxidation occurs over a solid catalyst:

$$\underset{\text{(IP)}}{CH_3\overset{\displaystyle OH}{\overset{\displaystyle |}{C}}HCH_3} + \underset{\text{(DO)}}{\tfrac{1}{2} O_2} = \underset{\text{(At)}}{CH_3\overset{\displaystyle O}{\overset{\displaystyle \|}{C}}CH_3} + \underset{\text{(W)}}{H_2O}$$

If acetone cannot adsorb and the rate of surface reaction between adsorbed isopropanol and adsorbed oxygen is the rate-determining step, a rate expression of the form:

$$r = \frac{k[IP][DO]^{\frac{1}{2}}}{(1 + K_1[IP] + K_2[W])(1 + K_3[DO]^{\frac{1}{2}})}$$

can be obtained from what reaction sequence?

6. For the reaction:

$$CO + Cl_2 = COCl_2$$

the rate expression given below can be obtained:

$$r = \frac{kP_{CO}P_{Cl_2}}{\left[1 + K_1P_{Cl_2} + K_2P_{COCl_2}\right]^2}$$

What does the exponent on the denominator imply and what does the lack of a K_3P_{CO} term in the denominator suggest?

7. For the reaction of A to form B over a solid catalyst, the reaction rate has the form:

$$r = \frac{kK_AP_A}{(1 + K_AP_A + K_BP_B)^2}$$

However, there is a large excess of inert in the reactant stream that is known to readily adsorb on the catalyst surface. How will this affect the reaction order with respect to A?

8. G. Thodor and C. F. Stutzman [*Ind. Eng. Chem.*, **50** (1958) 413] investigated the following reaction over a zirconium oxide-silica gel catalyst in the presence of methane:

$$C_2H_4 + HCl = C_2H_5Cl$$

If the equilibrium constant for the reaction is 35 at reaction conditions, find a reaction rate expression that describes the following data:

	Partial pressure (atm)			
$r \times 10^4$ (mol/h/(lb cat))	CH_4	C_2H_4	HCl	C_2H_5Cl
2.66	7.005	0.300	0.370	0.149
2.61	7.090	0.416	0.215	0.102
2.41	7.001	0.343	0.289	0.181
2.54	9.889	0.511	0.489	0.334
2.64	10.169	0.420	0.460	0.175
2.15	8.001	0.350	0.250	0.150
2.04	9.210	0.375	0.275	0.163
2.36	7.850	0.400	0.300	0.208
2.38	10.010	0.470	0.400	0.256
2.80	8.503	0.500	0.425	0.272

9. Ammonia synthesis is thought to take place on an iron catalyst according to the following sequence:

$$N_2 + 2* \rightleftharpoons 2N* \tag{1}$$

$$H_2 + 2* \rightleftharpoons 2H* \tag{2}$$

$$N* + H* \rightleftharpoons NH* + * \tag{3}$$

$$NH* + H* \rightleftharpoons NH_2* + * \tag{4}$$

$$NH_2* + H* \rightleftharpoons NH_3 + 2* \tag{5}$$

Obviously, step 2 must occur three times for every occurrence of step 1, and steps 3–5 each occur twice, to give the overall reaction of $N_2 + 3H_2 = 2NH_3$. Experimental evidence suggests that step 1 is the rate-determining step, meaning that all of the other steps can be represented by one pseudo-equilibrated overall reaction. Other independent evidence shows that nitrogen is the only surface species with any significant concentration on the surface (most abundant reaction intermediate). Thus, $[*]_0 = [*] + [N*]$. Equation (5.3.7) gives the rate expression consistent with the above assumptions.

(a) Now, assume that the rate-determining step is actually:

$$N_2 + * \rightleftharpoons N_2* \quad \text{(associative adsorption of } N_2)$$

with N_2* being the most abundant reaction intermediate. Derive the rate expression for the reversible formation of ammonia.

(b) Can the rate expression derived for part (a) and the one given in Equation (5.3.7) be discriminated through experimentation?

10. Nitrous oxide reacts with carbon monoxide in the presence of a ceria-promoted rhodium catalyst to form dinitrogen and carbon dioxide. One plausible sequence for the reaction is given below:

$$(1) \qquad N_2O + * \rightleftharpoons N_2O*$$
$$(2) \qquad N_2O* \longrightarrow N_2 + O*$$
$$(3) \qquad CO + * \rightleftharpoons CO*$$
$$(4) \qquad CO* + O* \longrightarrow CO_2 + 2*$$

$$\overline{\qquad N_2O + CO \Longrightarrow N_2 + CO_2 \qquad}$$

(a) Assume that the surface coverage of oxygen atoms is very small to derive a rate expression of the following form:

$$r = \frac{K_1 P_{N_2O}}{1 + K_2 P_{N_2O} + K_3 P_{CO}}$$

where the K's are collections of appropriate constants. Do not assume a rate-determining step.

(b) The rate expression in part (a) can be rearranged into a linear form with respect to the reactants. Use linear regression with the data in the following table to evaluate the kinetic parameters K_1, K_2, and K_3.

Rates of $N_2O + CO$ reaction at 543 K

P_{CO} (torr)	P_{N_2O} (torr)	Turnover rate (s^{-1})
30.4	7.6	0.00503
30.4	15.2	0.00906
30.4	30.4	0.0184
30.4	45.6	0.0227
30.4	76	0.0361
7.6	30.4	0.0386
15.2	30.4	0.0239
45.6	30.4	0.0117
76	30.4	0.00777

Source: J. H. Holles, M. A. Switzer, and R. J. Davis, *J. Catal.,* **190** (2000) 247.

(c) Use nonlinear regression with the data in part (b) to obtain the kinetic parameters. How do the answers compare?

(d) Which species, N_2O or CO, is present on the surface in greater amount?

11. The reaction of carbon monoxide with steam to produce carbon dioxide and dihydrogen is called the water gas shift (WGS) reaction and is an important process in the production of dihydrogen, ammonia, and other bulk chemicals. The overall reaction is shown below:

$$CO + H_2O = CO_2 + H_2$$

Iron-based solids that operate in the temperature range of 360 to 530°C catalyze the WGS reaction.

(a) One possible sequence is of the Rideal-Eley type that involves oxidation and reduction of the catalyst surface. This can be represented by the following steps:

$$H_2O + * \rightleftharpoons H_2 + O*$$
$$CO + O* \rightleftharpoons CO_2 + *$$

Derive a rate expression.

(b) Another possible sequence for the WGS reaction is of the Langmuir-Hinshelwood type that involves reaction of adsorbed surface species. For the following steps:

$$CO + * \rightleftharpoons CO* \qquad (1)$$
$$H_2O + 3* \rightleftharpoons 2H* + O* \qquad (2)$$
$$CO* + O* \rightleftharpoons CO_2 + 2* \qquad (3)$$
$$2H* \rightleftharpoons H_2 + 2* \qquad (4)$$

derive the rate expression. (Notice that step 2 is an overall equilibrated reaction.) Do not assume that one species is the most abundant reaction intermediate.

12. For the hydrogenation of propionaldehyde (CH_3CH_2CHO) to propanol ($CH_3CH_2CH_2OH$) over a supported nickel catalyst, assume that the rate-limiting step is the reversible chemisorption of propionaldehyde and that dihydrogen adsorbs dissociatively on the nickel surface.

(a) Provide a reasonable sequence of reaction steps that is consistent with the overall reaction.

(b) Derive a rate expression for the rate of consumption of propionaldehyde. At this point, do not assume a single *mari*.

(c) Under what conditions would the rate expression reduce to the experimentally observed function (where prop corresponds to propionaldehyde, and P represents partial pressure):

$$r = \frac{kP_{prop}}{P_{H_2}^{0.5}}$$

13. Some of the oxides of vanadium and molybdenum catalyze the selective oxidation of hydrocarbons to produce valuable chemical intermediates. In a reaction path proposed by Mars and van Krevelen (see Section 10.5), the hydrocarbon first reduces the surface of the metal oxide catalyst by reaction with lattice oxygen atoms. The resulting surface vacancies are subsequently re-oxidized by gaseous O_2. The elementary steps of this process are shown below. Electrons are added to the sequence to illustrate the redox nature of this reaction.

$$RH_2 \; + \; 2O^{-2} \longrightarrow RO + H_2O + 2\square \, (+\, 4e^-)$$

| hydrocarbon | lattice oxygen | product | surface vacancy |

$$\underline{O_2 + 2\square \, (+\, 4e^-) \longrightarrow 2O^{-2}}$$
$$RH_2 + O_2 \Longrightarrow RO + H_2O$$

Derive a rate expression consistent with the above sequence. For this problem, assume that the vacancies do not migrate on the surface. Thus, $2O^{-2}$ and $2\square$ can be considered as a single occupied and a single unoccupied surface site, respectively. Show that the expression can be transformed into the following form:

$$\frac{1}{r} = \frac{1}{K_1[RH_2]} + \frac{1}{K_2[O_2]}$$

where the K's are collections of appropriate constants.

CHAPTER 6

Effects of Transport Limitations on Rates of Solid-Catalyzed Reactions

6.1 | Introduction

To most effectively utilize a catalyst in a commercial operation, the reaction rate is often adjusted to be approximately the same order of magnitude as the rates of transport phenomena. If a catalyst particle in an industrial reactor were operating with an extremely low turnover frequency, diffusive transport of chemicals to and from the catalyst surface would have no effect on the measured rates. While this "reaction-limited" situation is ideal for the determination of intrinsic reaction kinetics, it is clearly an inefficient way to run a process. Likewise, if a catalyst particle were operating under conditions that normally give an extremely high turnover frequency, the overall observed reaction rate is lowered by the inadequate transport of reactants to the catalyst surface. A balance between reaction rate and transport phenomena is frequently considered the most effective means of operating a catalytic reaction. For typical process variables in industrial reactors, this balance is achieved by adjusting reaction conditions to give a rate on the order of 1 μmol/(cm^3-s) [P. B. Weisz, *CHEMTECH*, (July 1982) 424]. This reaction rate translates into a turnover frequency of about 1 s^{-1} for many catalysts (R. L. Burwell, Jr. and M. Boudart, in "Investigations of Rates and Mechanisms of Reactions," Part 1, Ch. 12, E. S. Lewis, Ed., John Wiley, New York, 1974).

Figure 6.1.1 depicts the concentration profile of a reactant in the vicinity of a catalyst particle. In region 1, the reactant diffuses through the stagnant boundary layer surrounding the particle. Since the transport phenomena in this region occur outside the catalyst particle, they are commonly referred to as *external,* or

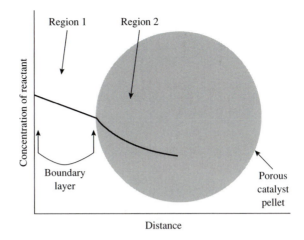

Figure 6.1.1 |
Concentration profile of a reacting species in the vicinity
of a porous catalyst particle. Distances are not to scale.

interphase, transport effects. In region 2, the reactant diffuses into the pores of the
particle, and transport phenomena in this region are called *internal*, or *intraphase*,
transport effects. Both external and internal transport effects may be important in a
catalytic reaction and are discussed separately in the following sections. In addition
to mass transfer effects, heat transfer throughout the catalyst particle and the stagnant boundary layer can dramatically affect observed reaction rates.

6.2 | External Transport Effects

For a solid-catalyzed reaction to take place, a reactant in the fluid phase must first
diffuse through the stagnant boundary layer surrounding the catalyst particle. This
mode of transport is described (in one spatial dimension) by the Stefan-Maxwell
equations (see Appendix C for details):

$$\nabla X_i = \sum_{\substack{j=1 \\ j \neq i}}^{n} \frac{1}{C\,D_{ij}}(X_i N_j - X_j N_i) \tag{6.2.1}$$

where X_i is the mole fraction of component i, C is the total concentration, N_i is the
flux of component i, and D_{ij} is the diffusivity of component i in j. The following
relationship for diffusion of A in a two component mixture at constant pressure
(constant total concentration) can be obtained from simplifying the Stefan-Maxwell
equations:

$$\nabla C_A = \frac{1}{D_{AB}}(X_A N_B - X_B N_A) \tag{6.2.2}$$

Since there are only two components in the mixture, $X_B = 1 - X_A$ and the above expression reduces to:

$$\nabla C_A = \frac{1}{D_{AB}} [X_A (N_A + N_B) - N_A] \tag{6.2.3}$$

Equimolar counterdiffusion ($N_A = -N_B$) can often be assumed and further simplification is thus possible to give:

$$\nabla C_A = \frac{-N_A}{D_{AB}} \tag{6.2.4}$$

The same equation can also be derived by assuming that concentrations are so dilute that $X_A (N_A + N_B)$ can be neglected. Equation (6.2.3) is known as Fick's First Law and can be written as:

$$N_A = -D_{AB} \nabla C_A \qquad \text{(equimolar counterdiffusion and/or} \atop \text{dilute concentration of } A) \tag{6.2.5}$$

The diffusivities of gases and liquids typically have magnitudes that are 10^{-1} and 10^{-5} cm^2 s^{-1}, respectively. The diffusivity of gases is proportional to $T^{1.5}$ and inversely proportional to P, whereas, the diffusivity of liquids is proportional to T and inversely proportional to viscosity $\bar{\mu}$ (may strongly depend on T).

To obtain the flux of reactant A through the stagnant boundary layer surrounding a catalyst particle, one solves Equation (6.2.5) with the appropriate boundary conditions. If the thickness of the boundary layer $\bar{\delta}$ is small compared to the radius of curvature of the catalyst particle, then the problem can be solved in one dimension as depicted in Figure 6.2.1. In this case, Fick's Law reduces to:

$$N_{Ax} = -D_{AB} \frac{dC_A}{dx} \tag{6.2.6}$$

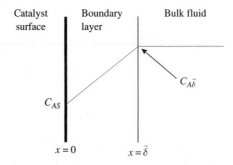

Figure 6.2.1 |
Concentration profile of reactant A in the vicinity of a catalyst particle.

Since the flux of A must be constant through the stagnant film (conservation of mass), the derivative of the flux with respect to distance in the film must vanish:

$$\frac{dN_{Ax}}{dx} = 0 \tag{6.2.7}$$

Differentiating Equation (6.2.6) (assuming constant diffusivity) and combining with Equation (6.2.7), yields the following differential equation that describes diffusion through a stagnant film:

$$\frac{d^2C_A}{dx^2} = 0 \tag{6.2.8}$$

with boundary conditions:

$$C_A = C_{AS} \quad \text{at } x = 0 \tag{6.2.9}$$

$$C_A = C_{A\bar{\delta}} \quad \text{at } x = \bar{\delta} \tag{6.2.10}$$

The solution of Equation (6.2.8) results in a linear concentration profile through the boundary layer:

$$C_A = C_{AS} + (C_{A\bar{\delta}} - C_{AS})\frac{x}{\bar{\delta}} \tag{6.2.11}$$

and the molar flux of A through the film is simply:

$$N_{Ax} = \frac{-D_{AB}}{\bar{\delta}} (C_{A\bar{\delta}} - C_{AS}) \tag{6.2.12}$$

Although diffusion of reacting species can be written in terms of the diffusivity and boundary layer thickness, the magnitude of $\bar{\delta}$ is unknown. Therefore, the mass-transfer coefficient is normally used. That is, the average molar flux from the bulk fluid to the solid surface is ($-x$ direction in Figure 6.2.1)

$$N_A = \bar{k}_c(C_{A\bar{\delta}} - C_{AS}) \tag{6.2.13}$$

where $\bar{k}_c$ is the mass transfer coefficient over the surface area of the particle. The mass transfer coefficient is obtained from correlations and is a function of the fluid velocity past the particle. If the fluid is assumed to be well mixed, the concentration of A at the edge of the stagnant boundary layer is equivalent to that in the bulk fluid, C_{AB}, and Equation (6.2.13) can therefore be written as:

$$N_A = \bar{k}_c(C_{AB} - C_{AS}) \tag{6.2.14}$$

At steady-state, the flux of A equals the rate of reaction thus preventing accumulation or depletion. For a simple first-order reaction, the kinetics depend on the surface rate constant, k_S, and the concentration of A at the surface:

$$r = k_S C_{AS} = \bar{k}_c\left(C_{AB} - C_{AS}\right) \tag{6.2.15}$$

Solving for C_{AS} yields:

$$C_{AS} = \frac{\bar{k}_c C_{AB}}{k_S + \bar{k}_c} \tag{6.2.16}$$

Substitution of the above expression for C_{AS} into Equation (6.2.15) gives a rate expression in terms of the measurable quantity, C_{AB}, the reactant concentration in the bulk fluid:

$$\mathbf{r} = \frac{k_S \bar{k}_c}{k_S + \bar{k}_c} C_{AB} = \frac{C_{AB}}{\dfrac{1}{k_S} + \dfrac{1}{\bar{k}_c}} \tag{6.2.17}$$

An overall, observed rate constant can be defined in terms of k_S and $\bar{k}_c$ as:

$$\frac{1}{k_{\text{obs}}} = \left(\frac{1}{k_S}\right) + \left(\frac{1}{\bar{k}_c}\right) \tag{6.2.18}$$

so that the rate expressed in terms of observable quantities can be written as:

$$\mathbf{r}_{\text{obs}} = k_{\text{obs}} C_{AB} \tag{6.2.19}$$

For rate laws that are noninteger or complex functions of the concentration, C_{AS} is found by trial and error solution of the flux expression equated to the reaction rate. The influence of diffusional resistance on the observed reaction rate is especially apparent for a very fast surface reaction. For that case, the surface concentration of reactant is very small compared to its concentration in the bulk fluid. The observed rate is then written according to Equation (6.2.15), but ignoring C_{AS}:

$$\mathbf{r}_{\text{obs}} = \bar{k}_c C_{AB} \tag{6.2.20}$$

The observed rate will appear to be first-order with respect to the bulk reactant concentration, regardless of the intrinsic rate expression applicable to the surface reaction. This is a clear example of how external diffusion can mask the intrinsic kinetics of a catalytic reaction. In a catalytic reactor operating under mass transfer limitations, the conversion at the reactor outlet can be calculated by incorporating Equation (6.2.20) into the appropriate reactor model.

Solution of a reactor problem in the mass transfer limit requires an estimation of the appropriate mass transfer coefficient. Fortunately, mass transfer correlations have been developed to aid the determination of mass transfer coefficients. For example, the Sherwood number, Sh, relates the mass transfer coefficient of a species A to its diffusivity and the radius of a catalyst particle, R_p:

$$Sh = \frac{\bar{k}_c (2R_p)}{D_{AB}} \tag{6.2.21}$$

For flow around spherical particles, the Sherwood number is correlated to the Schmidt number, Sc, and the Reynolds number, Re:

$$Sh = 2 + 0.6 Re^{1/2} Sc^{1/3} \tag{6.2.22}$$

$$Sc = \frac{\overline{\mu}}{\rho D_{AB}} \tag{6.2.23}$$

$$Re = \frac{u\rho(2R_p)}{\overline{\mu}} \tag{6.2.24}$$

where $\overline{\mu}$ is the viscosity (kg m^{-1} s^{-1}), ρ is the fluid density (kg m^{-3}), and u is the linear fluid velocity (m s^{-1}). However, most mass-transfer results are correlated in terms of Colburn J factors:

$$J = \frac{Sh}{Sc^{1/3} Re} \tag{6.2.25}$$

that are plotted as a function of the Reynolds number. These J factor plots are available in most textbooks on mass transfer. If one can estimate the fluid density, velocity, viscosity, diffusivity, and catalyst particle size, then a reasonable approximation of the mass-transfer coefficient can be found.

It is instructive to examine the effects of easily adjustable process variables on the mass-transfer coefficient. Combining Equations (6.2.21–6.2.24) gives the functional dependence of the mass-transfer coefficient:

$$\overline{k}_c \propto \frac{D_{AB}Sh}{R_p} \propto \frac{D_{AB}}{R_p} Re^{1/2} Sc^{1/3} \propto \frac{D_{AB}}{R_p} \left(\frac{R_p u \rho}{\overline{\mu}}\right)^{1/2} \left(\frac{\overline{\mu}}{\rho D_{AB}}\right)^{1/3} \propto \frac{\left(D_{AB}\right)^{2/3} u^{1/2} \rho^{1/6}}{\left(R_p\right)^{1/2} \left(\overline{\mu}\right)^{1/6}}$$

or

$$\overline{k}_c \propto \frac{\left(D_{AB}\right)^{2/3} u^{1/2} \rho^{1/6}}{\left(R_p\right)^{1/2} \left(\overline{\mu}\right)^{1/6}} \tag{6.2.26}$$

Equation (6.2.26) shows that decreasing the catalyst particle size and increasing the fluid velocity can significantly increase the mass-transfer coefficient. These simple variables may be used as process "handles" to decrease the influence of external mass-transfer limitations on the observed reaction rate.

To quickly estimate the importance of external mass-transfer limitations, the magnitude of the change in concentration across the boundary layer can be calculated from the observed rate and the mass-transfer coefficient:

$$r_{obs} = \overline{k}_c \left(C_{AB} - C_{AS}\right) \qquad \frac{r_{obs}}{\overline{k}_c} = \Delta C_A \tag{6.2.27}$$

If $\Delta C_A \ll C_{AB}$, then external mass-transfer limitations are not significantly affecting the observed rate.

The effects of heat transfer are completely analogous to those of mass transfer. The heat flux, q, across the stagnant boundary layer shown in Figure 6.2.1 is related to the difference in temperature and the heat-transfer coefficient, h_t, according to:

$$q = h_t \left(T_B - T_S\right) \tag{6.2.28}$$

Steady state requires that the heat flux is equivalent to the heat generated (or consumed) by reaction:

$$r_{obs}(\Delta H_r) = h_t(T_B - T_S) \tag{6.2.29}$$

where ΔH_r is the heat of reaction per mole of A converted. To estimate the influence of heat-transfer limitations on the observed rate, the change in temperature across the film is found by evaluating the observed rate of heat generated (or consumed) and the heat-transfer coefficient (obtained from J factor correlations, similar to the case of mass-transfer coefficients):

$$r_{obs}(\Delta H_r) = h_t(T_B - T_S), \qquad \frac{r_{obs}(\Delta H_r)}{h_t} = \Delta T \tag{6.2.30}$$

If $|\Delta T| \ll T_B$, then the effect of external heat-transfer limitations on the observed rate can be ignored. Equation (6.2.30) can also be used to find the maximum temperature change across the film. Using Equation (6.2.15) to eliminate the observed rate, the resulting equation relates the concentration change across the film to the temperature change:

$$\bar{k}_c(C_{AB} - C_{AS})\Delta H_r = h_t(T_B - T_S) \tag{6.2.31}$$

The maximum temperature change across the film will occur when C_{AS} approaches zero, which corresponds to the maximum observable rate. Solving Equation (6.2.31) for ΔT_{max} with $C_{AS} = 0$ gives the following expression:

$$\Delta T_{max} = \frac{\bar{k}_c \Delta H_r}{h_t} C_{AB} \tag{6.2.32}$$

that can always be calculated for a reaction, independent of an experiment. If both external heat and mass transfer are expected to affect the observed reaction rate, the balances must be solved simultaneously.

6.3 | Internal Transport Effects

Many solid catalysts contain pores in order to increase the specific surface area available for adsorption and reaction, sometimes up to 10^3 m^2 g^{-1}. Since nearly all of the catalytically active sites in highly porous solids are located in the pore network, diffusion of molecules in confined spaces obviously plays a critical role in the observed rate of reaction.

The preceding section assumed that the mass-transport mechanism in a fluid medium is dominated by molecule-molecule collisions. However, the mean free path of gases often exceeds the dimensions of small pores typical of solid catalysts. In this situation, called *Knudsen* diffusion, molecules collide more often with the pore walls than with other molecules. According to Equation (6.3.1), the Knudsen diffusivity of component A, D_{KA}, is proportional to $T^{1/2}$, but is independent of both pressure and the presence of other species:

$$D_{KA} = (9.7 \times 10^3) \cdot R_{\text{pore}} \cdot \left(\frac{T}{M_A}\right)^{1/2} \quad \text{cm}^2 \, \text{s}^{-1} \tag{6.3.1}$$

where R_{pore} is the pore radius in cm, T is the absolute temperature in Kelvins, and M_A is the molecular weight of A. Recall that the diffusivity D_{AB} for molecular diffusion depends on the pressure and the other species present but is independent of the pore radius. In cases where both molecule-molecule and molecule-wall collisions are important, neither molecular diffusivity nor Knudsen diffusivity alone can adequately describe the transport phenomena. Under the conditions of equimolar counterdiffusion of a binary mixture, a transition diffusivity of component A, D_{TA}, can be approximated by the Bosanquet equation (see Appendix C for derivation):

$$\frac{1}{D_{TA}} = \frac{1}{D_{AB}} + \frac{1}{D_{KA}} \tag{6.3.2}$$

VIGNETTE 6.3.1

Diffusion of molecules in porous catalysts is highly dependent on the dimensions of the pore network. Figure 6.3.1 shows typical values of the gas-phase diffusivity as a function of pore size. Transport of molecules in very large pores is essentially governed by molecular diffusion since collisions with other molecules are much more frequent than collisions with the pore wall. In the Knudsen regime, molecule-wall collisions are dominant and the diffusivity decreases with pore size. Further reduction in pore size to values typical of zeolites results in a dramatic decrease in diffusivity due to single file diffusion of molecules. This phenomenon is also called *configurational* diffusion [P. B. Weisz,

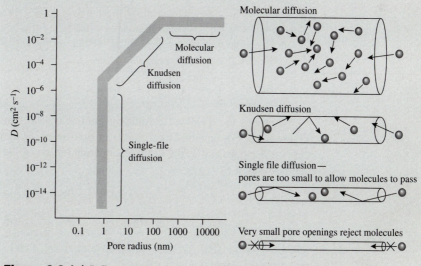

Figure 6.3.1 | Influence of pore size on diffusivity of gas-phase molecules.

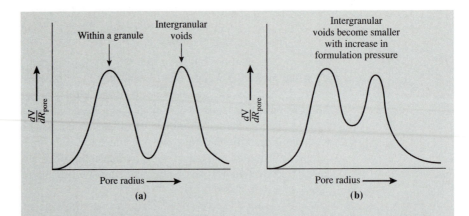

Figure 6.3.2 | Influence of pelletizing conditions on pore size distribution $\dfrac{dV}{dR_{pore}}$ in a bimodal pore network. **(a)** Small pores (micropores) are associated with the catalyst powder whereas large pores (macropores) result from the voids between granules in compressed catalyst pellets. **(b)** Decrease in pore size associated with intergranular voids as pellet in **(a)** is compressed as higher pressure.

CHEMTECH, (Aug. 1973) 498], and is described mathematically in a unique manner. For molecules exceeding the size of the pore aperture, the interior of the pore is inaccessible.

Small pores, called *micropores*, are often present in high surface area catalyst powders. These powders are compressed to form pellets so that the pressure drop in an industrial packed bed reactor is not too great. However, compression of the powders creates a new network of pores, called *macropores*, formed by the intergranular voids. If the original catalyst powder had a unimodal pore size distribution (PSD), the compressed catalyst pellet would have a bimodal PSD (see Figure 6.3.2.) The pore size associated with the voids is a function of the pressure used to form the pellet—high formulation pressures result in smaller void spaces. In contrast, the micropores of the catalyst powders are relatively unaffected by the compression process (see Figure 6.3.2).

Consider the idealized cylindrical pore in a solid catalyst slab, as depicted in Figure 6.3.3. For an isothermal, isobaric, first-order reaction of A to form B that occurs on the pore walls, the mole balance on a slice of the pore with thickness Δx can be written as:

$$(\text{Rate of input } A) - (\text{Rate of output } A) + (\text{Rate of generation } A) = 0 \qquad (6.3.3)$$

$$\pi R_{pore}^2 N_A\big|_x - \pi R_{pore}^2 N_A\big|_{x+\Delta x} - k_s C_A(2\pi R_{pore})(\Delta x) = 0 \qquad (6.3.4)$$

where N_A is the flux of A evaluated at both sides of the slice, k_s is the first-order rate constant expressed per surface area of the catalyst (volume/{surface area}/time), and $2\pi R_{pore}(\Delta x)$ is the area of the pore wall in the catalyst slice. Rearranging

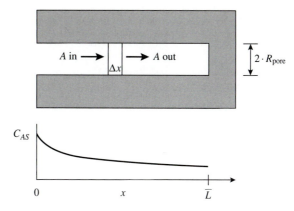

Figure 6.3.3 |
Schematic representation of component A diffusing and
reacting in an idealized cylindrical pore.

Equation (6.3.4) and taking the limit as Δx approaches zero gives the following
differential equation for the mole balance:

$$-\frac{dN_A}{dx} = \frac{2k_s}{R_{\text{pore}}} C_A \tag{6.3.5}$$

Recall that the Stefan-Maxwell equation relates the molar flux of A to its concentration gradient according to:

$$\nabla C_A = \sum_{\substack{j=1 \\ j \neq A}}^{n} \frac{1}{D_{Aj}} \left(X_A N_j - X_j N_A \right) \tag{6.3.6}$$

For diffusion in one dimension in the absence of bulk flow:

$$\frac{dC_A}{dx} = -\frac{N_A}{\left(\dfrac{1}{D_{AB}} + \dfrac{1}{D_{KA}} \right)^{-1}} \tag{6.3.7}$$

which is Fick's First Law that can be written as:

$$N_A = -D_{TA} \frac{dC_A}{dx} \tag{6.3.8}$$

where D_{TA} is defined by Equation (6.3.2). Substitution of Fick's Law into the
mole balance, Equation (6.3.5), yields the following second-order differential
equation:

$$\frac{d}{dx}\left(D_{TA} \frac{dC_A}{dx} \right) = \frac{2k_s}{R_{\text{pore}}} C_A \tag{6.3.9}$$

Assuming D_{TA} is constant:

$$\frac{d^2C_A}{dx^2} - \frac{2k_s}{D_{TA}R_{pore}}C_A = 0 \tag{6.3.10}$$

The surface rate constant can be rewritten on a volume basis by using the surface to volume ratio of a cylindrical pore:

$$\frac{\text{Area}}{\text{Volume}} = \frac{2\pi R_{pore}\overline{L}}{\pi R_{pore}^2 \overline{L}} = \frac{2}{R_{pore}} \tag{6.3.11}$$

and

$$k = \frac{2k_s}{R_{pore}} \tag{6.3.12}$$

To simplify the mole balance, let:

$$\chi = \frac{x}{\overline{L}} \tag{6.3.13}$$

$$\phi = \overline{L}\sqrt{\frac{k}{D_{TA}}} \tag{6.3.14}$$

and their substitution into Equation (6.3.10) gives:

$$\frac{d^2C_A}{d\chi^2} - \phi^2 C_A = 0 \tag{6.3.15}$$

Boundary conditions at each end of the pore are needed to solve the mole balance. At $x = 0$ (the pore mouth), the concentration of A is equal to C_{AS}. At the other end of the pore, the gradient in concentration is equal to zero. That is, there is no flux at the end of the pore. These conditions can be written as:

$$C_A = C_{AS} \quad \text{at } \chi = 0 \tag{6.3.16}$$

$$\frac{dC_A}{d\chi} = 0 \quad \text{at } \chi = 1 \tag{6.3.17}$$

The solution of Equation (6.3.15) using boundary conditions given in Equations (6.3.16) and (6.3.17) is:

$$C_A = C_{AS}\frac{\cosh\left[\phi\left(1 - \frac{x}{\overline{L}}\right)\right]}{\cosh[\phi]} \tag{6.3.18}$$

The term ϕ, also known as the *Thiele modulus,* is a dimensionless number composed of the square root of the characteristic reaction rate (kC_{AS}) divided by the characteristic diffusion rate $\left(\frac{D_{TA}C_{AS}}{\overline{L}^2}\right)$. The Thiele modulus indicates which process

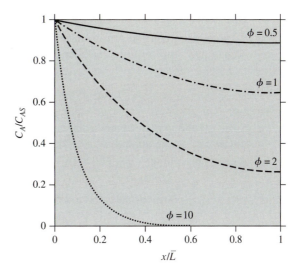

Figure 6.3.4 |
Effect of Thiele modulus on the normalized concentration
profiles in a catalyst pore with first-order surface reaction.

is rate-limiting. Figure 6.3.4 illustrates the concentration profile of reactant A along the pore for various values of the Thiele modulus. When ϕ is small, the diffusional resistance is insufficient to limit the rate of reaction and the concentration can be maintained near C_{AS} within the catalyst particle. However, when ϕ is large, a significant diffusional resistance prevents a constant concentration profile of A within the catalyst particle and thus lowers the observed rate.

Now consider a catalyst pellet with a random network of "zig-zag" pores. The surface of the pellet is composed of both solid material and pores. The flux equation derived earlier must be modified to account for the fact that the flux, N_A, is based only on the area of a pore. A parameter called the *porosity* of the pellet, or $\bar{\varepsilon}_p$, is defined as the ratio of void volume within the pellet to the total pellet volume (void + solid). The flux can be expressed in moles of A diffusing per unit pellet surface area (containing both solids and pores) by using $\bar{\varepsilon}_p$ as follows:

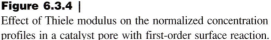

$$N_A = \bar{\varepsilon}_p N_A(\text{based on pore}) = -\bar{\varepsilon}_p D_{TA}\frac{dC_A}{dx}$$

Since the porosity of many solid catalysts falls between 0.3 and 0.7, a reasonable estimate of $\bar{\varepsilon}_p$ in the absence of experimental data is 0.5. The second parameter needed to modify the flux is the tortuosity, $\bar{\tau}$, which accounts for the deviations in the path length of the pores. Since the concentration gradient is based on the pellet geometry, the flux equation must be corrected to reflect the actual distance molecules travel in the pores. The tortuosity is the ratio of the "zig-zag length" to the "straight length" of the pore system. Obviously, $\bar{\tau}$ must be greater than or equal to one. For example, an ideal, cylindrical pore has $\bar{\tau}$ equal to 1, and a network of randomly oriented cylindrical pores has $\bar{\tau}$ equal to approximately 3. The flux equation can be written to take into account the "true" diffusion path length as:

$$N_A(\text{pellet}) = -\frac{\bar{\varepsilon}_p}{\bar{\tau}} D_{TA} \frac{dC_A}{dx}$$

Since the tortuosity of many solid catalysts falls between 2 and 7, a reasonable estimate of $\bar{\tau}$ in the absence of experimental data is 4. The diffusivity in a unimodal pore system can now be defined by the flux of reactant into the pellet according to:

$$N_A(\text{pellet}) = \frac{\bar{\varepsilon}_p}{\bar{\tau}} N_A(\text{pore}) = -\frac{\bar{\varepsilon}_p}{\bar{\tau}} D_{TA} \frac{dC_A}{dx} = -D_{TA}^e \frac{dC_A}{dx} \qquad (6.3.19)$$

where the superscript e refers to the *effective* diffusivity. Likewise, the following *effective* diffusivities can be written:

$$D_{KA}^e = \frac{\bar{\varepsilon}_p}{\bar{\tau}} D_{KA} \qquad (6.3.20)$$

$$D_{AB}^e = \frac{\bar{\varepsilon}_p}{\bar{\tau}} D_{AB} \qquad (6.3.21)$$

Now consider several ideal geometries of porous catalyst pellets shown in Figure 6.3.5. The first pellet is an infinite slab with thickness $2x_p$. However, since

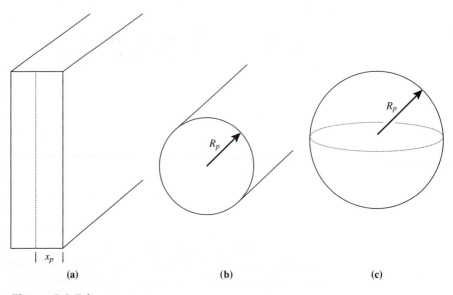

(a)   (b)   (c)

Figure 6.3.5 |
Schematic representations of ideal catalyst pellet geometries. (a) Infinite slab. (b) Infinite right cylinder. (c) Sphere.

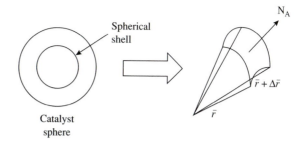

Figure 6.3.6 |
Schematic of the shell balance on a spherical catalyst
pellet.

pores can have openings on both faces of the slab because of symmetry, the characteristic length associated with the slab is half of the thickness, or x_p. The second pellet is an infinite right cylinder with radius R_p, and the third pellet is a sphere with radius R_p. End effects (or edge effects) are ignored in the cases of the slab and the cylinder.

As an example, simultaneous diffusion and reaction in a spherical catalyst pellet is described in detail below. The results are then generalized to other pellet shapes. The reaction is assumed to be isothermal, since the high thermal conductivity of most solid catalysts ensures a fairly constant temperature within a single pellet. In addition, the reaction is assumed to be isobaric, which implies negligible mole change upon reaction. For reactions with a significant mole change with conversion, the presence of a large excess of inert material can reduce the impact of reacting species on the total pressure. Isobaric conditions can therefore be achieved in a variety of catalytic reactions, regardless of reaction stoichiometry.

A diagram of the shell balance (material balance) for simultaneous diffusion and first order reaction of component A in a sphere is shown in Figure 6.3.6. The material balance in the spherical shell is given by:

(Rate of input A) $-$ (Rate of output A) $+$ (Rate of generation A) $= 0$

$$4\pi\bar{r}^2 N_A|_{\bar{r}} - 4\pi\bar{r}^2 N_A|_{\bar{r}+\Delta\bar{r}} - kC_A 4\pi\bar{r}^2\Delta\bar{r} = 0 \qquad (6.3.22)$$

The third term in Equation (6.3.22) is the rate of consumption of A in the differential volume defined between $\bar{r}$ and $\bar{r} + \Delta\bar{r}$. Simplifying Equation (6.3.22) and taking the limit as $\Delta\bar{r}$ approaches zero yields the following differential equation:

$$-\frac{d(\bar{r}^2 N_A)}{d\bar{r}} - \bar{r}^2 kC_A = 0 \qquad (6.3.23)$$

The flux of A can be expressed in terms of concentration for binary systems according to Fick's Law (in spherical coordinates):

$$N_A = -D_{TA}^e \frac{dC_A}{d\bar{r}} \qquad \text{(equimolar counterdiffusion)} \qquad (6.3.24)$$

Substitution of Equation (6.3.24) into (6.3.23) gives:

$$-\frac{d\left[\bar{r}^2\left(-D_{TA}^e\dfrac{dC_A}{d\bar{r}}\right)\right]}{d\bar{r}} - \bar{r}^2 kC_A = 0$$

$$D_{TA}^e\left[\bar{r}^2\frac{d^2C_A}{d\bar{r}^2} + 2\bar{r}\frac{dC_A}{d\bar{r}}\right] - \bar{r}^2 kC_A = 0$$

$$D_{TA}^e\left[\frac{d^2C_A}{d\bar{r}^2} + \frac{2}{r}\frac{dC_A}{d\bar{r}}\right] - kC_A = 0 \tag{6.3.25}$$

The above equation can be made dimensionless by the following substitutions:

$$\psi = \frac{C_A}{C_{AS}} \tag{6.3.26}$$

$$\omega = \frac{\bar{r}}{R_p} \tag{6.3.27}$$

where C_{AS} is the concentration of A on the external surface and R_p is the radius of the spherical particle. Rewriting Equation (6.3.25) in terms of dimensionless concentration and radius gives:

$$\frac{d^2\psi}{d\omega^2} + \frac{2}{\omega}\frac{d\psi}{d\omega} - \frac{(R_p)^2 k}{D_{TA}^e}\psi = 0 \tag{6.3.28}$$

The Thiele modulus, ϕ, for a sphere is defined as:

$$\phi = R_p\sqrt{\frac{k}{D_{TA}^e}} \tag{6.3.29}$$

so that Equation (6.3.28) becomes:

$$\frac{d^2\psi}{d\omega^2} + \frac{2}{\omega}\frac{d\psi}{d\omega} - \phi^2\psi = 0 \tag{6.3.30}$$

with boundary conditions:

$$\psi = 1 \qquad \text{at } \omega = 1 \quad \text{(surface of sphere)} \tag{6.3.31}$$

$$\frac{d\psi}{d\omega} = 0 \qquad \text{at } \omega = 0 \quad \text{(center of sphere)} \tag{6.3.32}$$

The zero-flux condition at the center results from the symmetry associated with the spherical geometry. The solution of the above differential equation with the stated boundary conditions is:

$$\psi = \frac{C_A}{C_{AS}} = \frac{\sinh(\phi\omega)}{\omega\sinh(\phi)} \tag{6.3.33}$$

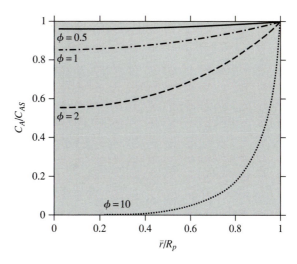

Figure 6.3.7 |

Effect of Thiele modulus on the normalized concentration profiles in a spherical catalyst particle with first-order reaction. The external surface of the particle is located at $\bar{r}/R_p = 1$.

Figure 6.3.7 illustrates the effect of the Thiele modulus on the concentration profile within a spherical catalyst pellet.

An *effectiveness factor*, η, can be defined as the ratio of the observed rate (r_{obs}) to the rate that would be observed in the absence of internal diffusional limitations (r_{max}):

$$\eta = \frac{r_{obs}}{r_{max}} = \frac{\int_0^{V_p} r(C_A)\,dV}{V_p r(C_{AS})} = \frac{\int_0^{R_p} r(C_A)4\pi\bar{r}^2\,d\bar{r}}{V_p r(C_{AS})} = \frac{S_p \int_0^{R_p} r(C_A)\left(\frac{\bar{r}}{R_p}\right)^2 d\bar{r}}{V_p r(C_{AS})} \qquad (6.3.34)$$

where V_p is the volume of the catalyst pellet, S_p is the external surface area of the pellet, and $r(C_A)$ is the reaction rate determined with concentration C_A. The denominator in Equation (6.3.34) is simply the rate of reaction in the catalyst pellet assuming the reactant concentration is equal to that on the external surface, C_{AS}. For a first-order reaction in a spherical particle, the rate observed in the absence of diffusional limitations is:

$$r_{max} = \frac{4}{3}\pi(R_p)^3 kC_{AS} \qquad (6.3.35)$$

At the steady state, the flux of A entering the pellet must be equivalent to the net rate of consumption of A in the pellet. Thus, the flux entering the sphere can be used to determine the observed rate of reaction in the presence of diffusional limitations.

$$r_{obs} = 4\pi (R_p)^2 D_{TA}^e \frac{dC_A}{d\bar{r}}\bigg|_{\bar{r}=R_p} \tag{6.3.36}$$

Substituting Equations (6.3.35) and (6.3.36) into (6.3.34) gives:

$$\eta = \frac{r_{obs}}{r_{max}} = \frac{4\pi (R_p)^2 D_{TA}^e \dfrac{dC_A}{d\bar{r}}\bigg|_{\bar{r}=R_p}}{\dfrac{4}{3}\pi (R_p)^3 kC_{AS}} = \frac{3}{R_p} \frac{D_{TA}^e}{k} \frac{\dfrac{dC_A}{d\bar{r}}\bigg|_{\bar{r}=R_p}}{C_{AS}} \tag{6.3.37}$$

The above equation is made nondimensional by the substitutions defined in Equations (6.3.26) and (6.3.27) and is:

$$\eta = \frac{3}{\phi^2} \frac{d\psi}{d\omega}\bigg|_{\omega=1} \tag{6.3.38}$$

The derivative is evaluated from the concentration profile, (6.3.33), to give:

$$\eta = \frac{3}{\phi^2} \frac{d\psi}{d\omega}\bigg|_{\omega=1} = \frac{3}{\phi^2} \frac{d}{d\omega}\left[\frac{\sinh(\phi\omega)}{\omega \sinh(\phi)}\right]\bigg|_{\omega=1}$$

$$\eta = \frac{3}{\phi^2}\left[\frac{1}{\sinh(\phi)}\left(\frac{\phi}{\omega}\cosh(\phi\omega) - \frac{1}{\omega^2}\sinh(\phi\omega)\right)\right]\bigg|_{\omega=1}$$

$$\eta = \frac{3}{\phi^2}\left[\frac{\phi}{\tanh(\phi)} - 1\right] = \frac{3}{\phi}\left[\frac{1}{\tanh(\phi)} - \frac{1}{\phi}\right] \tag{6.3.39}$$

Figure 6.3.8 illustrates the relationship between the effectiveness factor and the Thiele modulus for a spherical catalyst pellet.

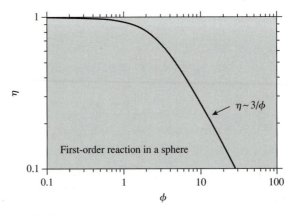

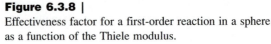

Figure 6.3.8 |
Effectiveness factor for a first-order reaction in a sphere as a function of the Thiele modulus.

Table 6.3.1 | Influence of catalyst particle geometry on concentration profile and effectiveness factor for a first-order, isothermal, isobaric reaction.

	Slab $(\omega = x/x_p)$	Cylinder[a] $(\omega = \bar{r}/R_p)$	Sphere $(\omega = \bar{r}/R_p)$
ϕ	$x_p\sqrt{\dfrac{k}{D_{TA}^e}}$	$R_p\sqrt{\dfrac{k}{D_{TA}^e}}$	$R_p\sqrt{\dfrac{k}{D_{TA}^e}}$
$\psi = C_A/C_{AS}$	$\dfrac{\cosh(\phi\omega)}{\cosh(\phi)}$	$\dfrac{I_0(\phi\omega)}{I_0(\phi)}$	$\dfrac{\sinh(\phi\omega)}{\omega\sinh(\phi)}$
η	$\dfrac{\tanh(\phi)}{\phi}$	$\dfrac{2I_1(\phi)}{\phi I_0(\phi)}$	$\dfrac{3}{\phi}\left[\dfrac{1}{\tanh(\phi)} - \dfrac{1}{\phi}\right]$

[a]I_i is a modified Bessel function of order i.

Table 6.3.2 | Characteristic length parameters of common pellet shapes.

	L_p
Slab	Length of pore, x_p
Cylinder	$R_p/2$
Sphere	$R_p/3$

A low value of the Thiele modulus results from a small diffusional resistance. For this case, the effectiveness factor is approximately 1 (values of ϕ typically less than 1). Large values of the Thiele modulus are characteristic of a diffusion-limited reaction with an effectiveness factor less than 1. For $\phi \gg 1$, the value of the effectiveness factor in a sphere approaches $3/\phi$, as illustrated in Figure 6.3.8.

The concentration profile and the effectiveness factor are clearly dependent on the geometry of a catalyst particle. Table 6.3.1 summarizes the results for catalyst particles with three common geometries.

Aris was the first to point out that the results for the effectiveness factor in different pellet geometries can be approximated by a single function of the Thiele modulus if the length parameter in ϕ is the ratio of the pellet volume, V_p, to the pellet external surface area, S_p [R. Aris, *Chem. Eng. Sci.*, **6** (1957) 262]. Thus, the length parameter, L_p, is defined by:

$$L_p = \frac{V_p}{S_p} \tag{6.3.40}$$

and the Thiele modulus is defined by:

$$\phi_0 = L_p\sqrt{\frac{k}{D_{TA}^e}} \tag{6.3.41}$$

where Table 6.3.2 summarizes the characteristic length parameter for common geometries.

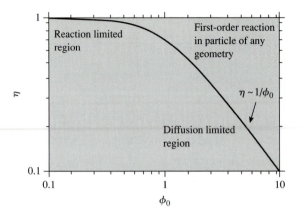

Figure 6.3.9 |
Effectiveness factor $[\eta = \tanh(\phi_0)/\phi_0]$ for a first-order
reaction in a catalyst as a function of the Thiele modulus
with generalized length parameter.

According to the above definitions, the effectiveness factor for any of the above
shapes can adequately describe simultaneous reaction and diffusion in a catalyst particle. The equation for the effectiveness factor in a slab is the simplest in Table 6.3.1
and will be used for all pellet shapes with the appropriate Thiele modulus:

$$\eta = \frac{\tanh(\phi_0)}{\phi_0} \qquad (6.3.42)$$

This relationship is plotted in Figure 6.3.9. The effectiveness factor for a severely
diffusion-limited reaction in a catalyst particle is approximated by the inverse of the
Thiele modulus.

EXAMPLE 6.3.1 |

The double bond isomerization of 1-hexene to form 2-hexene was studied in a laboratory
reactor containing rhodium particles supported on alumina at 150°C and atmospheric
pressure:

$$H_2C = CH - CH_2 - CH_2 - CH_2 - CH_3 \Rightarrow H_3C - CH = CH - CH_2 - CH_2 - CH_3$$

The reaction was found to be first order in 1-hexene with a rate constant of 0.14 s^{-1}. Find
the largest pellet size that can be used in an industrial reactor to achieve 70 percent of the
maximum rate. The pore radius of the alumina is 10 nm, and D_{AB} is 0.050 cm^2 s^{-1}.

■ Answer

It is desired to find the particle size that gives an internal effectiveness factor equal to 0.70.
For any geometry, the Thiele modulus is determined from:

$$\eta = \frac{\tanh(\phi_0)}{\phi_0} = 0.70$$

$$\phi_0 = 1.18$$

Assuming a spherical catalyst pellet with radius R_p, the Thiele modulus is:

$$\phi_0 = 1.18 = \frac{V_p}{S_p}\sqrt{\frac{k}{D_{TA}^e}} = \frac{R_p}{3}\sqrt{\frac{k}{D_{TA}^e}}$$

Since the rate constant is known, estimation of the effective diffusivity allows the calculation of particle radius. In the absence of experimental data, the porosity and tortuosity are assumed to be 0.5 and 4, respectively. Thus,

$$D_{AB}^e = \frac{\overline{\varepsilon}_p}{\tau}D_{AB} = \frac{0.5}{4} \cdot 0.050 = 0.0062 \text{ cm}^2 \text{ s}^{-1}$$

The Knudsen diffusivity is calculated from the temperature, pore radius, and molecular weight of hexene (84 g mol^{-1}) according to Equation (6.3.1):

$$D_{KA} = (9.7 \times 10^3) \cdot R_{\text{pore}} \cdot \sqrt{\frac{T}{M_A}} \quad (\text{cm}^2 \text{ s}^{-1})$$

$$D_{KA} = (9.7 \times 10^3)(1.0 \times 10^{-6}\text{cm})\sqrt{\frac{423K}{84}} = 0.022 \text{ cm}^2 \text{ s}^{-1}$$

$$D_{KA}^e = \frac{0.5}{4} \cdot 0.022 = 0.0027 \text{ cm}^2 \text{ s}^{-1}$$

The effective transition diffusivity is calculated from the Bosanquet equation assuming equimolar counter diffusion, which is what happens with isomerization reactions:

$$\frac{1}{D_{TA}^e} = \frac{1}{D_{AB}^e} + \frac{1}{D_{KA}^e}$$

$$D_{TA}^e = 0.0019 \text{ cm}^2 \text{ s}^{-1}$$

Substituting the necessary terms into the expression for the Thiele modulus yields the radius of the spherical catalyst pellet:

$$R_p = 3 \cdot 1.18 \cdot \sqrt{\frac{0.0019}{0.14}} = 0.41 \text{ cm}$$

Thus, spherical particles of about 1/3 in. diameter will have an effectiveness factor of 0.70.

It is worthwhile to examine how reasonably well a single characteristic length parameter describes reaction/diffusion in a finite cylinder, a very common catalyst pellet configuration. The pellet shown has a cylinder length ($2x_p$) and radius R_p:

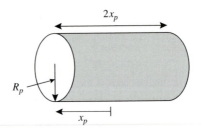

The material balance for a first-order reaction of A in the pellet is given by:

$$D_{TA}^e \left[\frac{\partial^2 C_A}{\partial \bar{r}^2} + \frac{1}{\bar{r}} \frac{\partial C_A}{\partial \bar{r}} + \frac{\partial^2 C_A}{\partial x^2} \right] = k C_A \tag{6.3.43}$$

with the flux equations being written for both the axial and radial directions:

$$N_A = -D_{TA}^e \frac{dC_A}{d\bar{r}} \quad \text{(radial)} \tag{6.3.44}$$

$$N_A = -D_{TA}^e \frac{dC_A}{dx} \quad \text{(axial)} \tag{6.3.45}$$

The solution of these equations, with appropriate boundary conditions, provides the concentration profile and the effectiveness factor for a finite cylinder. As discussed earlier, an approximation of the effectiveness factor:

$$\eta = \frac{\tanh(\phi_0)}{\phi_0} \tag{6.3.42}$$

can be used with any geometry as long as the Thiele modulus is based on the characteristic length defined by the volume-to-surface ratio. The volume and surface area of the finite cylindrical pellet are simply:

$$V_p = \pi (R_p)^2 \cdot 2x_p \tag{6.3.46}$$
$$S_p = 2\pi (R_p)^2 + \left(2\pi R_p \cdot 2x_p \right) \tag{6.3.47}$$

Thus, the characteristic length for the finite cylinder is:

$$\frac{V_p}{S_p} = \frac{\pi (R_p)^2 \cdot 2x_p}{2\pi (R_p)^2 + \left(2\pi R_p \cdot 2x_p \right)} = \frac{R_p}{\dfrac{R_p}{x_p} + 2} \tag{6.3.48}$$

and the Thiele modulus for use in Equation (6.3.42) is:

$$\phi_0 = \frac{V_p}{S_p} \sqrt{\frac{k}{D_{TA}^e}} = \left[\frac{R_p}{\dfrac{R_p}{x_p} + 2} \right] \sqrt{\frac{k}{D_{TA}^e}} \tag{6.3.49}$$

For comparison, the "radius" of an "equivalent" spherical particle, $R_{p(\text{sphere})}$, can be calculated by equating the volume-to-surface ratios:

$$\left(\frac{V_p}{S_p}\right)_{\text{finite cylinder}} = \left(\frac{V_p}{S_p}\right)_{\text{sphere}}$$

$$\frac{R_p}{\dfrac{R_p}{x_p} + 2} = \frac{R_{p(\text{sphere})}}{3}$$

Therefore, the radius of an "equivalent" sphere is:

$$R_{p(\text{sphere})} = 3\left(\frac{R_p}{\dfrac{R_p}{x_p} + 2}\right) \tag{6.3.50}$$

which can be used to evaluate the Thiele modulus and effectiveness factor. Figure 6.3.10 compares the effectiveness factor derived from the full solution of the material balance for a finite cylinder (individual points) to the approximate solution using the "equivalent" radius of a sphere based on the volume to surface ratio. Also shown is the solution for an infinite cylinder with equal (V_p/S_p). Clearly, the agreement among the sets of results confirms that substituting the characteristic

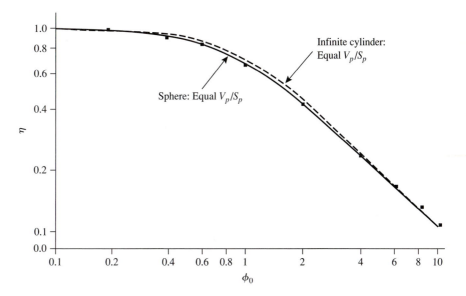

Figure 6.3.10 |
Effectiveness factors for sphere, infinite cylinder, and finite cylinder pellet geometries where the Thiele modulus is based on equal V_p/S_p. Individual points correspond to the numerical solutions of the material balance on a finite cylinder.

length in the Thiele modulus by (V_p/S_p) is an excellent approximation to the true solution.

EXAMPLE 6.3.2

The rate constant for the first-order cracking of cumene on a silica-alumina catalyst was measured to be 0.80 cm³/(s·gcat) in a laboratory reactor:

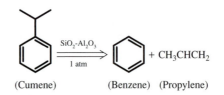

(Cumene)   (Benzene) (Propylene)

Is the observed rate constant the true rate constant or is there influence of pore diffusion? Additional data:

$$R_p = 0.25 \text{ cm}$$

$$\rho = 1.2 \text{ gcat cm}^{-3}$$

$$D_{TA}^e = 1.0 \times 10^{-3} \text{ cm}^2 \text{ s}^{-1}$$

■ **Answer**

Recall that the effectiveness factor can be approximated by:

$$\eta = \frac{\tanh(\phi_0)}{\phi_0}$$

when the Thiele modulus is defined in terms of the characteristic length of a pellet:

$$\phi_0 = L_p \sqrt{\frac{k}{D_{TA}^e}}$$

For the spherical particles in this problem:

$$\phi_0 = L_p \sqrt{\frac{k}{D_{TA}^e}} = \frac{R_p}{3}\sqrt{\frac{k}{D_{TA}^e}} = \left(\frac{(0.25)(\text{cm})}{3}\right)\sqrt{\frac{k\left(\frac{\text{cm}^3}{\text{s} \cdot \text{gcat}}\right) \cdot 1.2\left(\frac{\text{gcat}}{\text{cm}^3}\right)}{1.0 \times 10^{-3}\left(\frac{\text{cm}^2}{\text{s}}\right)}} = 2.9\sqrt{k}$$

Since the observed rate is first-order:

$$r_{obs} = k_{obs}C_{AS}$$

and, by definition, the effectiveness factor is:

$$\eta = \frac{\text{observed rate}}{\text{rate in absence of diffusion}} = \frac{k_{obs}C_{AS}}{kC_{AS}}$$

Substitute the Thiele modulus into the expression for the effectiveness factor to solve for k, the true rate constant, by trial and error:

$$\eta = \frac{\tanh(\phi_0)}{\phi_0} = \frac{\tanh(2.9\sqrt{k})}{2.9\sqrt{k}} = \frac{k_{obs}}{k} = \frac{0.80}{k}$$

$$k = 5.4 \frac{cm^3}{s \cdot gcat} \quad \eta = 0.15$$

Since η is small, there is a great influence of pore diffusion on the observed rate.

The material balance for simultaneous reaction and diffusion in a catalyst pellet can be extended to include more complex reactions. For example, the generalized Thiele modulus for an irreversible reaction of order n is:

$$\phi_0 = \frac{V_p}{S_p}\sqrt{\frac{n+1}{2}\frac{kC_{AS}^{n-1}}{D_{TA}^e}} \quad n > -1 \tag{6.3.51}$$

The generalized modulus defined in Equation (6.3.51) has been normalized so that the effectiveness factor is approximately $1/\phi_0$ at large values of ϕ_0, as illustrated in Figure 6.3.9.

The implications of severe diffusional resistance on observed reaction kinetics can be determined by simple analysis of this more general Thiele modulus. The observed rate of reaction can be written in terms of the intrinsic rate expression and the effectiveness factor as:

$$r_{obs} = \eta k C_{AS}^n \tag{6.3.52}$$

As discussed earlier, the effectiveness factor is simply the inverse of the Thiele modulus for the case of severe diffusional limitations (Figure 6.3.9.) Thus, the observed rate under strong diffusional limitations can be written as:

$$r_{obs} = \frac{1}{\phi_0} k C_{AS}^n \tag{6.3.53}$$

Substitution of the generalized Thiele modulus, Equation (6.3.51), into (6.3.53) gives the following expression for the observed rate:

$$r_{obs} = \frac{S_p}{V_p}\left(\frac{2}{n+1}D_{TA}^e k\right)^{\frac{1}{2}} C_{AS}^{(n+1)/2} = k_{obs}C_A^{n_{obs}} \tag{6.3.54}$$

The order of reaction observed under conditions of severe diffusional limitations, n_{obs}, becomes $(n+1)/2$ instead of the true reaction order n. The temperature dependence of the rate is also affected by diffusional limitations. Since the observed rate constant, k_{obs}, is proportional to $(D_{TA}^e k)^{\frac{1}{2}}$, the observed activation energy is $(E_D + E)/2$, where E_D is the activation energy for diffusion and E is the activation energy for reaction. Diffusional processes are weakly activated compared to chemical

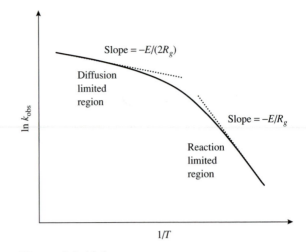

Figure 6.3.11 |
Temperature dependence of the observed rate constant of
a reaction occurring in a porous catalyst pellet.

reactions, and the value of E_D can often be neglected compared to E. Thus, the observed activation energy for a severely diffusion-limited reaction is approximately one half the true value. An Arrhenius plot of the *observed* rate constant, shown in Figure 6.3.11, illustrates the effect of diffusional resistances on the *observed* activation energy. At low temperatures, the reaction rate is not limited by diffusional resistances, and the observed activation energy is the true value. At high temperatures, the reaction rate is inhibited by diffusional resistances, and the activation energy is half the true value.

EXAMPLE 6.3.3 |

Develop expressions for the Thiele modulus and the concentration profile of A for the following *reversible* first-order reaction that takes place in a flat plate catalyst pellet:

$$A = B, \quad K = \frac{k_1}{k_{-1}}$$

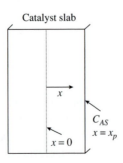

■ Answer

The material balance for diffusion/reaction in one dimension is given by:

$$\frac{d\left[-D_{TA}^e \dfrac{dC_A}{dx}\right]}{dx} = -r = -[k_1 C_A - k_{-1} C_B] \tag{6.3.55}$$

With C representing the total concentration, $C_A + C_B$, the rate expression can be rewritten in terms of C and C_A, thus eliminating the explicit dependence on C_B:

$$r = k_1 C_A - k_{-1}(C - C_A) = k_1\left(1 + \frac{1}{K}\right)C_A - \frac{k_1}{K}C = k_1\left(\frac{K+1}{K}\right)\left(C_A - \frac{C}{K+1}\right)$$

Assuming D_{TA}^e is constant, the following equation can be solved for the concentration profile:

$$D_{TA}^e \frac{d^2 C_A}{dx^2} = k_1\left(\frac{K+1}{K}\right)\left(C_A - \frac{C}{K+1}\right) \tag{6.3.56}$$

with boundary conditions:

$$C_A = C_{AS} \qquad \text{at } x = x_p \quad \text{(external surface of the slab)} \tag{6.3.57}$$

$$\frac{dC_A}{dx} = 0 \qquad \text{at } x = 0 \quad \text{(center line, point of symmetry)} \tag{6.3.58}$$

The following change of variables facilitates solution of the problem. Let $\overline{\psi} = C_A - C/(K+1)$ and $\chi = x/x_p$ so that the material balance can be written as:

$$\frac{d^2\overline{\psi}}{d\chi^2} = (x_p)^2 \frac{k_1}{D_{TA}^e}\left(\frac{1+K}{K}\right)\overline{\psi} = \phi^2\overline{\psi} \tag{6.3.59}$$

By expressing the material balance in this form, the Thiele modulus appears as the dimensionless constant ϕ:

$$\phi = x_p\sqrt{\frac{k_1}{D_{TA}^e}\left(\frac{1+K}{K}\right)} \tag{6.3.60}$$

The general solution of Equation (6.3.59), with arbitrary constants $\overline{\alpha}_1$ and $\overline{\alpha}_2$, is:

$$\overline{\psi} = \overline{\alpha}_1 \sinh(\phi\chi) + \overline{\alpha}_2 \cosh(\phi\chi) \tag{6.3.61}$$

The boundary conditions expressed in terms of the variables that are used to evaluate the constants $\overline{\alpha}_1$ and $\overline{\alpha}_2$ are:

$$\overline{\psi} = C_{AS} - \frac{C}{K+1} \qquad \text{at } \chi = 1 \tag{6.3.62}$$

$$\frac{d\overline{\psi}}{d\chi} = 0 \qquad \text{at } \chi = 0 \tag{6.3.63}$$

The constant $\bar{\alpha}_1$ vanishes due to the second boundary condition [Equation (6.3.63)]:

$$\frac{d\bar{\psi}}{d\chi} = \bar{\alpha}_1 \phi \cosh(0) + \bar{\alpha}_2 \phi \sinh(0) = 0$$

and

$$\cosh(0) = 1$$

so

$$\bar{\alpha}_1 = 0$$

The first boundary condition is used to evaluate $\bar{\alpha}_2$ and thus completes the solution of the problem as shown below:

$$C_{AS} - \frac{C}{K+1} = \bar{\alpha}_2 \cosh(\phi)$$

$$\bar{\alpha}_2 = \frac{C_{AS} - \dfrac{C}{K+1}}{\cosh(\phi)}$$

$$\bar{\psi} = \left(C_{AS} - \frac{C}{K+1}\right) \frac{\cosh(\phi\chi)}{\cosh(\phi)} \tag{6.3.64}$$

$$C_A - \frac{C}{K+1} = \left(C_{AS} - \frac{C}{K+1}\right) \frac{\cosh\left(\phi\dfrac{x}{x_p}\right)}{\cosh(\phi)}$$

$$C_A = \frac{C}{K+1} + \left(C_{AS} - \frac{C}{K+1}\right) \frac{\cosh\left(\phi\dfrac{x}{x_p}\right)}{\cosh(\phi)} \tag{6.3.65}$$

where the Thiele modulus is defined by:

$$\phi = x_p \sqrt{\frac{k_1}{D_{TA}^e}\left(\frac{1+K}{K}\right)} \tag{6.3.60}$$

EXAMPLE 6.3.4

Set up the equations necessary to calculate the effectiveness factor for a flat-plate catalyst pellet in which the following isothermal reaction takes place:

Benzene   Cyclohexane

■ Answer

To solve this problem, the Stefan-Maxwell relations for molecular diffusion in a multicomponent gas mixture (see Appendix C for details) should be used:

$$\nabla X_i = \sum_{\substack{j=1 \\ j \neq i}}^{n} \frac{1}{CD_{ij}} (X_i N_j - X_j N_i)$$

For the case of diffusion in one dimension within a porous medium, these equations yield the following expression, which is derived in Appendix C:

$$-\frac{1}{R_g T}\frac{dP_i}{dx} = \sum_{j \neq i}\left(\frac{X_j N_i - X_i N_j}{D_{ij}^e}\right) + \frac{N_i}{D_{Ki}^e} \tag{6.3.66}$$

The above equation reduces to a familiar form for two components if equimolar counterdiffusion of A and B ($N_A = -N_B$) at constant temperature is assumed:

$$-\frac{1}{R_g T}\frac{dP_A}{dx} = \frac{X_B N_A - X_A N_B}{D_{AB}^e} + \frac{N_A}{D_{KA}^e} = N_A\left(\frac{1}{D_{AB}^e} + \frac{1}{D_{KA}^e}\right)$$

$$N_A = -\left(\frac{1}{\dfrac{1}{D_{AB}^e} + \dfrac{1}{D_{KA}^e}}\right)\frac{d\left(\dfrac{P_A}{R_g T}\right)}{dx} = -D_{TA}^e \frac{dC_A}{dx} \tag{6.3.67}$$

Recall this equation is similar to Equation (6.3.7) for the flux of A in one dimension. To solve the multicomponent diffusion/reaction problem of benzene hydrogenation in one dimension, Equation (6.3.66) must instead be used. First, let:

Benzene = component 1

Dihydrogen = component 2

Cyclohexane = component 3

The following diffusivities (Knudsen and binary) need to be determined from tabulated data, handbooks, correlations, theoretical equations, etc.:

Benzene: D_{K1}, D_{12}, D_{13}

Dihydrogen: D_{K2}, D_{21}, D_{23}

Cyclohexane: D_{K3}, D_{31}, D_{32}

The porosity ($\bar{\varepsilon}_p$) and tortuosity ($\bar{\tau}$) of the flat plate catalyst pellet are then used to calculate the *effective* diffusivities associated with each component according to:

$$D_{Ki}^e = \frac{\bar{\varepsilon}_p}{\bar{\tau}} D_{Ki} \tag{6.3.68}$$

$$D_{ij}^e = \frac{\bar{\varepsilon}_p}{\bar{\tau}} D_{ij} \tag{6.3.69}$$

From the stoichiometry of the hydrogenation reaction, the ratios of the fluxes of the components are:

$$\frac{N_2}{N_1} = 3 \quad \text{and} \quad \frac{N_3}{N_1} = -1$$

Substitution of these relations into Equation (6.3.66) gives the appropriate flux equations:

$$-\frac{P}{R_g T}\frac{dX_1}{dx} = N_1\left[\left(\frac{X_2 - 3X_1}{D_{12}^e} + \frac{X_3 + X_1}{D_{13}^e}\right) + \frac{1}{D_{K1}^e}\right] \quad (6.3.70)$$

$$-\frac{P}{R_g T}\frac{dX_2}{dx} = N_1\left[\left(\frac{3X_1 - X_2}{D_{21}^e} + \frac{3X_3 + X_2}{D_{23}^e}\right) + \frac{3}{D_{K2}^e}\right] \quad (6.3.71)$$

Finally, the material balance on a slice of the catalyst pellet that is needed to completely specify the reaction/diffusion problem is:

$$-\frac{dN_1}{dx} = \text{Rate}(X_1, X_2) \quad (6.3.72)$$

Rate (X_1, X_2) is the rate expression for benzene hydrogenation that depends on X_1 and X_2. For example, the following rate equation could be used if the constants $\overline{\alpha}_A$ and $\overline{\alpha}_B$ were known at the reaction temperature:

$$\text{Rate}(X_1, X_2) = \frac{\overline{\alpha}_A X_1 X_2 P^2}{1 + \overline{\alpha}_B X_1 P} \quad (6.3.73)$$

The three equations representing the material balance and the flux relations can be solved simultaneously to determine the dependent variables (X_1, X_2, and N_1) as a function of the independent variable x. (Recall that X_3 can be expressed in terms of X_1 and X_2: $1 = X_1 + X_2 + X_3$.) The boundary conditions for these equations are:

$$X_1 = X_{1S} \quad \text{and} \quad X_2 = X_{2S} \qquad \text{at } x = x_s \quad \text{(external surface of pellet)}$$
$$N_1 = 0 \qquad \text{at } x = 0 \quad \text{(center line of the slab)}$$

To calculate the effectiveness factor, the actual reaction rate throughout the catalyst is divided by the rate determined at the conditions of the external surface, that is, Rate(X_{1S}, X_{2S}). The overall reaction rate throughout the particle is equivalent to the flux N_1 evaluated at the external surface of the catalyst. Thus, the final solution is:

$$\eta = \frac{N_1|_{x=x_s}}{\text{Rate}(X_{1S}, X_{2S})} \cdot \left(\frac{S_p}{V_p}\right) \quad (6.3.74)$$

The temperature profile in the catalyst pellet can be easily incorporated into this solution by including the energy balance in the system of equations.

The previous discussion focused on simultaneous diffusion and reaction in isothermal catalyst pellets. Since ΔH_r is significant for many industrially relevant reactions, it is necessary to address how heat transfer might affect solid-catalyzed reactions. For isothermal catalyst pellets, the effectiveness factor is less than or equal to unity, as illustrated in Figure 6.3.9. This is rationalized by examining the terms that comprise the effectiveness factor for the reaction of A in a flat plate catalyst pellet:

$$\eta = \frac{r_{obs}}{r_{max}} = \frac{S_p \int_0^{x_p} r(C_A)\,dx}{V_p\, r(C_{AS})} \tag{6.3.75}$$

For an isothermal pellet:

$$\int_0^{x_p} r(C_A)\,dx \leq r(C_{AS}) \tag{6.3.76}$$

If, as stated in Chapter 1, the rate is separable into two parts, one dependent on the temperature and the other dependent on the concentrations of reacting species, that is,

$$r = k(T) \cdot \overline{F}(C_A) \tag{6.3.77}$$

then the concentration of A inside the pellet is less than that on the external surface,

$$\overline{F}(C_A) \leq \overline{F}(C_{AS}) \tag{6.3.78}$$

for reactions with nonnegative reaction orders. Thus,

$$\int_0^{x_p} \overline{F}(C_A)\,dx \leq \overline{F}(C_{AS}) \tag{6.3.79}$$

and explains the upper limit of unity for the isothermal effectiveness factor in a catalyst pellet. The situation can be very different for a nonisothermal pellet. For example, the temperature dependence of the reaction rate constant, $k(T)$, is generally expressed in an Arrhenius form

$$k(T) = \overline{A} \exp\left(\frac{-E}{R_g T}\right) \tag{6.3.80}$$

For an endothermic reaction in the presence of significant heat-transfer resistance, the temperature at the surface of the pellet can exceed the temperature of the interior, which according to Equation (6.3.80), gives:

$$k(T) \leq k(T_S) \tag{6.3.81}$$

Since $\overline{F}(C_A) \leq \overline{F}(C_{AS})$ as discussed above, the effectiveness factor is always less than or equal to unity for an endothermic reaction. For an exothermic reaction, the opposite situation can occur. The temperature of the interior of the particle can exceed the surface temperature, $T > T_S$, which leads to:

$$k(T) \geq k(T_S) \tag{6.3.82}$$

Recall that $k(T)$ is a strong function of temperature. The effectiveness factor for an exothermic reaction can be less than, equal to, or greater than unity, depending on how $k(T)$ increases relative to $\overline{F}(C_A)$ within the particle. Thus, there are cases where the increase in $k(T)$ can be much larger than the decrease in $\overline{F}(C_A)$, for example,

$$k(T) \cdot \overline{F}(C_A) > k(T_S) \cdot \overline{F}(C_{AS})$$

To evaluate the effectiveness factor for a first-order, isobaric, nonisothermal, flat plate catalyst pellet, the material and energy balances must be solved simultaneously. As shown previously, the mole balance in a slab is given by:

$$\frac{dN_A}{dx} = -k(T) \cdot C_A \tag{6.3.83}$$

where the rate constant is of the Arrhenius form:

$$k(T) = \overline{A} \exp\left(\frac{-E}{R_g T}\right) \tag{6.3.84}$$

The flux of A can be written in terms of Fick's Law:

$$N_A = -D_{TA}^e \frac{dC_A}{dx} \tag{6.3.85}$$

and substituted into the mole balance to give:

$$D_{TA}^e \frac{d^2 C_A}{dx^2} = k(T) \cdot C_A \tag{6.3.86}$$

The energy balance is written in the same manner as the mole balance to give:

$$-\frac{dq}{dx} = (-\Delta H_r) \cdot k(T) \cdot C_A \tag{6.3.87}$$

where the flux is expressed in terms of the effective thermal conductivity of the fluid-solid system, λ^e, and the gradient in temperature:

$$q = -\lambda^e \frac{dT}{dx} \tag{6.3.88}$$

The heat of reaction, ΔH_r, is defined to be negative for exothermic reactions and positive for endothermic reactions. Substitution of Equation (6.3.88) into (6.3.87) results in:

$$\lambda^e \frac{d^2 T}{dx^2} = (-\Delta H_r) \cdot k(T) \cdot C_A \tag{6.3.89}$$

To render the material and energy balances dimensionless, let:

$$\chi = \frac{x}{x_p} \tag{6.3.90}$$

$$\psi = \frac{C_A}{C_{AS}} \tag{6.3.91}$$

$$\Gamma = \frac{T}{T_S} \tag{6.3.92}$$

The rate constant $k(T)$ is expressed in terms of T_S by first forming the ratio:

$$\frac{k(T)}{k(T_S)} = \exp\left[-\frac{E}{R_g}\left(\frac{1}{T} - \frac{1}{T_S}\right)\right] = \exp\left[-\gamma\left(\frac{1}{\Gamma} - 1\right)\right] \tag{6.3.93}$$

where:

$$\gamma = \frac{E}{R_g T_S} \tag{6.3.94}$$

The dimensionless group γ is known as the *Arrhenius* number. Substitution of the dimensionless variables into the material and energy balances gives:

$$\frac{d^2\psi}{d\chi^2} = \left[\frac{(x_p)^2 \cdot k(T_S)}{D_{TA}^e}\right] \cdot \exp\left[-\gamma\left(\frac{1}{\Gamma} - 1\right)\right] \cdot \psi \tag{6.3.95}$$

$$\frac{d^2\Gamma}{d\chi^2} = -\left[\frac{(x_p)^2 \cdot k(T_S) \cdot (-\Delta H_r) \cdot C_{AS}}{\lambda^e T_S}\right] \cdot \exp\left[-\gamma\left(\frac{1}{\Gamma} - 1\right)\right] \cdot \psi \tag{6.3.96}$$

Both equations can be expressed in terms of the Thiele modulus, ϕ, according to:

$$\frac{d^2\psi}{d\chi^2} = \phi^2 \cdot \exp\left[-\gamma\left(\frac{1}{\Gamma} - 1\right)\right] \cdot \psi \tag{6.3.97}$$

$$\frac{d^2\Gamma}{d\chi^2} = -\phi^2\left[\frac{(-\Delta H_r) \cdot D_{TA}^e \cdot C_{AS}}{\lambda^e T_S}\right] \cdot \exp\left[-\gamma\left(\frac{1}{\Gamma} - 1\right)\right] \cdot \psi \tag{6.3.98}$$

A new dimensionless grouping called the *Prater* number, β, appears in the energy balance:

$$\beta = \frac{(-\Delta H_r) \cdot D_{TA}^e \cdot C_{AS}}{\lambda^e T_S} \tag{6.3.99}$$

Thus, the energy balance is rewritten:

$$\frac{d^2\Gamma}{d\chi^2} = -\phi^2 \cdot \beta \cdot \exp\left[-\gamma\left(\frac{1}{\Gamma} - 1\right)\right] \cdot \psi \tag{6.3.100}$$

The material and energy balances are then solved simultaneously with the following boundary conditions:

$$\frac{d\psi}{d\chi} = \frac{d\Gamma}{d\chi} = 0 \qquad \text{at } \chi = 0 \tag{6.3.101}$$

$$\psi = \Gamma = 1 \qquad \text{at } \chi = 1 \tag{6.3.102}$$

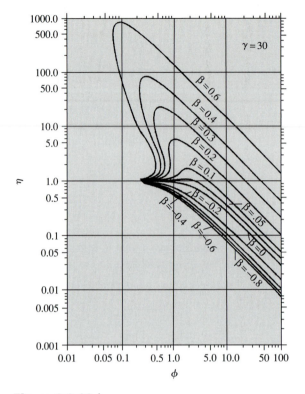

Figure 6.3.12 |
Effectiveness factors for a first-order reaction in a
spherical, nonisothermal catalysts pellet. (Reprinted
from P. B. Weisz and J. S. Hicks, "The Behavior of
Porous Catalyst Particles in View of Internal Mass and
Heat Diffusion Effects," *Chem. Eng. Sci.*, **17** (1962)
265, copyright 1962, with permission from Elsevier
Science.)

Since the equations are nonlinear, a numerical solution method is required. Weisz
and Hicks calculated the effectiveness factor for a first-order reaction in a spherical catalyst pellet as a function of the Thiele modulus for various values of the Prater
number [P. B. Weisz and J. S. Hicks, *Chem. Eng. Sci.*, **17** (1962) 265]. Figure 6.3.12
summarizes the results for an Arrhenius number equal to 30. Since the Arrhenius
number is directly proportional to the activation energy, a higher value of γ corresponds to a greater sensitivity to temperature. The most important conclusion to
draw from Figure 6.3.12 is that effectiveness factors for exothermic reactions (positive values of β) can exceed unity, depending on the characteristics of the pellet
and the reaction. In the narrow range of the Thiele modulus between about 0.1 and
1, three different values of the effectiveness factor can be found (but only two represent stable steady states). The ultimate reaction rate that is achieved in the pellet

depends on how the reaction is initiated. Effectiveness factors associated with the negative values of β on Figure 6.3.12 are all less than one, which is expected for endothermic reactions.

It is often useful to quickly estimate the maximum possible temperature rise, also known as the adiabatic temperature rise, in a catalyst pellet. Since no heat is transferred to the surroundings in this case, all energy generated (or consumed) by the reaction goes to heat (or cool) the pellet. The temperature difference between the surface and the pellet interior is directly related to the concentration difference. Dividing the material balance by the energy balance eliminates the reaction rate:

$$\frac{\left[\dfrac{d^2\psi}{d\chi^2}\right]}{\left[\dfrac{d^2\Gamma}{d\chi^2}\right]} = -\frac{1}{\beta} \tag{6.3.103}$$

$$d^2\Gamma = -\beta \cdot d^2\psi \tag{6.3.104}$$

Integrating once gives:

$$d\Gamma = -\beta \cdot d\psi + \overline{\alpha}_1 \tag{6.3.105}$$

The constant $\overline{\alpha}_1$ is evaluated by using the condition at the center of the pellet:

$$d\Gamma = d\psi = 0 \qquad \text{at } \chi = 0$$

$$\overline{\alpha}_1 = 0 \tag{6.3.106}$$

Integrating a second time gives:

$$\Gamma = -\beta \cdot \psi + \overline{\alpha}_2 \tag{6.3.107}$$

The constant $\overline{\alpha}_2$ is found by using the condition at the surface of the pellet:

$$\Gamma = \psi = 1 \qquad \text{at } \chi = 1$$

and

$$\overline{\alpha}_2 = 1 + \beta \tag{6.3.108}$$

Therefore, the relationship between temperature and concentration is:

$$\Gamma = 1 + \beta(1 - \psi)$$

$$\frac{T}{T_S} = 1 + \left[\frac{(-\Delta H_r)D_{TA}^e C_{AS}}{\lambda^e T_S}\right]\left(1 - \frac{C_A}{C_{AS}}\right)$$

$$T = T_S + \left[\frac{(-\Delta H_r)D_{TA}^e}{\lambda^e}\right](C_{AS} - C_A) \tag{6.3.109}$$

Equation (6.3.109) is called the *Prater* relation. From this relationship, the adiabatic temperature rise in a catalyst pellet can be calculated. The maximum temperature is reached when the reactant is completely converted in the pellet, that is, $C_A = 0$:

$$T_{\max} - T_S = \frac{(-\Delta H_r)D_{TA}^e C_{AS}}{\lambda^e} \tag{6.3.110}$$

$$\frac{T_{\max} - T_S}{T_S} = \frac{(-\Delta H_r)D_{TA}^e C_{AS}}{\lambda^e T_S} \tag{6.3.111}$$

Notice that the dimensionless maximum temperature rise in the catalyst pellet is simply the Prater number β:

$$\frac{T_{\max} - T_S}{T_S} = \beta$$

6.4 | Combined Internal and External Transport Effects

The previous two sections describe separately the significant role that diffusion through a stagnant film surrounding a catalyst pellet and transport through the catalyst pores can play in a solid-catalyzed chemical reaction. However, these two diffusional resistances must be evaluated simultaneously in order to properly interpret the observed rate of a catalytic reaction.

VIGNETTE 6.4.1

Kehoe and Butt studied the hydrogenation of benzene over a solid catalyst consisting of 58 wt. % Ni metal particles supported on kieselguhr powder [J. P. G. Kehoe and J. B. Butt, *AIChE J.*, **18** (1972) 347]. The catalyst had a surface area of 150 m^2 g^{-1} and an average pore radius of 3.7 nm. This powder was compressed into a cylindrical pellet (1.3 cm diameter by 5.8 cm height) large enough to incorporate four thermocouples along the radial direction. A second pellet was prepared by diluting the original Ni catalyst powder with alumina and graphite in order to increase the thermal conductivity of the catalyst by one order of magnitude. Figure 6.4.1 shows the temperature profile through the film and the catalyst for both pellets exposed to the same feed conditions and run at the same reaction rate. The undiluted catalyst pellet (upper figure) had a measurable intraparticle temperature gradient. Indeed, the center of the pellet was almost 30 K hotter than the external surface, whereas the temperature change over the external film was less than 10 K. The effectiveness factor for this pellet (1.9) was significantly greater than unity because the pellet was hotter than the surrounding bulk fluid. The temperature profile in the second pellet was virtually flat throughout the entire pellet because of its high thermal conductivity (Figure 6.4.1). The only significant temperature gradient for the second pellet was measured across the film. Since the second pellet was operating at a temperature higher than the bulk fluid, the effectiveness factor (1.2) also exceeded unity. However, it was not as high as that encountered with the first pellet because of its lower operating temperature.

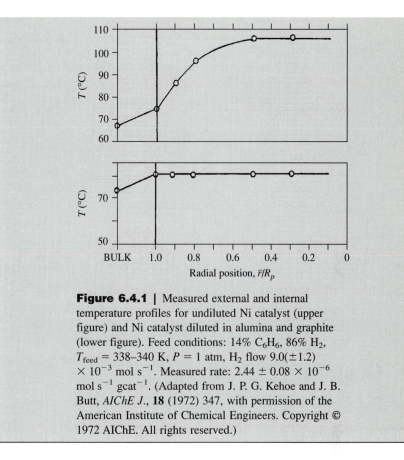

Figure 6.4.1 | Measured external and internal temperature profiles for undiluted Ni catalyst (upper figure) and Ni catalyst diluted in alumina and graphite (lower figure). Feed conditions: 14% C_6H_6, 86% H_2, T_{feed} = 338–340 K, P = 1 atm, H_2 flow 9.0($\pm$1.2) $\times 10^{-3}$ mol s^{-1}. Measured rate: 2.44 $\pm$ 0.08 $\times 10^{-6}$ mol s^{-1} gcat^{-1}. (Adapted from J. P. G. Kehoe and J. B. Butt, *AIChE J.*, **18** (1972) 347, with permission of the American Institute of Chemical Engineers. Copyright © 1972 AIChE. All rights reserved.)

Consider a first-order reaction occurring on a *nonporous* flat plate catalyst pellet. In Section 6.2, it was shown that the concentration of reactant A on the external surface of the catalyst is related to both the mass transfer coefficient, $\bar{k}_c$, and the surface rate constant, k_S:

$$C_{AS} = \frac{\bar{k}_c C_{AB}}{k_S + \bar{k}_c} \qquad (6.4.1)$$

A dimensionless parameter Da, called the *Damkohler* number, is defined to be the ratio of k_S and $\bar{k}_c$:

$$Da = \frac{k_S}{\bar{k}_c} \qquad (6.4.2)$$

so that the surface concentration can be written as:

$$C_{AS} = \frac{C_{AB}}{1 + Da} \qquad (6.4.3)$$

The Damkohler number indicates which characteristic first-order process is faster, external diffusion or reaction. For very large values of Da ($k_S \gg \bar{k}_c$), the surface concentration of reactant approaches zero, whereas for very small values of Da ($k_S \ll \bar{k}_c$), the surface concentration approaches the bulk fluid concentration. An *interphase effectiveness factor*, $\bar{\eta}$, is defined as the reaction rate based on surface conditions divided by the rate that would be observed in the absence of diffusional limitations:

$$\bar{\eta} = \frac{k_S C_{AS}}{k_S C_{AB}} = \frac{1}{1 + Da} \tag{6.4.4}$$

Now consider the first-order reaction in a porous flat plate catalyst pellet so that both external (interphase) and internal (intraphase) transport limitations are encountered. At steady state, the flux of A to the surface of the pellet is equal to the flux entering the pellet:

$$\bar{k}_c\left(C_{AS} - C_{AB}\right) = -D_{TA}^e \frac{dC_A}{dx}\bigg|_S \tag{6.4.5}$$

The energy balance is completely analogous:

$$h_t(T_S - T_B) = -\lambda^e \frac{dT}{dx}\bigg|_S \tag{6.4.6}$$

Rewriting these two equations in dimensionless form, using the usual substitution for distance:

$$\chi = \frac{x}{x_p} \tag{6.4.7}$$

yields:

$$\frac{d\left(\dfrac{C_A}{C_{AB}}\right)}{d\chi}\bigg|_{\chi=1} = Bi_m\left[1 - \frac{C_{AS}}{C_{AB}}\right] \tag{6.4.8}$$

$$\frac{d\left(\dfrac{T}{T_B}\right)}{d\chi}\bigg|_{\chi=1} = Bi_h\left[1 - \frac{T_S}{T_B}\right] \tag{6.4.9}$$

where:

$$Bi_m = \frac{x_p \bar{k}_c}{D_{TA}^e} \qquad \text{Biot number for mass} \tag{6.4.10}$$

$$Bi_h = \frac{x_p h_t}{\lambda^e} \qquad \text{Biot number for heat} \tag{6.4.11}$$

Since the concentration and temperature variables in Equations (6.4.8) and (6.4.9) are grouped with their respective bulk fluid values, new dimensionless parameters need to be defined. Let:

$$\Psi = \frac{C_A}{C_{AB}} \tag{6.4.12}$$

$$\overline{\Gamma} = \frac{T}{T_B} \tag{6.4.13}$$

The dimensionless material and energy balances, with associated boundary conditions, must be solved simultaneously to get the concentration and temperature profiles through the stagnant film and into the catalyst particle. Those relationships are given below:

$$\frac{d^2\Psi}{d\chi^2} = \phi^2 \Psi \exp\left[-\gamma\left(\frac{1}{\overline{\Gamma}} - 1\right)\right] \tag{6.4.14}$$

$$\frac{d^2\overline{\Gamma}}{d\chi^2} = -\beta\phi^2 \Psi \exp\left[-\gamma\left(\frac{1}{\overline{\Gamma}} - 1\right)\right] \tag{6.4.15}$$

where:

$$\phi^2 = \frac{(x_p)^2 \cdot k(T_B)}{D_{TA}^e} \qquad \beta = \frac{(-\Delta H_r)D_{TA}^e C_{AB}}{\lambda^e T_B} \qquad \gamma = \frac{E}{R_g T_B}$$

Notice that all of the parameters are based on bulk fluid values of the concentration and temperature. The boundary conditions are:

$$\frac{d\Psi}{d\chi} = \frac{d\overline{\Gamma}}{d\chi} = 0 \qquad \text{at } \chi = 0 \tag{6.4.16}$$

$$\frac{d\Psi}{d\chi} = Bi_m(1 - \Psi) \qquad \text{at } \chi = 1 \tag{6.4.17}$$

$$\frac{d\overline{\Gamma}}{d\chi} = Bi_h(1 - \overline{\Gamma}) \qquad \text{at } \chi = 1 \tag{6.4.18}$$

In general, solution of these equations requires a numerical approach.

EXAMPLE 6.4.1

Find an expression for the *overall* effectiveness factor of a first-order *isothermal* reaction in a flat plate catalyst pellet.

■ Answer
Since the reaction is isothermal, the energy balance can be ignored and the mass balance reduces to:

$$\frac{d^2\Psi}{d\chi^2} = \phi^2 \Psi \tag{6.4.19}$$

with boundary conditions:

$$\frac{d\Psi}{d\chi} = 0 \qquad\qquad \text{at } \chi = 0 \tag{6.4.20}$$

$$\frac{d\Psi}{d\chi} = Bi_m(1 - \Psi) \qquad \text{at } \chi = 1 \tag{6.4.21}$$

As discussed previously, the general solution of the differential equation is:

$$\Psi = \overline{\alpha}_1 \exp(\phi\chi) + \overline{\alpha}_2 \exp(-\phi\chi) \tag{6.4.22}$$

The constants are evaluated by using the appropriate boundary conditions. At the center of the pellet:

$$\left.\frac{d\Psi}{d\chi}\right|_{\chi=0} = 0 = \overline{\alpha}_1 \phi \exp\!\left(\phi \cdot 0\right) - \overline{\alpha}_2 \phi \exp\!\left(-\phi \cdot 0\right) \tag{6.4.23}$$

$$\overline{\alpha}_1 = \overline{\alpha}_2 \tag{6.4.24}$$

Substitution of Equation (6.4.24) into (6.4.22) eliminates one of the integration constants:

$$\Psi = \overline{\alpha}_1 \exp(\phi\chi) + \overline{\alpha}_1 \exp(-\phi\chi) \tag{6.4.25}$$

$$\Psi = 2\overline{\alpha}_1 \frac{\exp(\phi\chi) + \exp(-\phi\chi)}{2} \tag{6.4.26}$$

$$\Psi = 2\overline{\alpha}_1 \cosh(\phi\chi) \tag{6.4.27}$$

$$\frac{d\Psi}{d\chi} = 2\overline{\alpha}_1 \phi \sinh(\phi\chi) \tag{6.4.28}$$

The boundary condition at the external surface provides another relation for $d\Psi/d\chi$:

$$\left.\frac{d\Psi}{d\chi}\right|_{\chi=1} = Bi_m[1 - \Psi] = Bi_m[1 - 2\overline{\alpha}_1 \cosh(\phi)] \tag{6.4.29}$$

Equating Equations (6.4.28) and (6.4.29), at $\chi = 1$, enables the determination of $\overline{\alpha}_1$:

$$2\overline{\alpha}_1 \phi \sinh(\phi) = Bi_m[1 - 2\overline{\alpha}_1 \cosh(\phi)] \tag{6.4.30}$$

$$\overline{\alpha}_1 = \frac{Bi_m}{2[\phi\sinh(\phi) + Bi_m\cosh(\phi)]} \tag{6.4.31}$$

and therefore,

$$\Psi = \frac{Bi_m\cosh(\phi\chi)}{\phi\sinh(\phi) + Bi_m\cosh(\phi)} \tag{6.4.32}$$

Since the concentration profile is determined by Equation (6.4.32), evaluation of the *overall effectiveness factor*, η_o, is straightforward. By definition, η_o is the observed rate divided by the rate that would be observed at conditions found in the bulk fluid. Recall that the observed rate must equal the flux of A at the surface of the pellet at the steady state:

$$\eta_o = \frac{r_{obs}}{r_{max}} = \frac{(Area)D^e_{TA} \left.\frac{dC_A}{dx}\right|_{x=x_p}}{(Area)x_p k C_{AB}} = \frac{\left.\frac{d\Psi}{d\chi}\right|_{\chi=1}}{\phi^2} \tag{6.4.33}$$

$$\eta_o = \frac{\tanh(\phi)}{\phi\left[1 + \dfrac{\phi\tanh(\phi)}{Bi_m}\right]} \tag{6.4.34}$$

The overall effectiveness factor is actually comprised of the individual effectiveness factors for intraphase and interphase transport:

$$\eta_o = \eta_{\text{intraphase}} \cdot \eta_{\text{interphase}} = \eta \cdot \bar{\eta} \tag{6.4.35}$$

For example, an isothermal, first-order reaction in a flat plate catalyst pellet has individual effectiveness factors that are:

$$\eta = \frac{\tanh(\phi)}{\phi} \tag{6.4.36}$$

$$\bar{\eta} = \left[1 + \frac{\phi\tanh(\phi)}{Bi_m}\right]^{-1} \tag{6.4.37}$$

Common ranges of diffusivities, thermal conductivities, mass transfer coefficients, heat transfer coefficients, and catalyst pore sizes can be used to estimate the relative magnitude of artifacts in kinetic data obtained in industrial reactors. For gas-solid heterogeneous systems, the high thermal conductivity of solids compared to gases suggests that the temperature gradient in the film surrounding the catalyst particle is likely to be greater than the temperature gradient in the particle. Since the Knudsen diffusivity of gaseous molecules in a small pore of a catalyst particle is much lower than the molecular diffusivity in the stagnant film, intraphase gradients in mass are likely to be much greater than interphase gradients. For liquid-solid heterogeneous systems, internal temperature gradients are often encountered. A typical range of Bi_m/Bi_h is from 10 to 10^4 for gas-solid systems and from 10^{-4} to 10^{-1} for liquid solid systems.

VIGNETTE 6.4.2

One important process in the manufacturing of electronic materials and devices involves the deposition of thin films onto patterned surfaces. As feature sizes of silicon-based devices continue to shrink in order to increase memory capacity, deposition of uniform films onto surfaces with micron-scale trenches is required. One such film-growth process is called *chemical vapor deposition* (CVD) that involves the reaction of a vapor phase

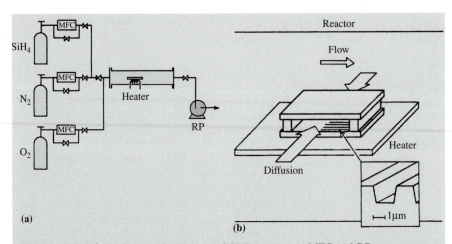

(a)

(b)

Figure 6.4.2 | **(a)** Schematic diagram of CVD apparatus. MFC and RP represent a mass flow controller and a rotary pump, respectively. **(b)** Schematic structure of the macrocavity reactor consisting of two silicon wafers, each patterned with microcavity trenches. (Reproduced from K. Watanabe and H. Komiyama, "Micro/Macrocavity Method Applied to the Study of the Step Coverage Formation Mechanism of SiO$_2$ Films by LPCVD," *J. Electrochem. Soc.*, **137** (1990) 1222, with permission of the Electrochemical Society, Inc.)

reactant with a surface. The deposition of reacting species on silicon wafers and in micron-sized trenches involves a balance between diffusion of reactants to the surface and the kinetics of surface deposition reactions. Watanabe and Komiyama analyzed the simultaneous diffusion/reaction phenomena involved with CVD of SiO$_2$ thin films by reaction of SiH$_4$ and O$_2$ [K. Watanabe and H. Komiyama, *J. Electrochem. Soc.*, **137** (1990) 1222]. Figures 6.4.2 (a) and (b) illustrate the CVD system used to study the deposition process. Micron-scale trenches were etched into silicon wafers by normal lithographic procedures. Two wafers were attached to each other with spacers between them, as illustrated in Figure 6.4.2(b). The openings between the two wafers were oriented parallel to the direction of flow. Thus, the reactor configuration has two regions that require diffusive transport of reactants, the macrocavity between the two wafers and the micron-sized trenches (microcavities) on each wafer.

The depth profile of SiO$_2$ film deposited in a macrocavity of width 0.55 mm at 673 K after 120 min of reaction is shown in Figure 6.4.3. The thickness of the film is greatest at the two open ends of the macrocavity and then rapidly decreases towards the center, presumably because of depletion of reactive species in the macrocavity interior.

Analysis of this system is rather straightforward since it is mathematically equivalent to a catalyst pellet in which reaction and diffusion occur simultaneously. The rate of reaction in this case is simply the growth rate of the film. The collision frequency of gas-phase molecules of A with the surface is given by gas kinetic theory as $(V_A/4)C_A$, where V_A = mean velocity of the gas phase A molecules $(\sqrt{8\bar{k}T/[\pi(M_A)]})$ and M_A is the

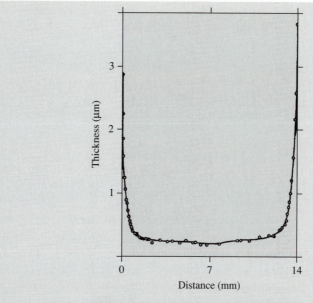

Figure 6.4.3 | Growth rate profile of SiO$_2$ in the macrocavity. (Reproduced from K. Watanabe and H. Komiyama, "Micro/Macrocavity Method Applied to the Study of the Step Coverage Formation Mechanism of SiO$_2$ Films by LPCVD," *J. Electrochem. Soc.*, **137** (1990) 1222, with permission of the Electrochemical Society, Inc.)

molecular weight of A. Thus, the growth rate of the film is simply the fraction of molecules that stick to the surface, sticking coefficient S_c, times the overall collision rate with the surface, $S_c(V_A/4)C_A$. The material balance on the macrocavity is analogous to Equation (6.3.10) and is:

$$\frac{d^2C_A}{dx^2} - \frac{S_c\left(\dfrac{V_A}{4}\right)}{D_{TA}\left(\dfrac{W_c}{2}\right)} C_A = 0 \qquad (6.4.38)$$

where W_c is the width of the macrocavity. Solving Equation (6.4.38) with the following boundary conditions:

$$C_A = C_A^0 \qquad \text{at } x = 0$$

$$\frac{dC_A}{dx} = 0 \qquad \text{at } x = L_c \quad \text{(where } L_c \text{ is the half length of the macrocavity)}$$

gives an analytical expression for C_A throughout the macrocavity and thus the growth rate of the deposited film, which is:

$$\text{Growth Rate} = S_c\left(\frac{V_A}{4}\right)C_A^0\left(\frac{\cosh[\phi'(L_c - x)/L_c]}{\cosh(\phi')}\right) \tag{6.4.39}$$

where ϕ' is the Thiele modulus. In this case, the Thiele modulus is the ratio of the characteristic deposition rate to the characteristic diffusion rate according to:

$$\phi' = L_c\sqrt{\frac{S_c\left(\dfrac{V_A}{4}\right)}{D_{TA}\left(\dfrac{W_c}{2}\right)}} \tag{6.4.40}$$

The growth rate profiles predicted by Equation (6.4.39) are given in Figure 6.4.4.

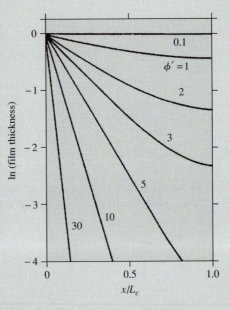

Figure 6.4.4 | Growth rate profiles in the macrocavity reactor predicted by Equation 6.4.39. Film thickness is normalized by the value at inlet of the cavity. (Adapted from K. Watanabe and H. Komiyama, "Micro/ Macrocavity Method Applied to the Study of the Step Coverage Formation Mechanism of SiO₂ Films by LPCVD," *J. Electrochem. Soc.*, **137** (1990) 1222, with permission of the Electrochemical Society, Inc.)

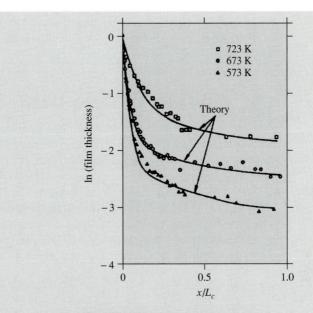

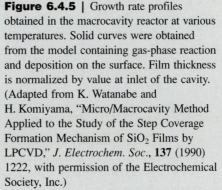

Figure 6.4.5 | Growth rate profiles obtained in the macrocavity reactor at various temperatures. Solid curves were obtained from the model containing gas-phase reaction and deposition on the surface. Film thickness is normalized by value at inlet of the cavity. (Adapted from K. Watanabe and H. Komiyama, "Micro/Macrocavity Method Applied to the Study of the Step Coverage Formation Mechanism of SiO_2 Films by LPCVD," *J. Electrochem. Soc.*, **137** (1990) 1222, with permission of the Electrochemical Society, Inc.)

Although the curves in Figure 6.4.4 represent the essential features of the simultaneous reaction/diffusion phenomena that occur in CVD processing, they do not reproduce the measured film thickness throughout the entire reactor as given in Figure 6.4.3. Watanabe and Komiyama derived a slightly more complex reaction/diffusion model in which the SiH_4 reacts in the gas phase to form a very dilute reactive intermediate which then deposits onto the wafer as SiO_2. This new model incorporates diffusion, gas-phase reaction, and surface deposition and fits the experimental data very well at three different temperatures, as illustrated in Figure 6.4.5.

Deposition of SiO_2 in the micron-sized trenches is also dependent on simultaneous diffusion and reaction. Watanabe and Komiyama found that the growth rate of the film at 573–723 K on the bottom of 1.2 μm deep trenches was less than that on the surface for trench widths less than about 3 μm. Clearly, solution of reaction/diffusion equations over many length scales is required to model the deposition of thin films during CVD processing of electronic materials.

6.5 | Analysis of Rate Data

To arrive at a rate expression that describes intrinsic reaction kinetics and is suitable for engineering design calculations, one must be assured that the kinetic data are free from artifacts that mask intrinsic rates. A variety of criteria have been proposed to guide kinetic analysis and these are thoroughly discussed by Mears [D. E. Mears, *Ind. Eng. Chem. Process Des. Develop.*, **10** (1971) 541].

A lack of significant intraphase diffusion effects (i.e., $\eta \geq 0.95$) on an irreversible, isothermal, first-order reaction in a spherical catalyst pellet can be assessed by the Weisz-Prater criterion [P. B. Weisz and C. D. Prater, *Adv. Catal.*, **6** (1954) 143]:

$$\frac{r_{obs}(R_p)^2}{D_{TA}^e C_{AS}} < 1 \tag{6.5.1}$$

where r_{obs} is the observed reaction rate per unit volume and R_p is the radius of a catalyst particle. An important aspect of this criterion is that it uses the *observed* rate and the reactant concentration at the *external* surface. The intrinsic rate and the concentration profile inside the pellet are not needed. For power law kinetics where n is the reaction order (other than 0), the following expression can be used:

$$\frac{r_{obs}(R_p)^2}{D_{TA}^e C_{AS}} < \frac{1}{n} \tag{6.5.2}$$

The influence of mass transfer through the film surrounding a spherical catalyst particle can also be examined with a similar expression. Satisfaction of the following inequality demonstrates that interphase mass transfer is not significantly affecting the measured rate:

$$\frac{r_{obs}R_p}{\bar{k}_c C_{AB}} < \frac{0.15}{n} \tag{6.5.3}$$

where $\bar{k}_c$ is the mass transfer coefficient and the reactant concentration is determined in bulk fluid. The above relationship is analogous to the modified Weisz-Prater criterion with $\bar{k}_c$ replacing D_{TA}^e/R_p.

Criteria have also been developed for evaluating the importance of intraphase and interphase heat transfer on a catalytic reaction. The Anderson criterion for estimating the significance of intraphase temperature gradients is [J. B. Anderson, *Chem. Eng. Sci.*, **18** (1963) 147]:

$$\frac{|\Delta H_r| r_{obs}(R_p)^2}{\lambda^e T_S} < 0.75 \frac{R_g T_S}{E} \tag{6.5.4}$$

where λ^e is the effective thermal conductivity of the particle and E is the true activation energy. Satisfying the above criterion guarantees that r_{obs} does not differ from the rate at constant temperature by more than 5 percent. Equation (6.5.4) is valid whether or not diffusional limitations exist in the catalyst particle. An analogous criterion for the lack of interphase temperature gradients has been proposed by Mears [D. E. Mears, *J. Catal.*, **20** (1971) 127]:

$$\frac{|\Delta H_r|\, r_{obs} R_p}{h_t T_B} < 0.15 \frac{R_g T_B}{E} \tag{6.5.5}$$

where h_t is the heat transfer coefficient and T_B refers to the bulk fluid temperature. The Mears criterion is similar to the Anderson criterion with h_t replacing λ^e/R_p. In addition, the Mears criterion is also valid in the presence of transport limitations in the catalyst particle.

While the above criteria are useful for diagnosing the effects of transport limitations on reaction rates of heterogeneous catalytic reactions, they require knowledge of many physical characteristics of the reacting system. Experimental properties like effective diffusivity in catalyst pores, heat and mass transfer coefficients at the fluid-particle interface, and the thermal conductivity of the catalyst are needed to utilize Equations (6.5.1) through (6.5.5). However, it is difficult to obtain accurate values of those critical parameters. For example, the diffusional characteristics of a catalyst may vary throughout a pellet because of the compression procedures used to form the final catalyst pellets. The accuracy of the heat transfer coefficient obtained from known correlations is also questionable because of the low flow rates and small particle sizes typically used in laboratory packed bed reactors.

Madon and Boudart propose a simple experimental criterion for the absence of artifacts in the measurement of rates of heterogeneous catalytic reactions [R. J. Madon and M. Boudart, *Ind. Eng. Chem. Fundam.*, **21** (1982) 438]. The experiment involves making rate measurements on catalysts in which the concentration of active material has been purposely changed. In the absence of artifacts from transport limitations, the reaction rate is directly proportional to the concentration of active material. In other words, the intrinsic turnover frequency should be independent of the concentration of active material in a catalyst. One way of varying the concentration of active material in a catalyst pellet is to mix inert particles together with active catalyst particles and then pelletize the mixture. Of course, the diffusional characteristics of the inert particles must be the same as the catalyst particles, and the initial particles in the mixture must be much smaller than the final pellet size. If the diluted catalyst pellets contain 50 percent inert powder, then the observed reaction rate should be 50 percent of the rate observed over the undiluted pellets. An intriguing aspect of this experiment is that measurement of the number of active catalytic sites is not involved with this test. However, care should be exercised when the dilution method is used with catalysts having a bimodal pore size distribution. Internal diffusion in the micropores may be important for both the diluted and undiluted catalysts.

Another way to change concentration of active material is to modify the catalyst loading on an inert support. For example, the number of supported transition metal particles on a microporous support like alumina or silica can easily be varied during catalyst preparation. As discussed in the previous chapter, selective chemisorption of small molecules like dihydrogen, dioxygen, or carbon monoxide can be used to measure the fraction of exposed metal atoms, or dispersion. If the turnover frequency is independent of metal loading on catalysts with identical metal dispersion, then the observed rate is free of artifacts from transport limitations. The metal particles on the support need to be the same size on the different catalysts to ensure that any observed differences in rate are attributable to transport phenomena instead of structure sensitivity of the reaction.

A minor complication arises when dealing with *exothermic* reactions, since the effectiveness factor for a catalyst pellet experiencing transport limitations can still equal one. To eliminate any ambiguity associated with this rare condition, the Madon-Boudart criterion for an exothermic reaction should be repeated at a different temperature.

The simplicity and general utility of the Madon-Boudart criterion make it one of the most important experimental tests to confirm that kinetic data are free from artifacts. It can be used for heterogeneous catalytic reactions carried out in batch, continuous stirred tank, and tubular plug flow reactors.

VIGNETTE 6.5.1

A good illustration of the Madon-Boudart criterion is the liquid-phase hydrogenation of cyclohexene to cyclohexane over supported Pt/SiO_2 catalysts that differ in Pt loading by a factor of 4 [R. J. Madon and M. Boudart, *Ind. Eng. Chem. Fundam.*, **21** (1982) 438].

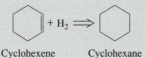

Cyclohexene   Cyclohexane

Two catalysts were prepared with 1.5 and 0.38 percent Pt supported on silica. Chemisorption experiments revealed that the percentage of metal exposed was 100 percent on both catalysts. The turnover frequency for liquid-phase cyclohexene hydrogenation (101.3 kPa H_2 pressure) was 2.67 s^{-1} and 2.51 s^{-1} at 275 K and 9.16 and 9.02 at 307 K. The similarity of the turnover frequencies at each of two different temperatures indicates that the measured rates were not influenced by transport limitations.

Development of rate expressions and evaluation of kinetic parameters require rate measurements free from artifacts attributable to transport phenomena. Assuming that experimental conditions are adjusted to meet the above-mentioned criteria for the lack of transport influences on reaction rates, rate data can be used to postulate a kinetic mechanism for a particular catalytic reaction.

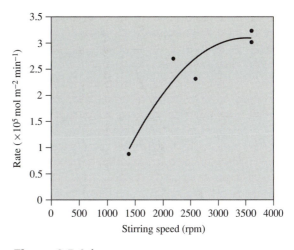

Figure 6.5.1 |
Effect of agitation on the rate of 2-propanol
dehydrogenation to acetone at 355 K over Ni catalysts.
[Rates are calculated at constant conversion level from
the data in D. E. Mears and M. Boudart, *AIChE J.*, **12**
(1966) 313.] In this case, increasing the stirring speed
increased the rate of acetone diffusion away from the
catalyst pellet and decreased product inhibition.

If mass and heat transfer problems are encountered in a catalytic reaction,
various strategies are employed to minimize their effects on observed rates. For
example, the mass transfer coefficient for diffusion through the stagnant film surrounding a catalyst pellet is directly related to the fluid velocity and the diameter
of the pellet according to Equation (6.2.26). When reactions are not mass transfer limited, the observed rate will be independent of process variables that affect
the fluid velocity around the catalyst pellets. Conversely, interphase transport limitations are indicated if the observed rate is a function of fluid flow. Consider the
results illustrated in Figure 6.5.1 for the dehydrogenation of 2-propanol to acetone over powdered nickel catalyst in a stirred reactor. The dependence of the rate
on stirring speed indicates that mass transfer limitations are important for stirring
speeds less than 3600 rpm. Additional experiments with different surface area catalysts confirmed that rates measured at the highest stirring speed were essentially
free of mass transfer limitations [D. E. Mears and M. Boudart, *AIChE J.*, **12** (1966)
313].

Both interphase and intraphase mass transfer limitations are minimized by
decreasing the pellet size of the catalyst. Since a packed bed of very small catalyst
particles can cause an unacceptably large pressure drop in a reactor, a compro-
mise between pressure drop and transport limitations is often required in
commercial reactors. Fortunately, laboratory reactors that are used to obtain

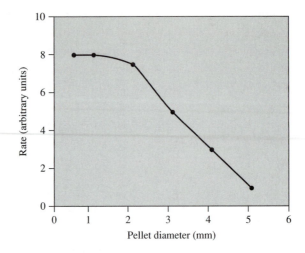

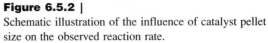

Figure 6.5.2 |
Schematic illustration of the influence of catalyst pellet
size on the observed reaction rate.

intrinsic reaction kinetics require relatively small amounts of catalyst and can be
loaded with very small particles. For the case presented in Figure 6.5.2, observed
rates measured on catalyst pellets larger than 1 mm are affected by transport
limitations.

Exercises for Chapter 6

1. The isothermal, first-order reaction of gaseous A occurs within the pores of a
    spherical catalyst pellet. The reactant concentration halfway between the
    external surface and the center of the pellet is equal to one-fourth the
    concentration at the external surface.

    (a) What is the relative concentration of A near the center of the pellet?

    (b) By what fraction should the pellet diameter be reduced to give an
    effectiveness factor of 0.7?

2. The isothermal, reversible, first-order reaction $A = B$ occurs in a flat plate
    catalyst pellet. Plot the dimensionless concentration of A (C_A/C_{AS}) as a
    function of distance into the pellet for various values of the Thiele modulus
    and the equilibrium constant. To simplify the solution, let $C_{AS} = 0.9(C_A + C_B)$
    for all cases.

3. A second-order, irreversible reaction with rate constant $k = 1.8 \text{ L mol}^{-1} \text{ s}^{-1}$
    takes place in a catalyst particle that can be considered to be a one-dimensional
    slab of half width $= 1$ cm.

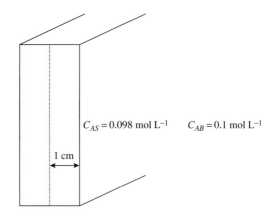

$C_{AS} = 0.098$ mol L^{-1} $C_{AB} = 0.1$ mol L^{-1}

1 cm

The concentration of reactant in the gas phase is 0.1 mol L^{-1} and at the surface is 0.098 mol L^{-1}. The gas-phase mass-transfer coefficient is 2 cm s^{-1}. Determine the intraphase effectiveness factor. (Contributed by Prof. J. L. Hudson, Univ. of Virginia.)

4. Consider the combustion of a coal particle occurring in a controlled burner. Assume the rate expression of the combustion reaction at the surface of the particle is given by:

$$r = 10^6 \exp\left[-100/(R_g T_S)\right] C_S \qquad E \text{ in kJ mol}^{-1}$$

where the rate is in units of moles O_2 reacted min^{-1} (m^2 external surface)$^{-1}$, T_S is the surface temperature of the coal particle in Kelvins, and C_S is the surface concentration of O_2. The estimated heat and mass-transfer coefficients from the gas phase to the particle are $h_t = 0.5$ kJ min^{-1} K^{-1} m^{-2} and $\bar{k}_c = 0.5$ m min^{-1}, and the heat of reaction is -150 kJ (mol O_2)$^{-1}$. Consider the bulk temperature of the gas to be $T_B = 500°C$ and the bulk concentration of O_2 to be 2 mol m^{-3}.

   (a) What is the maximum temperature difference between the bulk gas and the particle? What is the observed reaction rate under that condition?

   (b) Determine the *actual* surface temperature and concentration, and therefore the *actual* reaction rate.

5. The irreversible, first-order reaction of gaseous A to B occurs in spherical catalyst pellets with a radius of 2 mm. For this problem, the molecular diffusivity of A is 1.2×10^{-1} cm^2 s^{-1} and the Knudsen diffusivity is 9×10^{-3} cm^2 s^{-1}. The intrinsic first-order rate constant determined from detailed laboratory measurements was found to be 5.0 s^{-1}. The concentration of A in the surrounding gas is 0.01 mol L^{-1}. Assume the porosity and the tortuosity of the pellets are 0.5 and 4, respectively.

   (a) Determine the Thiele modulus for the catalyst pellets.

   (b) Find a value for the internal effectiveness factor.

(c) For an external mass-transfer coefficient of 32 s^{-1} (based on the external area of the pellets), determine the concentration of A at the surface of the catalyst pellets.

(d) Find a value for the overall effectiveness factor.

6. J. M. Smith (J. M. Smith, *Chemical Engineering Kinetics,* 2nd ed., McGraw-Hill, New York, 1970, p. 395) presents the following observed data for Pt-catalyzed oxidation of SO$_2$ at 480°C obtained in a differential fixed-bed reactor at atmospheric pressure and bulk density of 0.8 g/cm^3.

Mass velocity (g h^{-1} cm^{-2})	Bulk partial pressure (atm)			Observed rate (mol SO$_2$/h/gcat)
	SO$_2$	SO$_3$	O$_2$	
251	0.06	0.0067	0.2	0.1346
171	0.06	0.0067	0.2	0.1278
119	0.06	0.0067	0.2	0.1215
72	0.06	0.0067	0.2	0.0956

The catalyst pellets were 3.2 by 3.2 mm cylinders, and the Pt was superficially deposited upon the external surface. Compute both external mass and temperature gradients and plot ΔC_{SO_2} and ΔT versus the mass velocity. Can you draw any qualitative conclusions from this plot? If the reaction activation energy is 30 kcal/mol, what error in rate measurement attends neglect of an external ΔT? What error prevails if, assuming linear kinetics in SO$_2$, external concentration gradients are ignored?

Hints:

$$J = \frac{\bar{k}_c}{u} Sc^{2/3} = 0.817\, Re^{-1/2}$$

catalyst pellet is nonporous

reaction carried out with excess air

$\bar{\mu}_{air} = 1.339$ g/h/cm @ 480°C

$D_{SO_2-air} = 2.44$ ft^2/h @ 480°C

$\bar{\varepsilon}_B$ = void fraction of bed = 0.4

$Sc = 1.28$

Prandtl number $= \dfrac{\bar{\mu}\, C_p}{\lambda} = 0.686$

$C_p = 7.514$ cal/mol/K

$a = \dfrac{\text{surface area}}{\text{volume}} = 18.75$ mm^{-1}

$\Delta H_r = -30$ kcal/mol

7. The importance of diffusion in catalyst pellets can often be determined by measuring the effect of pellet size on the observed reaction rate. In this exercise, consider an irreversible first-order reaction occurring in catalyst pellets where the surface concentration of reactant A is $C_{AS} = 0.15$ M.

Data:

Diameter of sphere (cm)	0.2	0.06	0.02	0.006
r_{obs} (mol/h/cm^3)	0.25	0.80	1.8	2.5

(a) Calculate the intrinsic rate constant and the effective diffusivity.

(b) Estimate the effectiveness factor and the anticipated rate of reaction (r_{obs}) for a finite cylindrical catalyst pellet of dimensions 0.6 cm × 0.6 cm (diameter = length).

8. Isobutylene (A) reacts with water on an acidic catalyst to form t-butanol (B).

$$(CH_3)_2C = CH_2 + H_2O = (CH_3)_3COH$$

When the water concentration greatly exceeds that of isobutylene and t-butanol, the reversible hydration reaction is effectively first order in both the forward and reverse directions.

V. P. Gupta and W. J. M. Douglas [*AIChE J.*, **13** (1967) 883] carried out the isobutylene hydration reaction with excess water in a stirred tank reactor utilizing a cationic exchange resin as the catalyst. Use the following data to determine the effectiveness factor for the ion exchange resin at 85°C and 3.9 percent conversion.

Data:

Equilibrium constant @ 85°C = 16.6 = [B]/[A]

$D_{TA}^e = 2.0 \times 10^{-5}$ cm^2 s^{-1}

Radius of spherical catalyst particle = 0.213 mm

Density of catalyst = 1.0 g cm^{-3}

Rate of reaction at 3.9 percent conversion = 1.11×10^{-5} mol s^{-1} gcat^{-1}

$C_{AS} = 1.65 \times 10^{-2}$ M (evaluated at 3.9 percent conversion)

$C_A^0 = 1.72 \times 10^{-2}$ M (reactor inlet concentration)

(Problem adapted from C. G. Hill, Jr., *An Introduction to Chemical Engineering Kinetics and Reactor Design*, Wiley, NY, 1977.)

9. Ercan et al. studied the alkylation of ethylbenzene, *EB*, with light olefins (ethylene and propylene) over a commercial zeolite Y catalyst in a fixed-bed reactor with recycle [C. Ercan, F. M. Dautzenberg, C. Y. Yeh, and H. E. Barner, *Ind. Eng. Chem. Res.*, **37** (1998) 1724]. The solid-catalyzed liquid-phase reaction was carried out in excess ethylbenzene at 25 bar and 190°C. Assume

the reaction is pseudo-first-order with respect to olefin. The porosity of the catalyst was 0.5, the tortuosity was 5.0, and the density was 1000 kg m^{-3}. The observed rate (r_{obs}) and rate constant (k_{obs}) were measured for two different catalyst pellet sizes. Relevant results are given below:

R_p (mm)	$\bar{k}_c$ (m s^{-1})	External surface area of catalyst, S_p (m^2 kg^{-1})	r_{obs} (kmol (kgcat s)$^{-1}$)	k_{obs} (m^3 (kgcat s)$^{-1}$)
0.63	5.69×10^{-4}	4.62	8.64×10^{-6}	0.33×10^{-3}
0.17	1.07×10^{-3}	17.13	11.7×10^{-6}	1.06×10^{-3}

(a) Determine whether or not external and internal mass transfer limitations are significant for each case. Assume the diffusivity of olefins in ethylbenzene is $D_{AB} = 1.9 \times 10^{-4}$ cm^2 s^{-1}.

(b) Calculate the Thiele modulus, ϕ, and the internal effectiveness factor, η, for each case.

(c) Determine the overall effectiveness factor for each case.

10. Reaction rate expressions of the form:

$$r = \frac{kC_A}{1 + KC_A}$$

reveal zero-order kinetics when $KC_A \gg 1$. Solve the material balance (isothermal) equation for a slab catalyst particle using zero-order kinetics. Plot $C_A(x)/C_{AS}$ for a Thiele modulus of 0.1, 1.0, and 10.0. If the zero-order kinetics were to be used as an approximation for the rate form shown above when $KC_{AS} \gg 1$, would this approximation hold with the slab catalyst particle for the Thiele moduli investigated?

11. Kehoe and Butt [J. P. Kehoe and J. B. Butt, *AIChE J.*, **18** (1972) 347] have reported the kinetics of benzene hydrogenation of a supported, partially reduced Ni/kieselguhr catalyst. In the presence of a large excess of hydrogen (90 percent) the reaction is pseudo-first-order at temperatures below 200°C with the rate given by:

$$r = \left(P_H^0 k_1^0 K^0 \right) \exp\left[\frac{-E}{R_g T} \right] P_B$$

where

$P_B =$ benzene partial pressure, torr
$P_H^0 =$ dihydrogen partial pressure, torr
$K^0 = 4.22 \times 10^{-11}$ torr^{-1}
$k_1^0 = 4.22$ mol/gcat/s/torr
$E = -2.7$ kcal/mol

For the case of $P_H^0 = 685$ torr, $P_B = 75$ torr, and $T = 150°C$, estimate the effectiveness factor for this reaction carried out in a spherical catalyst particle of density 1.88 gcat/cm^3, $D_{TB}^e = 0.052$ cm^2/s, and $R_p = 0.3$ cm.

12. A first-order irreversible reaction is carried out on a catalyst of characteristic dimension 0.2 cm and effective diffusivity of 0.015 cm^2/s. At 100°C the intrinsic rate constant has been measured to be 0.93 s^{-1} with an activation energy of 20 kcal/mol.

    (a) For a surface concentration of 3.25×10^{-2} mol/L, what is the observed rate of reaction at 100°C?

    (b) For the same reactant concentration, what is the observed rate of reaction at 150°C? Assume that D_{TA}^e is independent of temperature.

    (c) What value of the activation energy would be observed?

    (d) Compare values of the Thiele modulus at 100°C and 150°C.

13. The catalytic dehydrogenation of cyclohexane to benzene was accomplished in an isothermal, differential, continuous flow reactor containing a supported platinum catalyst [L. G. Barnett et al., *AIChE J.*, **7** (1961) 211].

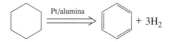

$$\begin{array}{c} C_6H_{12} = 0.002 \text{ mol s}^{-1} \\ \hline H_2 = 0.008 \text{ mol s}^{-1} \end{array} \quad \boxed{\begin{array}{c} T = 705 \text{ K} \\ P = 1.48 \text{ MPa} \\ 10.4 \text{ g catalyst} \end{array}} \quad f = 15.5\%$$

Dihydrogen was fed to the process to minimize deposition of carbonaceous residues on the catalyst. Assuming the reaction is first-order in cyclohexane and the diffusivity is primarily of the Knudsen type, estimate the tortuosity $\bar{\tau}$ of the catalyst pellets.

Additional Data:

    Diameter of catalyst pellet = 3.2 mm
    Pore volume of the catalyst = 0.48 cm^3 g^{-1}
    Surface area = 240 m^2 g^{-1}
    Pellet density $\rho_p = 1.332$ g cm^{-3}
    Pellet Porosity $\bar{\varepsilon}_p = 0.59$
    Effectiveness factor $\eta = 0.42$

14. For a slab with first-order kinetics:

$$\eta_o = \frac{\tanh\phi}{\phi[1 + \phi\tanh(\phi)/Bi_m]} \tag{1}$$

How important is the mass Biot number in Equation (1) with respect to its influence upon η_o for (a) $\phi = 0.1$, (b) $\phi = 1.0$, (c) $\phi = 5.0$, (d) $\phi = 10.0$?

Consider the effect of the Biot number significant if it changes η_o by more than 1 percent. Can you draw any qualitative conclusions from the behavior observed in parts (a)–(d)?

15. The liquid-phase hydrogenation of cyclohexene to cyclohexane (in an inert solvent) is conducted over solid catalyst in a semibatch reactor (dihydrogen addition to keep the total pressure constant).

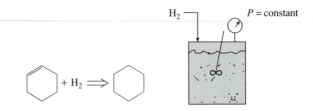

Draw a schematic representation of what is occurring at the microscopic level. Provide an interpretation for each of the following figures.

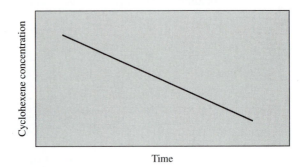

Figure 1 |
Linear relationship between cyclohexene concentration in the reactor and reaction time. Results apply for all conditions given in the figures provided below.

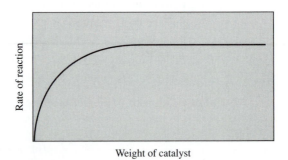

Figure 2 |
Effect of catalyst weight on reaction rate.

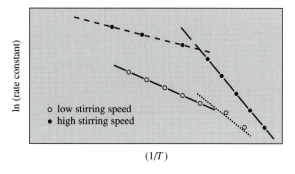

Figure 3 |
Evaluation of temperature effects on the reaction rate
constant.

CHAPTER 7

Microkinetic Analysis of Catalytic Reactions

7.1 | Introduction

A catalytic reaction consists of many elementary steps that comprise an overall mechanistic path of a chemical transformation. Although rigorous reaction mechanisms are known for simple chemical reactions, many catalytic reactions have been adequately described by including only the kinetically significant elementary steps. This approach was used in Chapter 5 to simplify complex heterogeneous catalytic reaction sequences to kinetically relevant surface reactions. The next step in furthering our understanding of catalytic reactions is to consolidate the available experimental data and theoretical principles that pertain to elementary steps in order to arrive at quantitative models. This is the role of microkinetic analysis, which is defined as an examination of catalytic reactions in terms of elementary steps and their relation with each other during a catalytic cycle (J. A. Dumesic, D. F. Rudd, L. M. Aparicio, J. E. Rekoske, and A. A. Trevino, *The Microkinetics of Heterogeneous Catalysis,* American Chemical Society, Washington, D.C., 1993, p. 1). In this chapter, three catalytic reactions will be examined in detail to illustrate the concept of microkinetic analysis and its relevance to chemical reaction engineering. The first example is asymmetric hydrogenation of an olefin catalyzed by a soluble organometallic catalyst. The second and third examples are ammonia synthesis and olefin hydrogenation, respectively, on heterogeneous transition metal catalysts.

7.2 | Asymmetric Hydrogenation of Prochiral Olefins

The use of soluble rhodium catalysts containing chiral ligands to obtain high stereoselectivity in the asymmetric hydrogenation of prochiral olefins represents one of the most important achievements in catalytic selectivity, rivaling the stereoselectivity of enzyme catalysts [J. Halpern, *Science,* **217** (1982) 401]. Many chiral ligands

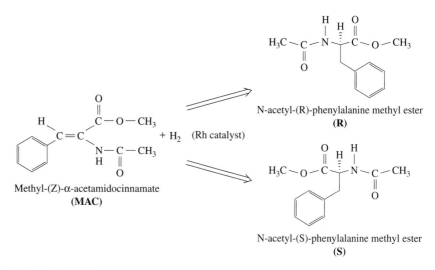

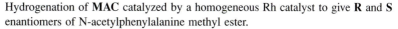

N-acetyl-(R)-phenylalanine methyl ester
(R)

Methyl-(Z)-α-acetamidocinnamate
(MAC)

N-acetyl-(S)-phenylalanine methyl ester
(S)

Figure 7.2.1 |
Hydrogenation of **MAC** catalyzed by a homogeneous Rh catalyst to give **R** and **S** enantiomers of N-acetylphenylalanine methyl ester.

have been used to create this class of new asymmetric hydrogenation catalysts. In addition, stereoselectivity has been observed with a variety of olefins during hydrogenation reactions in the presence of these chiral catalysts, demonstrating the general utility of the materials. As an example of microkinetic analysis, the asymmetric hydrogenation of methyl-(Z)-α-acetamidocinnamate, or **MAC,** to give the enantiomers (**R** and **S**) of N-acetylphenylalanine methyl ester will be discussed in detail. The overall reaction is illustrated in Figure 7.2.1.

C. R. Landis and J. Halpern found that cationic rhodium, Rh(I), with the chiral ligand R,R-1,2-bis[(phenyl-o-anisol)phosphino]ethane, or DIPAMP (see Figure 7.2.2), was very selective as a catalyst for the production of the **S** enantiomer of

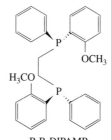

R,R-DIPAMP
R,R-1,2-bis[(phenyl-o-anisol)phosphino]ethane

Figure 7.2.2 |
Chiral ligand for enantioselective rhodium catalyst

N-acetylphenylalanine methyl ester during **MAC** hydrogenation in methanol solvent [C. R. Landis and J. Halpern, *J. Am. Chem. Soc.,* **109** (1987) 1746]. The DIPAMP coordinates to the rhodium through the phosphorus atoms and leaves plenty of space around the Rh cation for other molecules to bind and react. The unique feature of this system is that the chirality of the DIPAMP ligand induces the high stereoselectivity of the product molecules. An analogous nonchiral ligand on the Rh cation produces a catalyst that is not enantioselective.

To understand the origin of enantioselectivity in this system, the kinetics of the relevant elementary steps occurring during hydrogenation had to be determined. The catalytic cycle can be summarized as: (a) reversible binding of the olefin (**MAC**) to the Rh catalyst, (b) irreversible addition of dihydrogen to the Rh-olefin complex, (c) reaction of hydrogen with the bound olefin, and (d) elimination of the product into the solution to regenerate the original catalyst. Landis and Halpern have measured the kinetics of step (a) as well as the overall reactivity of the adsorbed olefin with dihydrogen [C. R. Landis and J. Halpern, *J. Am. Chem. Soc.,* **109** (1987) 1746]. Figure 7.2.3 shows two coupled catalytic cycles that occur in parallel for the production of **R** and **S** enantiomers from **MAC** in the presence of a common rhodium catalyst, denoted as *. Since the catalyst has a chiral ligand, **MAC** can add to the catalyst in two different forms, one that leads to the **R** product, called **MAC*R,** and one that leads to the **S** product, called **MAC*S**. The superscripts refer to the catalytic cycles that produce the two enantiomers.

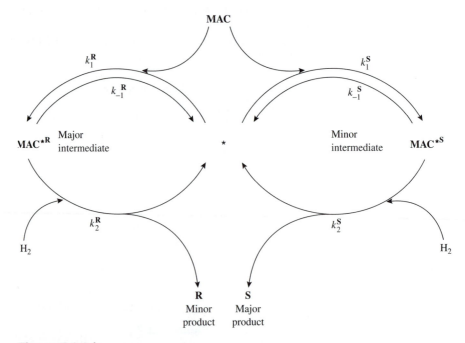

Figure 7.2.3 |
Scheme of the coupled catalytic cycles for the asymmetric hydrogenation of **MAC**.

Expressions for relative concentrations of the intermediates and products are developed from concepts discussed in earlier chapters, namely, the use of a total catalytic site balance and the application of the steady-state approximation. The total amount of rhodium catalyst in the reactor is considered constant, $[*]_0$, so that the site balance becomes:

$$[*]_0 = [\mathbf{MAC}^{*\mathbf{R}}] + [\mathbf{MAC}^{*\mathbf{S}}] + [*] \tag{7.2.1}$$

The steady-state approximation indicates that the concentrations of reactive intermediates remain constant with time. In other words, the net rate of **MAC** binding to the catalyst to form an intermediate must be the same as the rate of hydrogenation (or disappearance) of the intermediate. The steady-state approximation for the coupled catalytic cycles is expressed mathematically as:

$$\frac{d[\mathbf{MAC}^{*\mathbf{R}}]}{dt} = 0 = r_1^{\mathbf{R}} - r_{-1}^{\mathbf{R}} - r_2^{\mathbf{R}} \tag{7.2.2}$$

$$\frac{d[\mathbf{MAC}^{*\mathbf{S}}]}{dt} = 0 = r_1^{\mathbf{S}} - r_{-1}^{\mathbf{S}} - r_2^{\mathbf{S}} \tag{7.2.3}$$

or in terms of the rate expressions as:

$$0 = k_1^{\mathbf{R}}[\mathbf{MAC}][*] - k_{-1}^{\mathbf{R}}[\mathbf{MAC}^{*\mathbf{R}}] - k_2^{\mathbf{R}}[\mathbf{MAC}^{*\mathbf{R}}][H_2] \tag{7.2.4}$$

$$0 = k_1^{\mathbf{S}}[\mathbf{MAC}][*] - k_{-1}^{\mathbf{S}}[\mathbf{MAC}^{*\mathbf{S}}] - k_2^{\mathbf{S}}[\mathbf{MAC}^{*\mathbf{S}}][H_2] \tag{7.2.5}$$

Rearranging Equations (7.2.4 and 7.2.5) gives the concentrations of reactive intermediates:

$$[\mathbf{MAC}^{*\mathbf{R}}] = \frac{k_1^{\mathbf{R}}[\mathbf{MAC}][*]}{k_{-1}^{\mathbf{R}} + k_2^{\mathbf{R}}[H_2]} \tag{7.2.6}$$

$$[\mathbf{MAC}^{*\mathbf{S}}] = \frac{k_1^{\mathbf{S}}[\mathbf{MAC}][*]}{k_{-1}^{\mathbf{S}} + k_2^{\mathbf{S}}[H_2]} \tag{7.2.7}$$

that can be used to derive an equation for their relative concentration:

$$\frac{[\mathbf{MAC}^{*\mathbf{S}}]}{[\mathbf{MAC}^{*\mathbf{R}}]} = \left(\frac{k_1^{\mathbf{S}}}{k_1^{\mathbf{R}}}\right) \frac{k_{-1}^{\mathbf{R}} + k_2^{\mathbf{R}}[H_2]}{k_{-1}^{\mathbf{S}} + k_2^{\mathbf{S}}[H_2]} \tag{7.2.8}$$

The ratio of concentrations of the products **R** and **S** can be derived in a similar fashion:

$$r_{\mathbf{R}} = r_1^{\mathbf{R}} - r_{-1}^{\mathbf{R}} = r_2^{\mathbf{R}} \tag{7.2.9}$$

$$r_{\mathbf{S}} = r_1^{\mathbf{S}} - r_{-1}^{\mathbf{S}} = r_2^{\mathbf{S}} \tag{7.2.10}$$

$$\frac{[\mathbf{S}]}{[\mathbf{R}]} = \frac{r_{\mathbf{S}}}{r_{\mathbf{R}}} = \frac{r_2^{\mathbf{S}}}{r_2^{\mathbf{R}}} = \frac{k_2^{\mathbf{S}}[\mathbf{MAC}^{*\mathbf{S}}][H_2]}{k_2^{\mathbf{R}}[\mathbf{MAC}^{*\mathbf{R}}][H_2]} = \left(\frac{k_2^{\mathbf{S}}}{k_2^{\mathbf{R}}}\right) \frac{[\mathbf{MAC}^{*\mathbf{S}}]}{[\mathbf{MAC}^{*\mathbf{R}}]} \tag{7.2.11}$$

Table 7.2.1 | Kinetic parameters for the asymmetric hydrogenation of MAC catalyzed by Rh(R,R-DIPAMP) at 298 K.

Parameter	R Cycle	S Cycle
k_1 (L mmol^{-1} s^{-1})	5.3	11
k_{-1} (s^{-1})	0.15	3.2
K_1 (L mmol^{-1})	35	3.3
k_2 (L mmol^{-1} s^{-1})	1.1×10^{-3}	0.63

Source: C. R. Landis and J. Halpern, *J. Am. Chem. Soc.*, **109** (1987) 1746.

The relative concentration of reactive intermediates given in Equation (7.2.8) is then substituted into Equation (7.2.11) to give the ratio of final products:

$$\frac{[\mathbf{S}]}{[\mathbf{R}]} = \left(\frac{k_1^S \, k_2^S}{k_1^R \, k_2^R}\right) \frac{k_{-1}^R + k_2^R \, [\mathrm{H_2}]}{k_{-1}^S + k_2^S \, [\mathrm{H_2}]} \tag{7.2.12}$$

Notice that the relative concentrations of the intermediates and products depend only on temperature (through the individual rate constants) and the pressure of dihydrogen.

At sufficiently low dihydrogen pressures, the reversible binding of **MAC** to the catalyst is essentially equilibrated because:

$$k_{-1}^R \gg k_2^R \, [\mathrm{H_2}] \tag{7.2.13}$$

and

$$k_{-1}^S \gg k_2^S \, [\mathrm{H_2}] \tag{7.2.14}$$

In other words, the rate-determining step for the reaction at low dihydrogen pressures is the hydrogenation of bound olefin. Simplification of Equations (7.2.8) and (7.2.12) by assuming low dihydrogen pressure gives the following expressions for the relative concentrations of intermediates and products:

$$\frac{[\mathbf{MAC} *^S]}{[\mathbf{MAC} *^R]} = \left(\frac{k_1^S}{k_1^R}\right) \frac{k_{-1}^R}{k_{-1}^S} = \frac{K_1^S}{K_1^R} \tag{7.2.15}$$

$$\frac{[\mathbf{S}]}{[\mathbf{R}]} = \left(\frac{k_1^S \, k_2^S}{k_1^R \, k_2^R}\right) \frac{k_{-1}^R}{k_{-1}^S} = \frac{K_1^S}{K_1^R} \frac{k_2^S}{k_2^R} \tag{7.2.16}$$

Landis and Halpern have measured the relevant rate constants for this reaction system and they are summarized in Table 7.2.1 [C. R. Landis and J. Halpern, *J. Am. Chem. Soc.*, **109** (1987) 1746].

The values of the measured constants indicate that the **S** enantiomer is greatly favored over the **R** enantiomer at 298 K:

$$\frac{[\mathbf{S}]}{[\mathbf{R}]} = \frac{K_1^S \, k_2^S}{K_1^R \, k_2^R} = \frac{3.3}{35} \cdot \frac{0.63}{1.1 \times 10^{-3}} = 54 \tag{7.2.17}$$

This result is rather surprising since the reactive intermediate that leads to the **S** enantiomer is in the minority. Equation (7.2.15) illustrates the magnitude of this difference:

$$\frac{[\mathbf{MAC} *^{\mathbf{S}}]}{[\mathbf{MAC} *^{\mathbf{R}}]} = \frac{K_1^{\mathbf{S}}}{K_1^{\mathbf{R}}} = 0.094 \tag{7.2.18}$$

The reason that the *minor* reactive intermediate leads to the *major* product is due to the large rate constant for hydrogenation (k_2) associated with the **S** cycle compared to the **R** cycle. Clearly, the conventional "lock and key" analogy for the origin of enantioselectivity does not apply for this case since the selectivity is determined by kinetics of hydrogenation instead of thermodynamics of olefin binding.

A microkinetic analysis of this system also adequately explains the dependence of enantioselectivity on dihydrogen pressure. At sufficiently high pressures of dihydrogen,

$$k_{-1}^{\mathbf{R}} \ll k_2^{\mathbf{R}} [\mathbf{H}_2] \tag{7.2.19}$$

and

$$k_{-1}^{\mathbf{S}} \ll k_2^{\mathbf{S}} [\mathbf{H}_2] \tag{7.2.20}$$

Since the binding of olefin is not quasi-equilibrated at high pressure, the subsequent hydrogenation step cannot be considered as rate-determining. The relative concentrations of reactive intermediates and products are now given as:

$$\frac{[\mathbf{MAC} *^{\mathbf{S}}]}{[\mathbf{MAC} *^{\mathbf{R}}]} = \left(\frac{k_1^{\mathbf{S}}}{k_1^{\mathbf{R}}}\right)\frac{k_2^{\mathbf{R}}}{k_2^{\mathbf{S}}} = \left(\frac{11}{5.3}\right)\frac{1.1 \times 10^{-3}}{0.63} = 3.6 \times 10^{-3} \tag{7.2.21}$$

and

$$\frac{[\mathbf{S}]}{[\mathbf{R}]} = \left(\frac{k_1^{\mathbf{S}}}{k_1^{\mathbf{R}}}\right) = \left(\frac{11}{5.3}\right) = 2.1 \tag{7.2.22}$$

The effect of dihydrogen is to lower the enantioselectivity to the **S** product from $[\mathbf{S}]/[\mathbf{R}] = 54$ at low pressures to $[\mathbf{S}]/[\mathbf{R}] = 2.1$ at high pressures. The reason for the drop in selectivity is again a kinetic one. The *kinetic coupling* of the olefin binding and hydrogenation steps becomes important at high dihydrogen pressure [M. Boudart and G. Djega-Mariadassou, *Catal. Lett.,* **29** (1994) 7]. In other words, the rapid hydrogenation of the reactive intermediate prevents quasi-equilibration of the olefin binding step. The relative concentration of the *minor* intermediate **MAC***$^{\mathbf{S}}$ (that leads to the *major* product **S**) decreases significantly with increasing dihydrogen pressure since $k_2^{\mathbf{S}} \gg k_2^{\mathbf{R}}$. In this example, the kinetic coupling on the **S** cycle is much stronger than that on the **R** cycle, and that leads to an overall reduction in selectivity with increasing dihydrogen pressure.

It should be pointed out that selectivity does not depend on dihydrogen in the limit of very high or very low pressure. For intermediate dihydrogen pressures, Equations (7.2.8) and (7.2.12)—that depend on dihydrogen—should be used to calculate selectivity. The microkinetic methodology satisfactorily explains the surprising inverse relationship between intermediate and product selectivity and the unusual effect of dihydrogen pressure on product selectivity observed during asymmetric hydrogenation of prochiral olefins with a chiral catalyst.

7.3 | Ammonia Synthesis on Transition Metal Catalysts

The ammonia synthesis reaction is one of the most widely studied reactions and it has been discussed previously in this text. (See Example 1.1.1, Table 5.2.1, and Section 5.3.) In this section, results from the microkinetic analysis of ammonia synthesis over transition metal catalysts containing either iron or ruthenium will be presented.

A conventional ammonia synthesis catalyst, based on iron promoted with Al_2O_3 and K_2O, operates at high temperatures and pressures in the range of 673–973 K and 150–300 bars to achieve acceptable production rates. Metallic iron is the active catalytic component while Al_2O_3 and K_2O act as structural and chemical promoters, respectively. The goal of microkinetic analysis in this case is to study the conditions under which data collected on single crystals in ultrahigh vacuum or on model powder catalysts at ambient pressures can be extrapolated to describe the performance of a working catalyst under industrial conditions of temperature and pressure. Unfortunately, the kinetic parameters of all possible elementary steps are not known for most heterogeneous catalytic reactions. Thus, a strategy adopted by Dumesic et al. involves construction of a serviceable reaction path that captures the essential surface chemistry, and estimation of relevant parameters for kinetically significant elementary steps (J. A. Dumesic, D. F. Rudd, L. M. Aparicio, J. E. Rekoske, and A. A. Trevino, *The Microkinetics of Heterogeneous Catalysis,* American Chemical Society, Washington, D.C., 1993, p. 145).

In the case of ammonia synthesis on transition metal catalysts, a variety of reasonable paths with various levels of complexity can be proposed. For the sake of clarity, only one such path will be presented here. Stolze and Norskov successfully interpreted the high-pressure kinetics of ammonia synthesis based on a microscopic model established from fundamental surface science studies [P. Stolze and J. K. Norskov, *Phys. Rev. Lett.,* **55** (1985) 2502; P. Stolze and J. K. Norskov, *J. Catal.,* **110** (1988) 1]. Dumesic et al. re-analyzed that sequence of elementary steps, which is presented in Table 7.3.1, for ammonia synthesis over iron catalysts [J. A. Dumesic and A. A. Trevino, *J. Catal.,* **116** (1989) 119]. The rate constants for adsorption and desorption steps were estimated from results obtained on iron single crystal surfaces at ultrahigh vacuum conditions. The surface hydrogenation rates were estimated from the relative stabilities of surface NH_X species. In the following microkinetic analysis, an industrial plug flow reactor was modeled as a series of 10,000 mixing cells (see Example 3.4.3) in which the steady-state rate equations, the catalyst site balance, and the material balances for gaseous species were solved simultaneously. Thus, no assumptions with respect to a possible rate-determining step and a most abundant reaction intermediate were needed to complete the model.

The conditions chosen for the microkinetic analysis correspond to an industrial reactor with 2.5 cm^3 of catalyst operating at 107 bar with a stoichiometric feed of

Table 7.3.1 | A proposed mechanism of ammonia synthesis over iron catalysts.[a]

Number	Forward rate	Elementary step	Reverse rate
1	$2 \times 10^1 P_{N_2} \theta_*$	$N_2 + * \rightleftharpoons N_2*$	$2 \times 10^{14} e^{-43/(R_gT)} \theta_{N_2}$
2	$4 \times 10^9 e^{-29/(R_gT)} \theta_{N_2} \theta_*$	$N_2* + * \rightleftharpoons 2N*$	$1 \times 10^9 e^{-155/(R_gT)} (\theta_N)^2$
3	$2 \times 10^9 e^{-81/(R_gT)} \theta_H \theta_N$	$H* + N* \rightleftharpoons NH* + *$	$1 \times 10^7 e^{-23/(R_gT)} \theta_{NH} \theta_*$
4	$1 \times 10^{13} e^{-36/(R_gT)} \theta_{NH} \theta_H$	$NH* + H* \rightleftharpoons NH_2* + *$	$1 \times 10^{12} \theta_{NH_2} \theta_*$
5	$4 \times 10^{13} e^{-39/(R_gT)} \theta_{NH_2} \theta_H$	$NH_2* + H* \rightleftharpoons NH_3* + *$	$2 \times 10^{13} \theta_{NH_3} \theta_*$
6	$4 \times 10^{12} e^{-39/(R_gT)} \theta_{NH_3}$	$NH_3* \rightleftharpoons NH_3 + *$	$2 \times 10^3 P_{NH_3} \theta_*$
7	$7 \times 10^1 P_{H_2} (\theta_*)^2$	$H_2 + 2* \rightleftharpoons 2H*$	$3 \times 10^{13} e^{-94/(R_gT)} (\theta_H)^2$

[a] Rates are in units of molecules per second per site, pressures (P) are in pascals, and activation energies are in kilojoules per mole. Concentrations of surface species are represented by fractional surface coverages as discussed in Chapter 5. Kinetic parameters are adapted from the data of Stolze and Norskov [P. Stolze and J. K. Norskov, *Phys. Rev. Lett.*, **55** (1985) 2502; P. Stolze and J. K. Norskov, *Surf. Sci. Lett.*, **197** (1988) L230; P. Stolze and J. K. Norskov, *J. Catal.*, **110** (1988) 1].

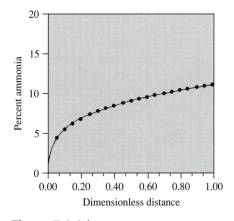

Figure 7.3.1 |
Ammonia concentration calculated from microkinetic model versus longitudinal distance from reactor inlet. (Figure adapted from "Kinetic Simulation of Ammonia Synthesis Catalysis" by J. A. Dumesic and A. A. Trevino, in *Journal of Catalysis*, Volume 116:119, copyright © 1989 by Academic Press, reproduced by permission of the publisher and the authors.)

dinitrogen and dihydrogen. The space velocity was 16,000 h^{-1}, the catalyst bed density was 2.5 g cm^{-3}, and the catalyst site density was 6×10^{-5} mol g^{-1}. The ammonia concentration in the effluent of the reactor operating at 723 K was measured experimentally to be 13.2 percent. Figure 7.3.1 illustrates the ammonia concentration calculated from the microkinetic model as a function of axial distance along

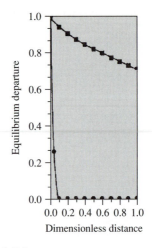

Figure 7.3.2 |
Departure from equilibrium for the two slowest
elementary steps in ammonia synthesis. Squares are for
step 2 and circles are for step 3. (Figure adapted from
"Kinetic Simulation of Ammonia Synthesis Catalysis"
by J. A. Dumesic and A. A. Trevino, in *Journal of
Catalysis*, Volume 116:119, copyright © 1989 by
Academic Press, reproduced by permission of the
publisher and the authors.)

the reactor. The effluent concentration approached the experimental value of 13.2
percent, which illustrates the consistency of the model with observation. The utility of a microkinetic model, however, is that it allows for a detailed understanding
of the reaction kinetics. For example, a rate-determining step can be found, if one
exists, by examining the departure of the elementary steps from equilibrium under
simulated reaction conditions. Figure 7.3.2 presents the departure from equilibrium,
(forward rate − reverse rate)/forward rate, for the slowest steps as the reaction
proceeds through the reactor. Step 2, the dissociation of dinitrogen, is clearly the
rate-determining step in the mechanism throughout the reactor. All of the other steps
in Table 7.3.1 are kinetically insignificant. Thus, the turnover frequency of ammonia synthesis depends critically on the kinetics of dissociative adsorption of dinitrogen (steps 1 and 2), as discussed earlier in Section 5.3. As long as the individual
quasi-equilibrated elementary steps can be summed to one overall quasi-equilibrated
reaction with a thermodynamically consistent equilibrium constant, small errors in
the individual pre-exponential factors and activation energies are irrelevant. A microkinetic model also allows for the determination of relative coverages of various
intermediates on the catalyst surface. Figure 7.3.3 illustrates the change in fractional
surface reaction of N* and H* through the reactor. Adsorbed nitrogen is the most
abundant reaction intermediate throughout the iron catalyst bed except at the entrance.

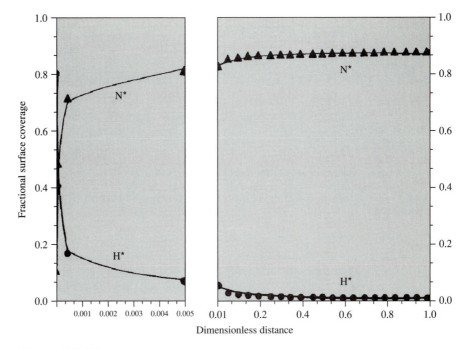

Figure 7.3.3 |
Fractional surface coverages of predominant adsorbed species versus dimensionless
distance from the reactor inlet. (Figure adapted from "Kinetic Simulation of Ammonia
Synthesis Catalysis" by J. A. Dumesic and A. A. Trevino, in *Journal of Catalysis*,
Volume 116:119, copyright © 1989 by Academic Press, reproduced by permission of the
publisher and the authors.)

Since the formation of N* is the rate-determining step, its high coverage during
steady-state reaction on iron results from the equilibrium of H_2 and NH_3. Thus, the
high coverage of H* instead of N* at the entrance to the reactor (Figure 7.3.3) results
from the vanishingly low pressure of ammonia at that point.

Another test of validity is to check the performance of the model against experimental rate data obtained far from equilibrium. The microkinetic model presented in Table 7.3.1 predicts within a factor of 5 the turnover frequency of ammonia synthesis on magnesia-supported iron particles at 678 K and an ammonia
concentration equal to 20 percent of the equilibrium value. This level of agreement
is reasonable considering that the catalyst did not contain promoters and that the
site density may have been overestimated. The model in Table 7.3.1 also predicts
within a factor of 5 the rate of ammonia synthesis over an Fe(111) single crystal at
20 bar and 748 K at ammonia concentrations less than 1.5 percent of the equilibrium
value.

It should be emphasized that since the rate of ammonia synthesis on iron depends critically on the dissociative adsorption of dinitrogen (steps 1 and 2), any

Table 7.3.2 | A proposed mechanism of ammonia synthesis over ruthenium catalysts.[a]

Number	Forward rate	Elementary step	Reverse rate
1	$6 \times 10^{-2} e^{-33/(R_gT)} P_{N_2} (\theta_*)^2$	$N_2 + 2* \rightleftharpoons 2N*$	$2 \times 10^{10} e^{-137/(R_gT)} (\theta_N)^2$
2	$6 \times 10^{13} e^{-86/(R_gT)} \theta_H \theta_N$	$H* + N* \rightleftharpoons NH* + *$	$3 \times 10^{14} e^{-41/(R_gT)} \theta_{NH} \theta_*$
3	$5 \times 10^{13} e^{-60/(R_gT)} \theta_{NH} \theta_H$	$NH* + H* \rightleftharpoons NH_2* + *$	$2 \times 10^{13} e^{-9/(R_gT)} \theta_{NH_2} \theta_*$
4	$3 \times 10^{13} e^{-17/(R_gT)} \theta_{NH_2} \theta_H$	$NH_2* + H* \rightleftharpoons NH_3* + *$	$9 \times 10^{12} e^{-65/(R_gT)} \theta_{NH_3} \theta_*$
5	$6 \times 10^{13} e^{-84/(R_gT)} \theta_{NH_3}$	$NH_3* \rightleftharpoons NH_3 + *$	$2 \times 10^3 P_{NH_3} \theta_*$
6	$6 \times 10^2 P_{H_2} (\theta_*)^2$	$H_2 + 2* \rightleftharpoons 2H*$	$2 \times 10^{13} e^{-89/(R_gT)} (\theta_H)^2$

[a] Rates are in units of molecules per second per site, pressures (P) are in pascals, and activation energies are in kilojoules per mole. Concentrations of surface species are represented by fractional surface coverages as discussed in Chapter 5. Kinetics parameters are adapted from the data of Hinrichsen et al. [O. Hinrichsen, F. Rosowski, M. Muhler, and G. Ertl, *Chem. Eng. Sci.*, **51** (1996) 1683.]

microkinetic model that properly accounts for those steps and that is thermodynamically consistent with the overall reaction will effectively describe the kinetics of ammonia synthesis. This feature explains the success of the microkinetic model used to describe an industrial reactor even though the kinetic parameters for dinitrogen adsorption/desorption match those obtained from studies on iron single crystals in ultrahigh vacuum.

Ammonia synthesis on supported ruthenium catalysts has also been the subject of microkinetic analysis. Table 7.3.2 presents the kinetic parameters for one model of ammonia synthesis catalyzed by cesium-promoted ruthenium particles supported on MgO [O. Hinrichsen, F. Rosowski, M. Muhler, and G. Ertl, *Chem. Eng. Sci.*, **51** (1996) 1683]. In contrast to the model in Table 7.3.1, the dissociative adsorption of dinitrogen is represented by a single step. Many of the rate constants in Table 7.3.2 were determined from independent steady-state and transient experiments on the same catalyst. The experiments included temperature-programmed adsorption and desorption of dinitrogen, isotopic exchange of labeled and unlabeled dinitrogen, and temperature-programmed reaction of adsorbed nitrogen atoms with gaseous dihydrogen. The unknown rate constants were estimated by regression analysis of the experimental data and checked to ensure that they were within physically reasonable limits. As in the previous example with the iron catalyst, the steady-state reactor containing the supported Ru catalyst was modeled as a series of mixing cells in which the steady-state rate equations, the catalyst site balance and the material balances for gaseous species were solved simultaneously.

Similar to the results found with iron catalysts, the rate-determining step during ammonia synthesis on ruthenium catalysts is the dissociative adsorption of dinitrogen. However, the overall reaction rate is strongly inhibited by dihydrogen, indicating that adsorbed hydrogen is the most abundant reaction intermediate that covers most of the surface [B. C. McClaine, T. Becue, C. Lock, and R. J. Davis,

J. Mol. Catal. A: Chem., **163** (2000) 105]. In the example using iron catalysts, inhibition by dihydrogen was observed only at extremely low pressures of ammonia. The reaction on ruthenium is quite different, since the equilibrium involving dihydrogen and ammonia on the catalyst surface continues to favor hydrogen atoms instead of nitrogen atoms, even at reasonable pressures of ammonia. The microkinetic model in Table 7.3.2 accounts for the ammonia effluent concentration from a reactor operating at both atmospheric pressure and elevated pressure, over a wide temperature range and at exit conditions near and far from equilibrium. Figure 7.3.4 compares

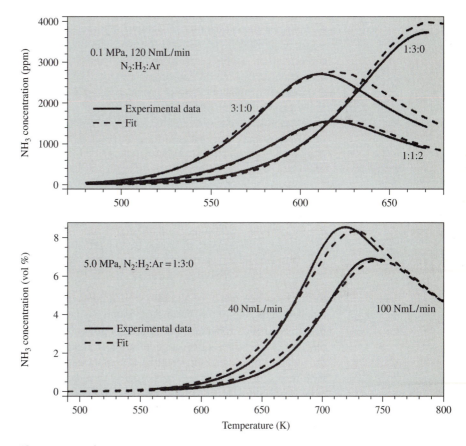

Figure 7.3.4 |
Comparison of calculated and measured ammonia concentrations at the effluent of a steady-steady ammonia synthesis reactor containing ruthenium particles supported on magnesia and promoted by cesium. [Adapted from O. Hinrichsen, F. Rosowski, M. Muhler, and G. Ertl, "The Microkinetics of Ammonia Synthesis Catalyzed by Cesium-Promoted Supported Ruthenium," *Chem. Eng. Sci.,* **51** (1996) 1683, copyright 1996, with permission from Elsevier Science.]

the results from the model to those measured experimentally. The decrease in ammonia effluent concentration at the highest temperatures indicates that equilibrium was reached in the reactor. The essential features of the reaction (i.e., proper temperature dependence of the rate appropriate level of ammonia produced near and far from equilibrium, and inhibition of the rate by dihydrogen) were well reproduced by the kinetic model.

7.4 | Ethylene Hydrogenation on Transition Metals

Hydrogenation is an important industrial reaction that often requires the presence of a heterogeneous catalyst to achieve commercial yields. Ethylene, C_2H_4, is the smallest olefin that can be used to investigate the addition of hydrogen atoms to a carbon-carbon double bond. Even though many experiments and theoretical studies have been carried out on this simple system, the reaction is still not completely understood. Microkinetic analysis provides insights into the relevant elementary steps in the catalytic cycle.

A simple mechanism that has been proposed for ethylene hydrogenation on metal catalysts is that of Horiuti and Polanyi [J. Horiuti and M. Polanyi, *J. Chem. Soc., Faraday Trans.*, **30** (1934) 1164]:

Step 1	$H_2 + 2* \rightleftarrows 2H*$
Step 2	$C_2H_4 + 2* \rightleftarrows *C_2H_4*$
Step 3	$*C_2H_4* + H* \rightleftarrows *C_2H_5 + 2*$
Step 4	$*C_2H_5* + H* \rightleftarrows *C_2H_6 + 2*$
Overall	$C_2H_4 + H_2 = C_2H_6$

In this sequence, ethylene and dihydrogen compete for active sites on the transition metal surface. While this basic mechanism has been used to describe olefin hydrogenation kinetics for many years, it does not adequately account for the observed reaction characteristics over a wide range of conditions. For example, ethylene hydrogenation on a platinum catalyst at temperatures less than 300 K is zero-order with respect to ethylene and half-order with respect to dihydrogen. The zero-order dependence on C_2H_4 indicates that the surface is nearly covered with carbon containing species, whereas the half-order dependence on H_2 suggests that step 1, dissociative adsorption of dihydrogen, is an equilibrated process. If the adsorption steps were truly competitive, then increasing the ethylene pressure should decrease the vacant site concentration and thus the hydrogen surface coverage. In other words, the rate of hydrogenation should decrease with increasing ethylene pressure, which is not observed at low temperature.

One way to explain the observed reaction orders is to also allow for a *noncompetitive* dihydrogen adsorption step in the sequence. This added complexity makes sense because more surface sites are available to dihydrogen than ethylene because of the very small size of a H_2 molecule. The catalytic cycle for ethylene hydrogenation with both *competitive* and *noncompetitive* adsorption is now:

Step 1	$H_2 + 2* \rightleftharpoons 2H*$
Step 1a	$H_2 + 2*' \rightleftharpoons 2H*'$
Step 2	$C_2H_4 + 2* \rightleftharpoons *C_2H_4*$
Step 3	$*C_2H_4* + H* \rightleftharpoons *C_2H_5 + 2*$
Step 3a	$*C_2H_4* + H*' \rightleftharpoons *C_2H_5 + * + *'$
Step 4	$*C_2H_5 + H* \rightleftharpoons C_2H_6 + 2*$
Step 4a	$*C_2H_5 + H*' \rightleftharpoons C_2H_6 + * + *'$
Overall	$C_2H_4 + H_2 = C_2H_6$

where $*'$ represents an active site that is accessible to dihydrogen but not to ethylene. Rekoske et al. have studied an analogous sequence of steps for ethylene hydrogenation on Pt, but their analysis also included a hydrogen activation step [J. E. Rekoske, R. D. Cortright, S. A. Goddard, S. B. Sharma, and J. A. Dumesic, *J. Phys. Chem.*, **96** (1992) 1880]. The additional activation step was needed to reconcile results from microkinetic analysis to the observed isotopic distribution of products when deuterium was added to the system. Nevertheless, their analysis showed that the main features of Pt-catalyzed ethylene hydrogenation could be captured over a very wide range of conditions by the coupled competitive and noncompetitive mechanism of Horiuti and Polanyi. Figure 7.4.1 compares the measured and calculated turnover frequencies over an order of magnitude range

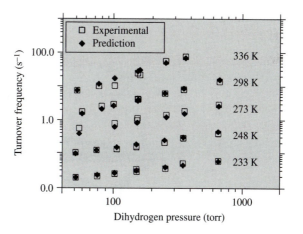

Figure 7.4.1 |

Comparison of results from microkinetic model and experimental observation. [Adapted with permision from J. E. Rekoske, R. D. Cortright, S. A. Goddard, S. B. Sharma, and J. A. Dumesic, *J. Phys. Chem.*, **96** (1992) 1880. Copyright 1992 American Chemical Society.]

of dihydrogen pressure and a 100 K span in temperature. The observed order of reaction with respect to dihydrogen increases from 0.47 at 223 K to 1.10 at 336 K. Clearly, the kinetic model reproduces the observed rates at all of the conditions tested. The model also predicted the order of reaction with respect to ethylene (not shown) quite well. This microkinetic analysis suggests that ethylene hydrogenates mostly through the noncompetitive route at low temperatures whereas the competitive route dominates at high temperatures.

The rapid increase in computing power and the advent of new quantum chemical methods over the last decade allow kinetic parameters for surface reactions to be estimated from *ab initio* quantum chemical calculations and kinetic simulation schemes. Indeed, the adsorption enthalpies of gas-phase species can be calculated, the activation barriers to form transition states from surface bound species can be predicted, and the evolution of surface species with time can be simulated. Figure 7.4.2 shows, for example, the transition state for the hydrogenation of adsorbed ethylene (step 3) on a model palladium surface consisting of 7 Pd atoms arranged in a plane. The activation energy associated with ethyl formation on a Pd(111) metal surface was calculated from first principles quantum mechanics to be about 72 kJ mol^{-1} [M. Neurock and R. A. van Santen, *J. Phys. Chem. B,* **104** (2000) 11127]. The configuration of the adsorbed ethyl species, the product of step 3, is illustrated for two different Pd surfaces in Figure 7.4.3. The energies for chemisorption of this reactive intermediate on a Pd$_{19}$ cluster and a semi-infinite Pd(111) slab are –130 and –140 kJ mol^{-1}, respectively, and are essentially the same within the error of the calculation. Results involving adsorbed ethylene, adsorbed hydrogen and adsorbed ethyl indicate

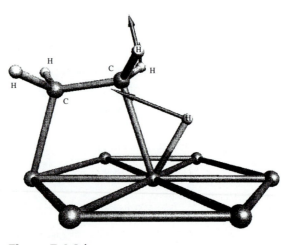

Figure 7.4.2 |
The isolated transition state structure for the formation of ethyl from ethylene and hydrogen on a model Pd$_7$ cluster. The vectors correspond to the motion along the reaction coordinate. [Adapted with permission from M. Neurock and R. A. van Santen, *J. Phys. Chem. B,* **104** (2000) 11127. Copyright 2000 American Chemical Society.]

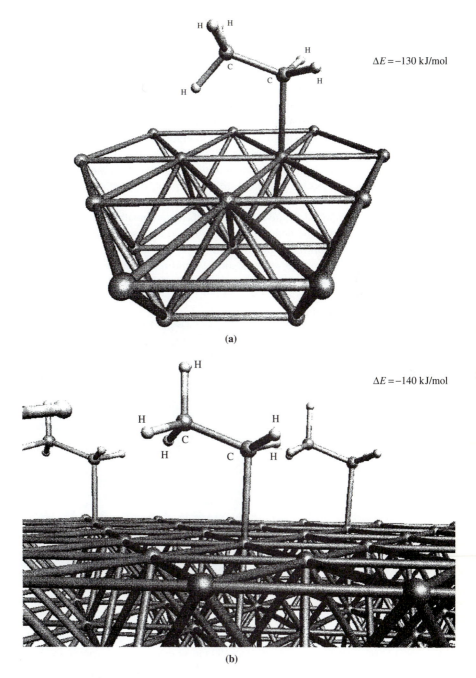

$\Delta E = -130 \, \text{kJ/mol}$

(a)

$\Delta E = -140 \, \text{kJ/mol}$

(b)

Figure 7.4.3 |
Structures and energies for the chemisorption of ethyl on (**a**) a Pd_{19} cluster model and
(**b**) a model Pd(111) surface. [Reproduced with permission from M. Neurock and R. A. van
Santen, *J. Phys. Chem. B,* **104** (2000) 11127. Copyright 1992 American Chemical Society.]

that the overall heat of reaction for step 3 is about 3 kJ mol^{-1} endothermic on a Pd surface. In principle, these types of calculations can be performed on all of the species in the Horiuti-Polanyi mechanism, and the results can be used in a subsequent kinetic simulation to ultimately give a reaction rate. Rates obtained in this fashion are truly from first principles since regression of experimental data is not required to obtain kinetic parameters.

Hansen and Neurock have simulated the hydrogenation of ethylene on a Pd catalyst using a Monte Carlo algorithm to step through the many reactions that occur on a model surface [E. W. Hansen and M. Neurock, *J. Catal.*, **196** (2000) 241]. In essence, the reaction is simulated by initializing a grid of Pd atoms and allowing all possible reactions, like adsorption of reactants, elementary surface reactions, and desorption of products, to take place. The simulation moves forward in time, event by event, updating the surface composition as the reaction proceeds. The details of the method are too complicated to describe here but can be found in E. W. Hansen and M. Neurock, *J. Catal.*, **196** (2000) 241. The input to the simulation involves a complete set of adsorption and reaction energies. An important feature of this simulation is that it also accounts for interactions among the species on the surface.

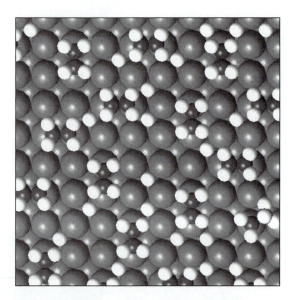

Figure 7.4.4 |
Snapshot of a Pd(100) surface during a simulation of ethylene hydrogenation at 298 K, 25 torr of ethylene and 100 torr of dihydrogen. [Figure from "First-Principles-Based Monte Carlo Simulation of Ethylene Hydrogenation Kinetics on Pd," by E. W. Hansen and M. Neurock, in *Journal of Catalysis,* Volume 196:241, copyright © 2000 by Academic Press, reproduced by permission of the publisher and authors.]

Although many kinetic models assume that the catalyst is an ideal Langmuir surface (all sites have identical thermodynamic properties and there are no interactions among surface species), modern surface science has proven that ideality is often not the case.

Results from the simulation reproduced the experimentally observed kinetics (activation energy, turnover frequency, and orders of reaction) for ethylene hydrogenation on a Pd catalyst. A snapshot of the Pd surface during a simulation of ethylene hydrogenation is given in Figure 7.4.4. At this point in the simulation, highly mobile hydrogen atoms moved rapidly around the surface and reacted with fairly stationary hydrocarbons. In this model, no distinction was made between the competitive and noncompetitive adsorption steps in the mechanism. The results suggest that lateral interactions among species can explain some of the experimental observations not easily accounted for in the Horiuti-Polanyi mechanism.

7.5 | Concluding Remarks

This chapter illustrated the concepts involved in microkinetic analysis of catalytic reactions. The first example involving asymmetric hydrogenation of prochiral olefins with a chiral homogeneous catalyst illustrated how precise rate measurements of critical elementary steps in a catalytic cycle can yield a kinetic model that describes the overall performance of a reaction in terms of enantioselectivity and response to reaction conditions. The second example regarding ammonia synthesis showed how kinetic parameters of elementary steps measured on single crystals under ultrahigh vacuum conditions can be incorporated into a kinetic model that applies to reactors operating under industrial conditions of high pressure. The last example involving ethylene hydrogenation revealed how quantum chemical calculations provide estimates of the energies associated with elementary steps on catalytic surfaces that can be subsequently used in reaction simulations. Finally, microkinetic analysis is clearly moving toward the routine use of quantum chemical calculations as inputs to kinetic models, with the ultimate goal of describing industrial chemical reactors.

Exercises for Chapter 7

1.   In Section 7.2, the reaction paths of the Rh-catalyzed asymmetric hydrogenation of **MAC** were described in detail. The ratio of the **S** product to the **R** product, [**S**]/[**R**], was expressed in the limit of low [Equation (7.2.16)] and high [Equation (7.2.22)] pressure (or concentration) of dihydrogen. Use the rate constants in Table 7.2.1 to plot [**S**]/[**R**] as a function of H_2 concentration. At what concentrations are Equation (7.2.16) and Equation (7.2.22) valid?

2.   Landis and Halpern studied the temperature dependence of the various rate constants in Table 7.2.1 [C. R. Landis and J. Halpern, *J. Am. Chem. Soc.,* **109** (1987) 1746]. The values of the individual activation energies and pre-exponential factors are:

Kinetic parameters for the asymmetric hydrogenation of MAC catalyzed by Rh(R,R-DIPAMP).

Parameter	R cycle		S cycle	
	E^a	$\overline{A}^b$	E^a	$\overline{A}^b$
k_1 (L mmol^{-1} s^{-1})	4.9	2.32×10^4	6.9	1.21×10^6
k_{-1} (s^{-1})	13.3	8.53×10^8	13.0	9.94×10^9
k_2 (L mmol^{-1} s^{-1})	10.7	7.04×10^4	7.5	1.84×10^5

^{a} Activation energy, kcal mol^{-1}.
^{b} Pre-exponential factor in the units of the rate constant.
Source: C. R. Landis and J. Halpern, *J. Am. Chem. Soc.,* **109** (1987) 1746.

Determine the ratio of the **S** product to the **R** product, [**S**]/[**R**], as a function of temperature, over the range of 0 to 37°C. Keep the H_2 concentration at a constant value between the limits of high and low pressure as discussed in Exercise 1.

3. A microkinetic analysis of ammonia synthesis over transition metals is presented in Section 7.3. Use the results of that analysis to explain how adsorbed nitrogen atoms (N*) can be the most abundant reaction intermediate on iron catalysts even though dissociative chemisorption of N_2 is considered the rate-determining step.

4. Describe the main differences in the kinetics of ammonia synthesis over iron catalysts compared to ruthenium catalysts.

5. The Horiuti-Polanyi mechanism for olefin hydrogenation as discussed in Section 7.4 involves 4 steps:

Step 1	$H_2 + 2* \rightleftharpoons 2H*$
Step 2	$C_2H_4 + 2* \rightleftharpoons *C_2H_4*$
Step 3	$*C_2H_4* + H* \rightleftharpoons *C_2H_5 + 2*$
Step 4	$*C_2H_5 + H* \rightleftharpoons C_2H_6 + 2*$
Overall	$C_2H_4 + H_2 = C_2H_6$

Derive a rate expression for the hydrogenation of ethylene on Pt assuming steps 1, 2, and 3 are quasi-equilibrated, step 4 is virtually irreversible, and $*C_2H_5$ is the most abundant reaction intermediate covering almost the entire surface $([*]_0 \sim [*C_2H_5])$. Discuss why the rate expression cannot properly account for the experimentally observed half order dependence in H_2 and zero-order dependence in ethylene. Could the observed reaction orders be explained if adsorbed ethylene $(*C_2H_4*)$ were the most abundant reaction intermediate? Explain your answer.

6. Read the paper entitled "Microkinetics Modeling of the Hydroisomerization of n-Hexane," by A. van de Runstraat, J. van Grondelle, and R. A. van Santen, *Ind. Eng. Chem. Res.,* **36** (1997) 3116. The isomerization of n-hexane is a classic

example of a reaction involving a bifunctional catalyst, where a transition metal component facilitates hydrogenation/dehydrogenation and an acidic component catalyzes structural rearrangement. Section 5.3 illustrates the important reactions involved in hydrocarbon isomerization over bifunctional catalysts. Summarize the key findings of the microkinetic analysis by van de Runstraat et al.

CHAPTER 8

Nonideal Flow in Reactors

8.1 | Introduction

In Chapter 3, steady-state, isothermal ideal reactors were described in the context of their use to acquire kinetic data. In practice, conditions in a reactor can be quite different than the ideal requirements used for defining reaction rates. For example, a real reactor may have nonuniform flow patterns that do not conform to the ideal PFR or CSTR mixing patterns because of corners, baffles, nonuniform catalyst packings, etc. Additionally, few real reactors are operated at isothermal conditions; rather they may be adiabatic or nonisothermal. In this chapter, techniques to handle nonideal mixing patterns are outlined. Although most of the discussion will center around common reactor types found in the petrochemicals industries, the analyses presented can be employed to reacting systems in general (e.g., atmospheric chemistry, metabolic processes in living organisms, and chemical vapor deposition for microelectronics fabrication). The following example illustrates how the flow pattern within the same reaction vessel can influence the reaction behavior.

EXAMPLE 8.1.1 |

In order to approach ideal PFR behavior, the flow must be turbulent. For example, with an open tube, the Reynolds number must be greater than 2100 for turbulence to occur. This flow regime is attainable in many practical situations. However, for laboratory reactors conducting liquid-phase reactions, high flow rates may not be achievable. In this case, laminar flow will occur. Calculate the mean outlet concentration of a species A undergoing a first-order reaction in a tubular reactor with laminar flow and compare the value to that obtained in a PFR when $(kL)/u = 1$ (u = average linear flow velocity).

■ Answer

The material balance on a PFR reactor accomplishing a first-order reaction at constant density is:

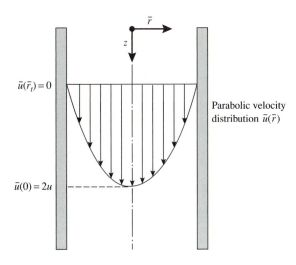

Figure 8.1.1 | Schematic representation of laminar velocity profile in a circular tube.

$$\frac{dF_A}{dV_R} = v\frac{dC_A}{dV_R} = \frac{uA_C}{A_C}\frac{dC_A}{dz} = u\frac{dC_A}{dz} = -kC_A$$

Integration of this equation with $C_A = C_A^0$ at the entrance of the reactor ($z = 0$) gives:

$$C_A = C_A^0 \exp\left[-\frac{kz}{u}\right]$$

For laminar flow:

$$\bar{u}(\bar{r}) = 2u\left[1 - \left(\bar{r}/\bar{r}_t\right)^2\right]$$

where $\bar{r}_t$ is the radius of the tubular reactor (see Figure 8.1.1).

The material balance on a laminar-flow reactor with negligible mass diffusion (discussed later in this chapter) is:

$$\bar{u}(\bar{r})\frac{\partial C_A}{\partial z} = -kC_A$$

Since $\bar{u}(\bar{r})$ is not a function of z, this equation can be solved to give:

$$C_A(\bar{r}) = C_A^0 \exp\left[-\frac{kz}{\bar{u}(\bar{r})}\right]$$

To obtain the mean concentration, $\overline{C}_A$, $C_A(\bar{r})$ must be integrated over the radial dimension as follows:

$$\overline{C}_A = \frac{\int_0^{\overline{r}^t} C_A(\overline{r})\overline{u}(\overline{r})2\pi\overline{r}d\overline{r}}{\int_0^{\overline{r}^t} \overline{u}(\overline{r})2\pi\overline{r}d\overline{r}}$$

Thus, the mean outlet concentration of A, $\overline{C}_A^L$, can be obtained by evaluating $\overline{C}_A$ at $z = L$. For $(kL)/u = 1$ the outlet value of C_A from the PFR, C_A^P, is $0.368 \, C_A^0$ while for the laminar-flow reactor $\overline{C}_A^L = 0.443 \, C_A^0$. Thus, the deviation from PFR behavior can be observed in the outlet conversion of A: 63.2 percent for the PFR versus 55.7 percent for the laminar-flow reactor.

8.2 | Residence Time Distribution (RTD)

In Chapter 3, it was stated that the ideal PFR and CSTR are the theoretical limits of fluid mixing in that they have no mixing and complete mixing, respectively. Although these two flow behaviors can be easily described, flow fields that deviate from these limits are extremely complex and become impractical to completely model. However, it is often not necessary to know the details of the entire flow field but rather only how long fluid elements reside in the reactor (i.e., the distribution of residence times). This information can be used as a diagnostic tool to ascertain flow characteristics of a particular reactor.

The "age" of a fluid element is defined as the time it has resided within the reactor. The concept of a fluid element being a small volume relative to the size of the reactor yet sufficiently large to exhibit continuous properties such as density and concentration was first put forth by Danckwerts in 1953. Consider the following experiment: a tracer (could be a particular chemical or radioactive species) is injected into a reactor, and the outlet stream is monitored as a function of time. The results of these experiments for an ideal PFR and CSTR are illustrated in Figure 8.2.1. If an impulse is injected into a PFR, an impulse will appear in the outlet because there is no fluid mixing. The pulse will appear at a time $t_1 = t_0 + \tau$, where τ is the space time ($\tau = V/v$). However, with the CSTR, the pulse emerges as an exponential decay in tracer concentration, since there is an exponential distribution in residence times [see Equation (3.3.11)]. For all nonideal reactors, the results must lie between these two limiting cases.

In order to analyze the residence time distribution of the fluid in a reactor the following relationships have been developed. Fluid elements may require differing lengths of time to travel through the reactor. The distribution of the exit times, defined as the $E(t)$ curve, is the residence time distribution (RTD) of the fluid. The exit concentration of a tracer species $C(t)$ can be used to define $E(t)$. That is:

$$E(t) = \frac{C(t)}{\int_0^\infty C(\overline{t})d\overline{t}} \tag{8.2.1}$$

such that:

$$\int_0^\infty E(\overline{t})d\overline{t} = 1 \tag{8.2.2}$$

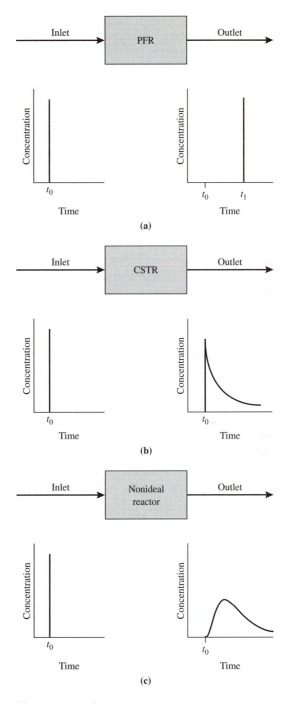

Figure 8.2.1 |
Concentrations of tracer species using an impulse input.
(a) PFR ($t_1 = t_0 + \tau$). **(b)** CSTR. **(c)** Nonideal reactor.

With this definition, the fraction of the exit stream that has residence time (age) between t and $t + dt$ is:

$$E(t)dt \tag{8.2.3}$$

while the fraction of fluid in the exit stream with age less than t_1 is:

$$\int_0^{t_1} E(\bar{t})d\bar{t} \tag{8.2.4}$$

EXAMPLE 8.2.1

Calculate the RTD of a perfectly mixed reactor using an impulse of n moles of a tracer.

■ **Answer**

The impulse can be described by the Dirac delta function, that is:

$$\delta(t - t_0) \begin{cases} = 0, & t \neq t_0 \\ \neq 0, & t = t_0 \end{cases}$$

such that:

$$\int_{-\infty}^{\infty} \delta(t - t_0)dt = 1$$

The unsteady-state mass balance for a CSTR is:

$$V\frac{dC}{dt} = n\delta(t) - vC$$

$$\text{accumulation}\quad\text{input}\qquad\text{output}$$

where t_0 in the Dirac delta is set to zero and:

$$\int_{-\infty}^{\infty} \delta(t)dt = 1$$

Integration of this differential equation with $C(0) = 0$ gives:

$$\text{(a)}\quad \tau\frac{dC}{dt} + C = \left(\frac{n}{v}\right)\delta(t)$$

$$\text{(b)}\quad \int_0^t d[C\exp(\bar{t}/\tau)] = \left(\frac{n}{v}\right)\int_0^t \frac{\delta(\bar{t})}{\tau}\exp(\bar{t}/\tau)d\bar{t}$$

$$\text{(c)}\quad C\exp(\bar{t}/\tau)\Big|_0^t = \left(\frac{n}{v}\right)\int_0^t \frac{\delta(\bar{t})}{\tau}\exp(\bar{t}/\tau)d\bar{t}$$

(d) $\qquad C(t)\exp(t/\tau) - C(0)\exp(0) = \left(\dfrac{n}{v}\right)\displaystyle\int_0^t \dfrac{\delta(\bar{t})}{\tau}\exp(\bar{t}/\tau)d\bar{t}$

(e) $\qquad C(t) = \left(\dfrac{n}{v}\right)\exp(-t/\tau)\displaystyle\int_0^t \dfrac{\delta(\bar{t})}{\tau}\exp(\bar{t}/\tau)d\bar{t}$

(f) $\qquad$ another property of the Dirac delta function is:

$$\int_{-\infty}^{\infty}\delta(t - t_0)f(t)dt = f(t_0)$$

(g) $\qquad C(t) = \left(\dfrac{n}{v}\right)\dfrac{\exp(-t/\tau)}{\tau}\exp(0/\tau)$

(h) $\qquad C(t) = \left(\dfrac{n}{v}\right)\dfrac{\exp(-t/\tau)}{\tau}$

(i) $\qquad E(t) = \dfrac{C(t)}{\left(\dfrac{n}{v}\right)\displaystyle\int_0^{\infty}\dfrac{\exp(-\bar{t}/\tau)}{\tau}d\bar{t}}$

(j) $\qquad E(t) = \dfrac{\dfrac{\exp(-t/\tau)}{\tau}}{-\exp(-t/\tau)\big|_0^{\infty}}$

(k) $\qquad E(t) = \dfrac{\exp(-t/\tau)}{\tau}$

Thus, for a perfectly mixed reactor (or often called completely backmixed), the RTD is an exponential curve.

VIGNETTE 8.2.1

The concept of using a "tracer" species to measure the mixing characteristics is not limited to chemical reactors. In the area of pharmacokinetics, the time course of renal excretion of species originating from intravenous injections in many ways resembles the input of a pulse of tracer into a chemical reactor. Normally, a radioactive labeled (^{2}H, ^{14}C, ^{32}P, etc.) version of a drug is used to follow the pharmacokinetics of the drug in animals and humans. Analyses like those presented in this chapter and in other portions of the text are used to ascertain clearance times and other parameters of importance. Good coverage of these topics can be found in M. Rowland and T. N. Tozer, *Clinical Pharmacokinetics, Concepts and Application,* 3rd ed., Lippincott, Williams and Wilkins, Philadelphia, 1995.

Two types of tracer experiments are commonly employed and they are the input of a pulse or a step function. Figure 8.2.1 illustrates the exit concentration curves and thus the shape of the $E(t)$-curves (same shape as exit concentration curve) for an impulse input. Figure 8.2.2 shows the exit concentration for a step input of tracer. The $E(t)$-curve for this case is related to the time derivative of the exit concentration.

By knowing the $E(t)$-curve, the mean residence time can be obtained and is:

$$\langle t \rangle = \frac{\displaystyle\int_0^\infty \bar{t} E(\bar{t}) d\bar{t}}{\displaystyle\int_0^\infty E(\bar{t}) d\bar{t}} = \int_0^\infty \bar{t} E(\bar{t}) d\bar{t} \tag{8.2.5}$$

EXAMPLE 8.2.2

Calculate the mean residence time for a CSTR.

■ **Answer**

The exit concentration profile from a step decrease in the inlet concentration is provided in Equation (3.3.11) and using this function to calculate the $E(t)$-curve gives:

$$E(t) = \frac{\exp(-t/\tau)}{\tau}$$

Therefore application of Equation (8.2.5) to this $E(t)$-curve yields the following expression:

$$\langle t \rangle = \frac{1}{\tau} \int_0^\infty \bar{t} \exp(-\bar{t}/\tau) d\bar{t}$$

Since:

$$\int_0^\infty x \exp(-x) dx = 1$$

$$\langle t \rangle = \frac{1}{\tau} \int_0^\infty \tau^2 (\bar{t}/\tau) \exp(-\bar{t}/\tau) d(\bar{t}/\tau) = \tau$$

As was shown in Chapter 3, the mean exit time of any reactor is the space time, τ.

The RTD curve can be used as a diagnostic tool for ascertaining features of flow patterns in reactors. These include the possibilities of bypassing and/or regions of stagnant fluid (i.e., dead space). Since these maldistributions can cause unpredictable conversions in reactors, they are usually detrimental to reactor operation. Thus, experiments to determine RTD curves can often point to problems and suggest solutions, for example, adding or deleting baffles and repacking of catalyst particles.

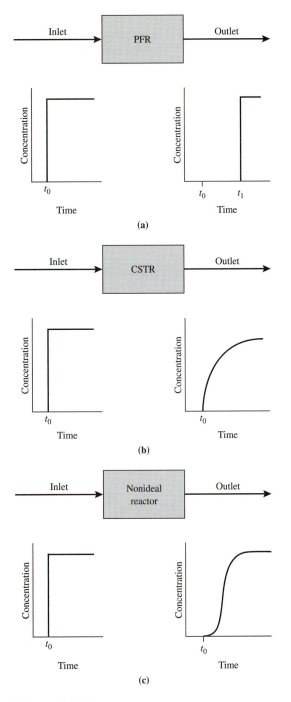

Figure 8.2.2 |
Concentrations of tracer species using a step input.
(a) PFR ($t_1 = t_0 + \tau$). **(b)** CSTR. **(c)** Nonideal reactor.

EXAMPLE 8.2.3

In Section 3.5 recycle reactors and particularly a Berty reactor were described. At high impeller rotation speed, a Berty reactor should behave as a CSTR. Below are plotted the dimensionless exit concentrations, that is, $C(t)/C^0$, of cis-2-butene from a Berty reactor containing alumina catalyst pellets that is operated at 4 atm pressure and 2000 rpm impeller rotation speed at temperatures of 298 K and 427 K. At these temperatures, the cis-2-butene is not isomerized over the catalyst pellets. At $t = 0$, the feed stream containing 2 vol % cis-2-butene in helium is switched to a stream of pure helium at the same total flow rate. Reaction rates for the isomerization of cis-2-butene into 1-butene and trans-2-butene are to be measured at higher temperatures in this reactor configuration. Can the CSTR material balance be used to ascertain the rate data?

■ **Answer**

The exit concentrations from an ideal CSTR that has experienced a step decrease in feed concentration are [from Equation (3.3.11)]:

$$C/C^0 = \exp[-t/\tau]$$

If the RTD is that of an ideal CSTR (i.e., perfect mixing), then the decline in the exit concentration should be in the form of an exponential decay. Therefore, a plot of $\ln(C/C^0)$ versus time should be linear with a slope of $-\tau^{-1}$. Using the data from the declining portions of the concentration profiles shown in Figure 8.2.3, excellent linear fits to the data are obtained (see Figure 8.2.4) at both temperatures indicating that the Berty reactor is behaving as a CSTR at 298 K $\leq T \leq$ 427 K. Since the complete backmixing is achieved over such a large temperature range, it is most likely that the mixing behavior will also occur at slightly higher temperatures where the isomerization reaction will occur over the alumina catalyst.

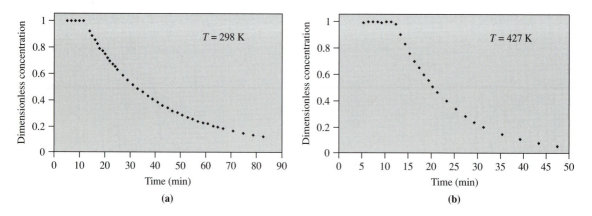

Figure 8.2.3 | Dimensionless concentration (C/C^0) of cis-2-butene in exit stream of Berty reactor as a function of time. See Example 8.2.3 for additional details.

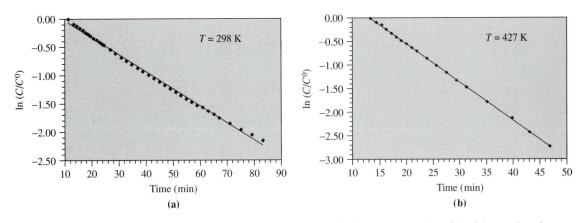

Figure 8.2.4 | Logarithm of the dimensionless concentration of cis-2-butene in exit stream of Berty reactor as a function of time. See Example 8.2.3 for additional details.

8.3 | Application of RTD Functions to the Prediction of Reactor Conversion

The application of the RTD to the prediction of reactor behavior is based on the assumption that each fluid element (assume constant density) behaves as a batch reactor, and that the total reactor conversion is then the average conversion of all the fluid elements. That is to say:

$$\begin{bmatrix} \text{mean concentration} \\ \text{of reactant in} \\ \text{reactor outlet} \end{bmatrix} = \sum \begin{bmatrix} \text{concentration of} \\ \text{reactant remaining in} \\ \text{a fluid element of age} \\ \text{between } t \text{ and } t + dt \end{bmatrix} \begin{bmatrix} \text{fraction of exit stream} \\ \text{that consists of fluid} \\ \text{elements of age} \\ \text{between } t \text{ and } t + dt \end{bmatrix} \qquad (8.3.1)$$

where the summation is over all fluid elements in the reactor exit stream. This equation can be written analytically as:

$$\langle C_A \rangle = \int_0^\infty C_A(\bar{t}) E(\bar{t}) d\bar{t} \qquad (8.3.2)$$

where $C_A(t)$ depends on the residence time of the element and is obtained from:

$$\frac{dC_A}{dt} = -v_A r(C_A) \qquad (8.3.3)$$

with

$$C_A(0) = C_A^0$$

For a first-order reaction:

$$\frac{dC_A}{dt} = -kC_A \tag{8.3.4}$$

or

$$C_A = C_A^0 \exp[-kt] \tag{8.3.5}$$

Insertion of Equation (8.3.5) into Equation (8.3.2) gives:

$$\langle C_A \rangle = \int_0^\infty C_A^0 \exp[-k\bar{t}]E(\bar{t})d\bar{t} \tag{8.3.6}$$

Take for example the ideal CSTR. If the $E(t)$-curve for the ideal CSTR is used in Equation (8.3.6) the result is:

$$\langle C_A \rangle = \frac{C_A^0}{\tau} \int_0^\infty \exp(-k\bar{t}) \exp(-\bar{t}/\tau)d\bar{t}$$

or

$$\frac{\langle C_A \rangle}{C_A^0} = \frac{1}{\tau} \int_0^\infty \exp\left[-\left(k + \frac{1}{\tau}\right)\bar{t}\right]d\bar{t}$$

that gives after integration:

$$\frac{\langle C_A \rangle}{C_A^0} = \frac{1}{\tau}\left[-\left(\frac{1}{k + \frac{1}{\tau}}\right)\exp\left[-\left(k + \frac{1}{\tau}\right)t\right]\Bigg|_0^\infty\right] = \frac{1}{k\tau + 1} \tag{8.3.7}$$

Notice that the result shown in Equation (8.3.7) is precisely that obtained from the material balance for an ideal CSTR accomplishing a first-order reaction. That is:

$$vC_A^0 = vC_A + VkC_A$$

or

$$\frac{C_A}{C_A^0} = \frac{1}{k\tau + 1} \tag{8.3.8}$$

Unfortunately, if the reaction rate is not first-order, the RTD cannot be used so directly to obtain the conversion. To illustrate why this is so, consider the two reactor schemes shown in Figure 8.3.1.

Froment and Bischoff analyze this problem as follows (G. F. Froment & K. B. Bischoff, *Chemical Reactor Analysis and Design,* Wiley, 1979). Let the PFR and CSTR have space times of τ_1 and τ_2, respectively. The overall RTD for either system will be that of the CSTR but with a delay caused by the PFR. Thus, a tracer experiment cannot distinguish configuration (I) from (II) in Figure 8.3.1.

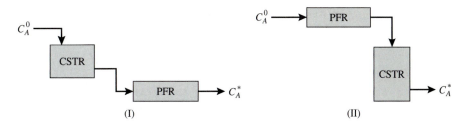

Figure 8.3.1 |
PFR and CSTR in series. (I) PFR follows the CSTR, (II) CSTR follows the PFR.

A first-order reaction occurring in either reactor configuration will give for the two-reactor network:

$$\frac{C_A^*}{C_A^0} = \frac{e^{-k\tau_1}}{1 + k\tau_2} \qquad (8.3.9)$$

This is easy to see; for configuration (I):

$$C_A = \frac{C_A^0}{1 + k\tau_2} \qquad \text{(CSTR)}$$

$$\frac{C_A^*}{C_A} = \exp[-k\tau_1] \qquad \text{(PFR)}$$

or

$$\frac{C_A^*}{C_A^0} = \frac{e^{-k\tau_1}}{1 + k\tau_2}$$

and for configuration (II):

$$\frac{C_A}{C_A^0} = \exp[-k\tau_1] \qquad \text{(PFR)}$$

$$\frac{C_A^*}{C_A} = \frac{1}{1 + k\tau_2} \qquad \text{(CSTR)}$$

or

$$\frac{C_A^*}{C_A^0} = \frac{e^{-k\tau_1}}{1 + k\tau_2}$$

Now with second-order reaction rates, configuration (I) gives:

$$\frac{C_A^*}{C_A^0} = \frac{-1 + \sqrt{\left(\dfrac{1 + 4(kC_A^0\tau_2)}{1 + kC_A^0\tau_1}\right)}}{2kC_A^0\tau_2} \qquad (8.3.10)$$

while configuration (II) yields:

$$\frac{C_A^*}{C_A^0} = \frac{-1 + \sqrt{1 + 4kC_A^0\tau_2}}{2kC_A^0\tau_2 + kC_A^0\tau_1\left(-1 + \sqrt{1 + 4kC_A^0\tau_2}\right)} \tag{8.3.11}$$

If $kC_A^0 = 1$ and $\tau_2/\tau_1 = 4$, configurations (I) and (II) give outlet dimensionless concentrations (C_A^*/C_A^0) of 0.25 and 0.28, respectively. Thus, while first-order kinetics (linear) yield the same outlet concentrations from reactor configurations (I) and (II), the second-order kinetics (nonlinear) do not. The reasons for these differences are as follows. First-order processes depend on the length of time the molecules reside in the reactors but not on exactly where they are located during their trajectory through the reactors. Nonlinear processes depend on the encounter of more than one set of molecules (fluid elements), so they depend both on residence time and also what they experience at each time. The RTD measures *only* the time that fluid elements reside in the reactor but provides *no information* on the details of the mixing. The terms *macromixing* and *micromixing* are used for the RTD and mixing details, respectively. For a given state of perfect macromixing, two extremes in micromixing can occur: complete segregation and perfect micromixing. These types of mixing schemes can be used to further refine the reactor analysis. These methods will not be described here because they lack the generality of the procedure discussed in the next section.

In addition to the problems of using the RTD to predict reactor conversions, the analysis provided above is only strictly applicable to isothermal, single-phase systems. Extensions to more complicated behaviors are not straightforward. Therefore, other techniques are required for more general predictive and design purposes, and some of these are discussed in the following section.

8.4 | Dispersion Models for Nonideal Reactors

There are numerous models that have been formulated to describe nonideal flow in vessels. Here, the axial dispersion or axially-dispersed plug flow model is described, since it is widely used. Consider the situation illustrated in Figure 8.4.1. (The steady-state PFR is described in Chapter 3 and the RTD for a PFR discussed in Section 8.2.)

The transient material balance for flow in a PFR where no reaction is occurring can be written as:

$$\underset{\text{(accumulation)}}{A_C \partial z \frac{\partial C_i}{\partial t}} = \underset{\text{(in)}}{uA_C C_i} - \underset{\text{(out)}}{\left[uA_C C_i + \partial(uA_C C_i)\right]} \tag{8.4.1}$$

or

$$\frac{\partial C_i}{\partial t} = -\frac{\partial}{\partial z}(uC_i) \tag{8.4.2}$$

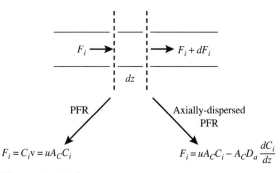

Figure 8.4.1 |
Descriptions for the molar flow rate of species i in a
PFR and an axially-dispersed PFR. A_C: cross-sectional
diameter of tube, u: linear velocity, D_a: axial dispersion
coefficient.

where A_C is the cross-sectional area of the tube. If u = constant, then:

$$\frac{\partial C_i}{\partial t} = -u \frac{\partial C_i}{\partial z} \tag{8.4.3}$$

Now if diffusion/dispersion processes that mix fluid elements are superimposed on
the convective flow in the axial direction (z direction), then the total flow rate can
be written as:

$$F_i = uA_C C_i - A_C D_a \frac{dC_i}{dz} \tag{8.4.4}$$

$$\text{(convection)} \quad \text{(dispersion)}$$

Note that D_a is called the *axial-dispersion coefficient*, and that the dispersion term
of the molar flow rate is formulated by analogy to molecular diffusion. Fick's First
Law states that the flux of species A (moles/area/time) can be formulated as:

$$N_A = -D_{AB} \frac{dC_A}{dz} + uC_A \tag{8.4.5}$$

for a binary mixture, where D_{AB} is the molecular diffusion coefficient. Since axial
dispersion processes will occur by molecular diffusion during laminar flow, at this
condition the dispersion coefficient will be the molecular diffusion coefficient. However, with turbulent flow, the processes are different and D_a must be obtained from
correlations. Since D_a is the molecular diffusion coefficient during laminar flow, it
is appropriate to write the form of the dispersion relationship as in Equation (8.4.4)
and then obtain D_a from correlations assuming this form of the molar flow rate expression. Using Equation (8.4.4) to develop the transient material balance relationship for the axially-dispersed PFR gives:

$$A_C \partial z \frac{\partial C_i}{\partial t} = \left(uA_C C_i - A_C D_a \frac{\partial C_i}{\partial z} \right) - \left[\left(uA_C C_i - A_C D_a \frac{\partial C_i}{\partial z} \right) \right.$$

(accumulation)   (in)   (out)

$$\left. + \partial \left(uA_C C_i - A_C D_a \frac{\partial C_i}{\partial z} \right) \right] \tag{8.4.6}$$

(out)

or for constant u and D_a:

$$\frac{\partial C_i}{\partial t} + u \frac{\partial C_i}{\partial z} = D_a \frac{\partial^2 C_i}{\partial z^2} \tag{8.4.7}$$

If $\theta = t/\langle t \rangle = (tu)/L$ (L: length of the reactor), $Z = z/L$ and $Pe_a = (Lu)/D_a$ (axial Peclet number), then Equation (8.4.7) can be written as:

$$\frac{\partial C_i}{\partial \theta} + \frac{\partial C_i}{\partial Z} = \frac{1}{Pe_a} \frac{\partial^2 C_i}{\partial Z^2} \tag{8.4.8}$$

The solution of Equation (8.4.8) when the input (i.e., C_i at $t = 0$) is an impulse is:

$$C_i = \left(\frac{Pe_a}{4\pi\theta} \right)^{\frac{1}{2}} \exp\left[\frac{-(1-\theta)^2 Pe_a}{4\theta} \right] \tag{8.4.9}$$

Thus, for the axially-dispersed PFR the RTD is:

$$E(\theta) = \left(\frac{Pe_a}{4\pi\theta} \right)^{\frac{1}{2}} \exp\left[\frac{-(1-\theta)^2 Pe_a}{4\theta} \right] \tag{8.4.10}$$

A plot of $E(\theta)$ versus θ is shown in Figure 8.4.2 for various amounts of dispersion. Notice that as $Pe_a \to \infty$ (no dispersion), the behavior is that of a PFR while as $Pe_a \to 0$ (maximum dispersion), it is that of a CSTR. Thus, the axially-dispersed reactor can simulate all types of behaviors between the ideal limits of no back-mixing (PFR) and complete backmixing (CSTR).

The dimensionless group Pe_a is a ratio of convective to dispersive flow:

$$Pe_a = \frac{Lu}{D_a} = \frac{\text{convective flow}}{\text{dispersive flow}} \text{ in the axial direction} \tag{8.4.11}$$

The Peclet number is normally obtained via correlations, and Figure 8.4.3 illustrates data from Wilhelm that are plotted as a function of the Reynolds number for packed beds (i.e., tubes packed with catalyst particles). Notice that both the Pe_a and Re numbers use the particle diameter, d_p, as the characteristic length:

$$Pe_a = \frac{d_p u}{D_a}, \qquad Re = \frac{d_p u \rho}{\overline{\mu}}$$

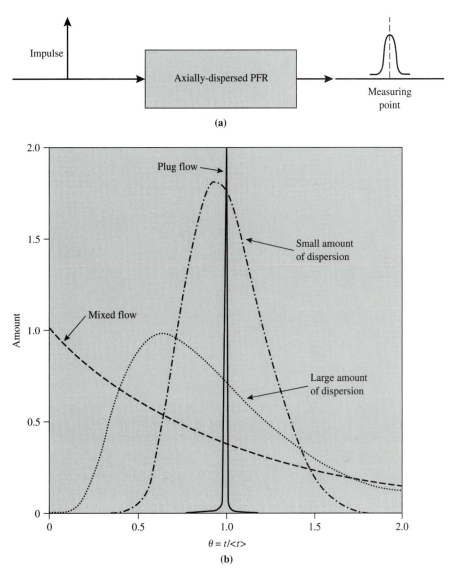

(a)

(b)

Figure 8.4.2 |
(a) Configuration illustrating pulse input to an axially-dispersed PFR. (b) Results observed at measuring point.

It is always prudent to check the variables used in each dimensionless group prior to their application. This is especially true with Peclet numbers, since they can have many different characteristic lengths.

Notice that in packed beds, $Pe_a = 2$ for gases with turbulent flow $Re = (d_p u \rho)/\overline{\mu} > 40$, while for liquids Pe_a is below 1. Additionally, for unpacked tubes,

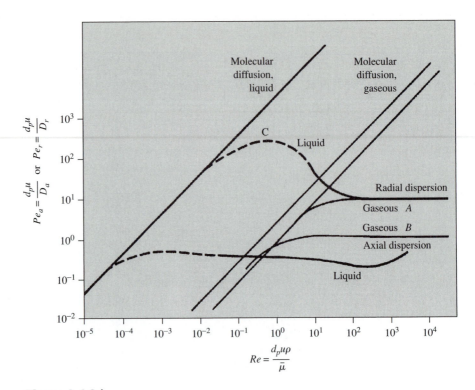

Figure 8.4.3 |
Axial and radial Peclet numbers as a function of Reynolds number for packed-beds.
[Adapted from R. H. Wilhelm, *Pure App. Chem.*, **5** (1962) 403, with permission of the
International Union of Pure and Applied Chemistry.]

$Pe_a(d_t u/D_a)$ is about 10 with turbulent flow ($Re = (d_t u\rho)/\overline{\mu}$ greater than 2100)
(not shown). Thus, all real reactors will have some effects of dispersion. The question is, how much? Consider again Equation (8.4.7) but now define $Pe_a = d_e u/D_a$
where d_e is an effective diameter and could be either d_p for a packed bed or d_t for
an open tube. Equation (8.4.7) can be then written as:

$$\frac{\partial C_i}{\partial \theta} + \frac{\partial C_i}{\partial Z} = \frac{1}{Pe_a}\left(\frac{d_e}{L}\right)\frac{\partial^2 C_i}{\partial Z^2} \qquad (8.4.12)$$

If the flow rate is sufficiently high to create turbulent flow, then Pe_a is a constant and
the magnitude of the right-hand side of the equation is determined by the aspect ratio, L/d_e. By solving Equation, (8.4.12) and comparing the results to the solutions
of the PFR [Equation (8.4.3)], it can be shown that for open tubes, $L/d_t > 20$ is sufficient to produce PFR behavior. Likewise, for packed beds, $L/d_p > 50$ (isothermal)
and $L/d_p > 150$ (nonisothermal) are typically sufficient to provide PFR characteristics. Thus, the effects of axial dispersion are minimized by turbulent flow in long
reactors.

B. G. Anderson et al. [*Ind. Eng. Chem. Res.*, **37** (1998) 815] obtained *in situ* images of pulses of ^{11}C-labeled alkanes that were passing through packed beds of zeolites by using positron emission tomography (PET). PET is a technique developed primarily for nuclear medicine that is able to create three-dimensional images of gamma-ray emitting species within various organs of the human body. By using PET, Anderson et al. could obtain complete concentration profiles of ^{11}C-labeled alkanes as a function of time in a packed-bed reactor upon introduction of a pulse of the tracer alkane. Using analyses similar to those illustrated in this chapter, the following data were obtained:

Zeolite	T (°C)	Axial dispersion coeff. (m²/s)
H-Mordenite	230	1.1×10^{-4}
H-Beta	230	2.1×10^{-4}
H-ZSM-22	170	1.2×10^{-4}
H-Ferrierite	170	1.0×10^{-4}

The measured axial dispersion coefficients are around 1×10^{-4} m²/s. The reactor was 4 mm in diameter and the volumetric flow rate was 150 mL/min. Assuming a void volume of 0.5 and $Pe = 2$, D_a is calculated to be approximately 1×10^{-4} m²/s. Thus, the value calculated with the information presented in Figure 8.4.3 is in good agreement with the experimental findings.

8.5 | Prediction of Conversion with an Axially-Dispersed PFR

Consider (so that an analytical solution can be obtained) an isothermal, axially-dispersed PFR accomplishing a first-order reaction. The material balance for this reactor can be written as:

$$D_a \frac{d^2C_A}{dz^2} - u \frac{dC_A}{dz} - kC_A = 0 \qquad (8.5.1)$$

If $y = C_A/C_A^0$, $Z = z/L$, and $Pe_a = uL/D_a$, then Equation (8.5.1) can be put into dimensionless form as:

$$\frac{1}{Pe_a} \frac{d^2y}{dZ^2} - \frac{dy}{dZ} - \left(\frac{kL}{u}\right)y = 0 \qquad (8.5.2)$$

The proper boundary conditions used to solve Equation (8.5.2) have been exhaustively discussed in the literature. Consider the reactor schematically illustrated in Figure 8.5.1. The conditions for the so-called "open" configuration are:

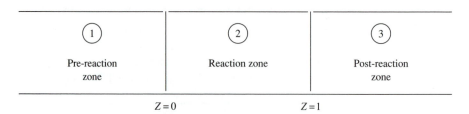

Figure 8.5.1 |
Schematic of hypothetical reactor.

$$\left. \begin{array}{ll} Z = -\infty, & y = 1 \\ Z = +\infty, & y = \text{is finite} \\ \\ Z = 0, & y(0_-) = y(0_+) = y(0) \\ Z = 1, & y(1_-) = y(1_+) \end{array} \right\} \tag{8.5.3}$$

Note that the use of these conditions specifies that the flux continuity:

$$\left[uC_A - D_{a_1} \frac{dC_A}{dz} \right]\bigg|_{0_-} = \left[uC_A - D_{a_2} \frac{dC_A}{dz} \right]\bigg|_{0_+}$$

gives:

$$D_{a_1} \frac{dC_A}{dz} = D_{a_2} \frac{dC_A}{dz}$$

That is to say that if the dispersion coefficients in zones 1 and 2 are not the same, then there will be a discontinuity in the concentration gradient at $z = 0$. Alternatively, Danckwerts formulated conditions for the so-called "closed" configuration that do not allow for dispersion in zones 1 and 3 and they are:

$$uC_A|_{0_-} = \left[uC_A - D_a \frac{dC_A}{dz} \right]\bigg|_{0_+}$$

$$\frac{dC_A}{dz}\bigg|_{L_-} = 0 \tag{8.5.4}$$

The Danckwerts boundary conditions are used most often and force discontinuities in both concentration and its gradient at $z = 0$.

EXAMPLE 8.5.1 |

Consider an axially-dispersed PFR accomplishing a first-order reaction. Compute the dimensionless concentration profiles for $L/d_p = 5$ and 50 and show that at isothermal conditions the values for $L/d_p = 50$ are nearly those from a PFR. Assume $Pe = d_p u/D_a = 2$, $d_p = 0.004$ m and $k/u = 25$ m^{-1}.

■ Answer

The material balance for the axially-dispersed PFR is:

$$\frac{d^2y}{dZ^2} - Pe_a\,(L/d_p)\,\frac{dy}{dZ} - \left(\frac{L^2kPe_a}{ud_p}\right)y = 0$$

or

$$\frac{d^2y}{dZ^2} - (500L)\,\frac{dy}{dZ} - (12500L^2)y = 0$$

The solution to this equation using the Danckwerts boundary conditions of:

$$1 = y - \frac{1}{Pe_a}\left(\frac{d_p}{L}\right)\frac{dy}{dZ} \quad \text{at } Z = 0$$

$$\frac{dy}{dZ} = 0 \qquad\qquad \text{at } Z = 1$$

gives the desired form of y as a function of Z and the result is:

$$y = \overline{\alpha}_1 \exp(524LZ) + \overline{\alpha}_2 \exp(-23.9LZ)$$

where $\overline{\alpha}_1$ and $\overline{\alpha}_2$ vary with L/d_p. The material balance equation for the PFR is:

$$-u\frac{dC_A}{dz} = kC_A$$

or

$$\frac{dy}{dZ} = -\left(\frac{Lk}{u}\right)y$$

with

$$y = 1 \quad \text{at } Z = 0$$

The solution to the PFR material balance gives:

$$y_{PF} = \exp\left[-(kLZ)/u\right] = \exp\left[-25LZ\right]$$

Note that the second-term of y is nearly (but not exactly) that of the expression for y_{PF} and that the first-term of y is a strong function of L. Therefore, it is clear that L will significantly affect the solution y to a much greater extent than y_{PF} and that $y \neq y_{PF}$ even for very long L. However, as shown in Figure 8.5.2, at $L/dp = 50$, $y \approx y_{PF}$ for all practical matters. Notice that $y \neq 1$ at $Z = 0$ because of the dispersion process. There is a forward movement of species A because of the concentration gradient within the reaction zone. The dispersion always produces a lower conversion at the reactor outlet than that obtained with no mixing (PFR)—recall conversion comparisons between PFR and CSTR.

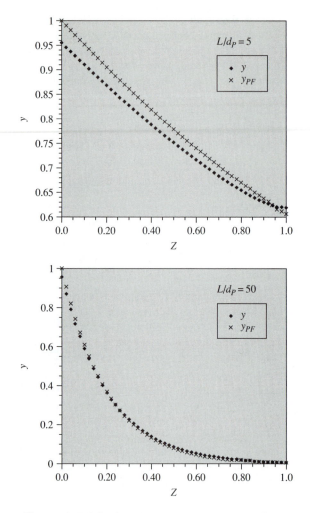

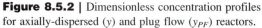

Figure 8.5.2 | Dimensionless concentration profiles
for axially-dispersed (y) and plug flow (y_{PF}) reactors.

VIGNETTE 8.5.1

Y. Park et al. [*Biotech. Bioeng.*, **26** (1984) 457] analyzed a fixed-film bioreactor for the
continuous production of penicillin. The bioreactor (Figure 8.5.3, left) was modeled in
the two extremes of contacting patterns (Figure 8.5.3, right). Notice from the results
shown in Figure 8.5.4 that the productivity of penicillin at any substrate feed concentration, So, is higher for the CSTR configuration. Thus, it would be important to know
the exact mixing pattern within the bioreactor and to modify it to resemble more closely
a CSTR reactor.

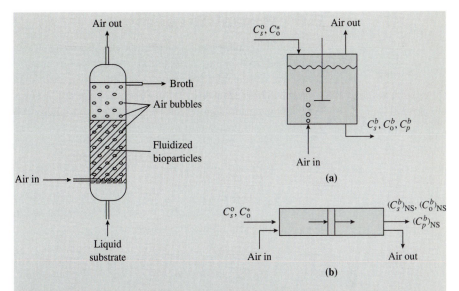

Figure 8.5.3 | Schematic of bioreactor and two extremes used for modeling the contacting pattern. **(a)** Complete backmix. **(b)** Plug flow. [Reproduced from "Analysis of a Continuous, Aerobic Fixed-Film Bioreactor. I. Steady-State Behavior," by Y. Park, M. E. Davis, and D. A. Wallis, *Biotech. Bioeng.*, **26** (1984) 457, copyright © 1984 Wiley-Liss, Inc., a subsidiary of John Wiley and Sons, Inc.]

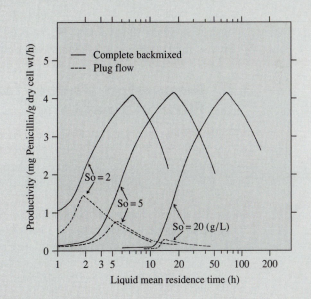

Figure 8.5.4 | Comparison of performance of completely backmixed and plug flow contacting patterns. [Reproduced from "Analysis of a Continuous, Aerobic Fixed-Film Bioreactor. I. Steady-State Behavior," by P. Park, M. E. Davis, and D. A. Wallis, *Biotech. Bioeng.*, **26** (1984) 457, copyright © 1984, Wiley-Liss, Inc., a subsidiary of John Wiley and Sons, Inc.]

8.6 | Radial Dispersion

Like axial dispersion, radial dispersion can also occur. Radial-dispersion effects normally arise from radial thermal gradients that can dramatically alter the reaction rate across the diameter of the reactor. Radial dispersion can be described in an analogous manner to axial dispersion. That is, there is a radial dispersion coefficient. A complete material balance for a transient tubular reactor could look like:

$$\frac{\partial C}{\partial t} + u\frac{\partial C}{\partial z} = D_a \frac{\partial^2 C}{\partial z^2} + D_r \left[\frac{\partial^2 C}{\partial \bar{r}^2} + \frac{1}{r}\frac{\partial C}{\partial \bar{r}} \right] \tag{8.6.1}$$

If $\theta = t/\langle t \rangle = (tu)/L$, $Z = z/L$, $\bar{R} = \bar{r}/d_e$ (d_e is d_p for packed beds, d_t for unpacked tubes), $Pe_a = (ud_e)/D_a$ and $Pe_r = (ud_e)/D_r$ (D_r is the *radial-dispersion coefficient*), then Equation (8.6.1) can be written as:

$$\frac{\partial C}{\partial \theta} + \frac{\partial C}{\partial Z} = \frac{1}{Pe_a}\left(\frac{L}{d_e}\right)^{-1}\frac{\partial^2 C}{\partial Z^2} + \frac{1}{Pe_r}\left(\frac{L}{d_e}\right)\left[\frac{\partial^2 C}{\partial \bar{R}^2} + \frac{1}{\bar{R}}\frac{\partial C}{\partial \bar{R}}\right] \tag{8.6.2}$$

The dimensionless group Pe_r is a ratio of convective to dispersive flow in the radial direction:

$$Pe_r = \frac{d_e u}{D_r} = \frac{\text{convective flow}}{\text{dispersive flow}} \text{ in radial direction} \tag{8.6.3}$$

Referring to Figure 8.4.3, for packed beds with turbulent flow, $Pe_r = 10$ if $d_e = d_p$. For unpacked tubes, $d_e = d_t$ and $Pe_r \approx 1000$ with turbulent flow (not shown). Solution of Equation (8.6.1) is beyond the level of this text.

8.7 | Dispersion Models for Nonideal Flow in Reactors

As illustrated above, dispersion models can be used to described reactor behavior over the entire range of mixing from PFR to CSTR. Additionally, the models are *not* confined to single-phase, isothermal conditions or first-order, reaction-rate functions. Thus, these models are very general and, as expected, have found widespread use. What must be kept in mind is that as far as reactor performance is normally concerned, radial dispersion is to be maximized while axial dispersion is minimized.

The analysis presented in this chapter can be used to describe reaction containers of any type—they need not be tubular reactors. For example, consider the situation where blood is flowing in a vessel and antibodies are binding to cells on the vessel wall. The situation can be described by the following material balance:

$$u\frac{\partial C}{\partial z} = D_a \frac{\partial^2 C}{\partial z^2} + D_r \left(\frac{\partial^2 C}{\partial \bar{r}^2} + \frac{1}{\bar{r}}\frac{\partial C}{\partial \bar{r}}\right) \tag{8.7.1}$$

with

$$uC_0 = uC - D_a \frac{\partial C}{\partial z} \qquad \text{at } z = 0, \text{ all } \bar{r}$$

$$\frac{\partial C}{\partial z} = 0 \qquad \text{at } z = L, \text{ all } \bar{r}$$

$$\frac{\partial C}{\partial \bar{r}} = 0 \qquad \text{at } \bar{r} = 0, \text{ all } z$$

$$-D_r \frac{\partial C}{\partial \bar{r}} = \overline{AB} \qquad \text{at } \bar{r} = \bar{r}_t, \text{ all } z$$

where C is the concentration of antibody, $\bar{r}_t$ is the radius of the blood vessel and $\overline{AB}$ is the rate of antibody binding to the blood vessel cells. Thus, the use of the dispersion model approach to describing flowing reaction systems is quite robust.

Exercises for Chapter 8

1. Find the residence time distribution, that is, the effluent concentration of tracer A after an impulse input at $t = 0$, for the following system of equivolume CSTRs with a volumetric flow rate of liquid into the system equal to v:

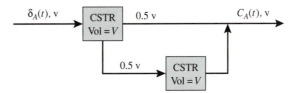

How does the RTD compare to that of a single CSTR with volume $2V$?

2. Sketch the RTD curves for the sequence of plug flow and continuous stirred tank reactors given in Figure 8.3.1.

3. Consider three identical CSTRs connected in series according to the diagram below.

(a) Find the RTD for the system and plot the E curve as a function of time.

(b) How does the RTD compare to the result from an axially-dispersed PFR (Figure 8.4.2)? Discuss you answer in terms of the axial Peclet number.

(c) Use the RTD to calculate the exit concentration of a reactant undergoing first-order reaction in the series of reactors. Confirm that the RTD method gives the same result as a material balance on the system (see Example 3.4.3).

4. Calculate the mean concentration of A at the outlet ($z = L$) of a laminar flow, tubular reactor ($\overline{C}_A^L$) accomplishing a second-order reaction (kC_A^2), and compare the result to that obtained from a PFR when $[(C_A^0 kL)/u] = 1$. Referring to Example 8.1.1, is the deviation from PFR behavior a strong function of the reaction rate expression (i.e., compare results from first- and second-order rates)?

5. Referring to Example 8.2.3, compute and plot the dimensionless exit concentration from the Berty reactor as a function of time for decreasing internal recycle ratio to the limit of PFR behavior.

6. Consider the axially-dispersed PFR described in Example 8.5.1. How do the concentration profiles change from those illustrated in the Example if the second boundary condition is changed from:

$$\frac{dy}{dZ} = 0 \qquad \text{at} \quad Z = 1$$

to

$$y = 0 \qquad \text{at} \quad Z = \infty$$

7. Write down in dimensionless form the material balance equation for a laminar flow tubular reactor accomplishing a first-order reaction and having both axial and radial diffusion. State the necessary conditions for solution.

8. Falch and Gaden studied the flow characteristics of a continuous, multistage fermentor by injecting an impulse of dye to the reactor [E. A. Falch and E. L. Gaden, Jr., *Biotech. Bioengr.*, **12** (1970) 465]. Given the following RTD data from the four-stage fermentor, calculate the Pe_a that best describes the data.

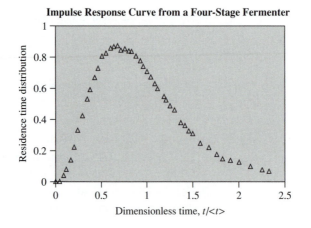

Impulse Response Curve from a Four-Stage Fermenter

$t/<t>$	RTD	$t/<t>$	RTD
0.000	0.000	0.950	0.745
0.050	0.001	0.990	0.710
0.090	0.040	1.035	0.675
0.120	0.080	1.080	0.630
0.170	0.140	1.110	0.600
0.210	0.220	1.180	0.550
0.245	0.330	1.200	0.525
0.295	0.420	1.240	0.485
0.340	0.530	1.290	0.460
0.370	0.590	1.365	0.380
0.420	0.670	1.410	0.360
0.460	0.730	1.450	0.325
0.500	0.810	1.490	0.310
0.540	0.830	1.580	0.250
0.590	0.860	1.670	0.220
0.630	0.870	1.760	0.180
0.670	0.875	1.820	0.150
0.710	0.850	1.910	0.140
0.750	0.855	2.000	0.125
0.790	0.845	2.120	0.100
0.825	0.840	2.250	0.080
0.870	0.810	2.320	0.070
0.920	0.780		

9. Using the value of the Pe_a determined in Exercise 8, compute the concentration profile in the reactor for the reaction of catechol to L-dopa catalyzed by whole cells according to the following rate of reaction:

$$r = -\frac{dC_S}{dt} = \frac{r_{max}C_S}{K_m + C_S}$$

where C_S is the concentration of substrate catechol. This reaction is discussed in Example 4.2.4. The values of the various parameters are the same as those determined by the nonlinear regression analysis in Example 4.2.5, that is, $C_S^0 = 0.027$ mol L^{-1}, $r_{max} = 0.0168$ mol L^{-1} h^{-1}, and $K_m = 0.00851$ mol L^{-1}. Assume that the mean residence time of the reactor is 2 h.

10. Using the same rate expression and parameter values given in Exercise 9, compute the concentration profile assuming PFR behavior and compare to the results in Exercise 9.

CHAPTER 9

Nonisothermal Reactors

9.1 | The Nature of the Problem

In Chapter 3, the isothermal material balances for various ideal reactors were derived (see Table 3.5.1 for a summary). Although isothermal conditions are most useful for the measurement of kinetic data, real reactor operation is normally nonisothermal. Within the limits of heat exchange, the reactor can operate isothermally (maximum heat exchange) or adiabatically (no heat exchange); recall the limits of reactor behavior given in Table 3.1.1. Between these bounds of heat transfer lies the most common form of reactor operation—the nonisothermal regime (some extent of heat exchange). The three types of reactor operations yield different temperature profiles within the reactor and are illustrated in Figure 9.1.1 for an exothermic reaction.

If a reactor is operated at nonisothermal or adiabatic conditions then the material balance equation must be written with the temperature, T, as a variable. For example with the PFR, the material balance becomes:

$$\frac{dF_i}{dV_R} = v_i \mathrm{r}(F_i, T) \tag{9.1.1}$$

Since the reaction rate expression now contains the independent variable T, the material balance cannot be solved alone. The solution of the material balance equation is only possible by the simultaneous solution of the energy balance. Thus, for nonisothermal reactor descriptions, an energy balance must accompany the material balance.

9.2 | Energy Balances

Consider a generalized flow reactor as illustrated in Figure 9.2.1. Applying the first law of thermodynamics to the reactor shown in Figure 9.2.1, the following is obtained:

$$Q - (\overline{W}_s + P_{\text{out}} V_{\text{out}} - P_{\text{in}} V_{\text{in}}) = \overline{U}_{\text{out}} + \overline{m}_{\text{out}}(\overline{z}_{\text{out}}\,(g/g_c) + \tfrac{1}{2} u_{\text{out}}^2/g_c) \\ - \overline{U}_{\text{in}} - \overline{m}_{\text{in}}\,(\overline{z}_{\text{in}}\,(g/g_c) + \tfrac{1}{2} u_{\text{in}}^2/g_c) \tag{9.2.1}$$

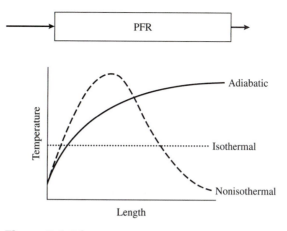

Figure 9.1.1 |
Temperature profiles in a PFR accomplishing an
exothermic reaction.

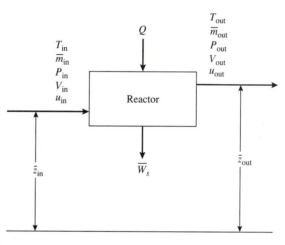

Figure 9.2.1 |
Schematic of flow reactor and energy terms.

where $\overline{W}_s$ is "shaft work," that is, pump or turbine work, $\overline{U}_i$ is the internal energy of
stream i, $\overline{m}_i$ is the mass of stream i, P_i is the pressure of stream i, V_i is the volume
of mass i, u_i is the velocity of stream i, $\overline{z}_i$ is the height of stream i above a datum
plane, and g_c is the gravitational constant, for i denoting either the outlet or inlet
stream. For most normal circumstances:

$$\overline{W}_s = 0 \qquad (9.2.2)$$

and

$$\overline{m}_{out}\left(\overline{z}_{out}(g/g_c) + \tfrac{1}{2}\,u_{out}^2/g_c\right) \cong \overline{m}_{in}\left(\overline{z}_{in}(g/g_c) + \tfrac{1}{2}\,u_{in}^2/g_c\right)$$

Using these assumptions to simplify Equation (9.2.1) yields:

$$Q = (\overline{U}_{out} + P_{out}\,V_{out}) - (\overline{U}_{in} + P_{in}\,V_{in}) \qquad (9.2.3)$$

Recall that the enthalpy, H, is:

$$H = \overline{U} + PV \qquad (9.2.4)$$

Thus, Equation (9.2.3) can be written as:

$$Q = H_{out} - H_{in} \qquad (9.2.5)$$

However, since it is more typical to deal with rates of energy transfer, $\dot{Q}$, rather than energy, Q, when dealing with reactors, Equation (9.2.5) can be differentiated with respect to time to give:

$$\dot{Q} = h_{out}\,\dot{m}_{out} - h_{in}\,\dot{m}_{in} \qquad (9.2.6)$$

where $\dot{Q}$ is the rate of heat transfer, h_i is the enthalpy per unit mass of stream i, and $\dot{m}$ is the mass flow rate of stream i. Generalizing Equation (9.2.6) to multi-input, multi-output reactors yields a generalized energy balance subject to the assumptions stated above:

$$\dot{Q} = \sum_{\substack{\text{outlet}\\\text{streams}}} h_{out}\,\dot{m}_{out} - \sum_{\substack{\text{inlet}\\\text{streams}}} h_{in}\,\dot{m}_{in} \qquad \left(\begin{smallmatrix}\text{conservation}\\\text{of energy}\end{smallmatrix}\right) \qquad (9.2.7)$$

$$0 = \sum_{\substack{\text{outlet}\\\text{streams}}} \dot{m}_{out} - \sum_{\substack{\text{inlet}\\\text{streams}}} \dot{m}_{in} \qquad \left(\begin{smallmatrix}\text{conservation}\\\text{of mass}\end{smallmatrix}\right) \qquad (9.2.8)$$

Note that the enthalpy includes sensible heat and heat of reaction effects as will be illustrated below.

For a closed reactor (e.g., a batch reactor), the potential and kinetic energy terms in Equation (9.2.1) are not relevant. Additionally, $\overline{W}_s \cong 0$ for most cases (including the work input from the stirring impellers). Since most reactions carried out in closed reactors involve condensed phases, $\Delta(PV)$ is small relative to $\Delta\overline{U}$, and for this case is:

$$Q = \Delta\overline{U} = \overline{U}_{products} - \overline{U}_{reactants} \qquad (9.2.9)$$

or from Equations (9.2.9) and (9.2.4):

$$Q = \Delta H, \; PV = \text{constant} \qquad \text{(conservation of energy)} \qquad (9.2.10)$$

9.3 | Nonisothermal Batch Reactor

Consider the batch reactor schematically illustrated in Figure 9.3.1. Typically, reactants are charged into the reactor from point (I), the temperature of the reactor

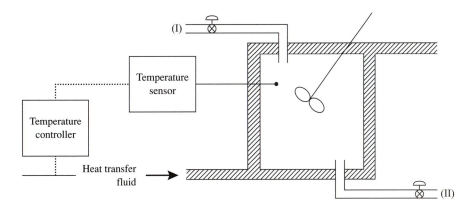

Figure 9.3.1 |
Schematic of a batch reactor.

is increased by elevating the temperature in the heat transfer fluid, a temperature maximum is reached, the reactor is then cooled by decreasing the temperature of the heat transfer fluid and products discharged via point (II).

To describe this process, material and energy balances are required. Recall that the mass balance on a batch reactor can be written as [refer to Equation (3.2.1)]:

$$\frac{1}{V}\frac{dn_i}{dt} = v_i r\left(n_i, T\right) \tag{9.3.1}$$

If in the rare case that the reactor is accomplishing a constant pressure gas-phase reaction at nonisothermal conditions:

$$V = V_0 \left(1 + \varepsilon_i f_i\right)\left(\frac{T}{T^0}\right) \tag{9.3.2}$$

To solve the mass balance, it must be accompanied by the simultaneous solution of the energy balance (i.e., the solution of Equation (9.2.9)). To do this, Equation (9.2.9) can be written in more convenient forms. Consider that the enthalpy contains both sensible heat and heat of reaction effects. That is to say that Equation (9.2.10) can be written as:

$$
\begin{aligned}
\text{heat exchange} = &\ (\text{change in sensible heat effects}) \\
&+ (\text{the energy consumed or released} \\
&\quad \text{by reaction})
\end{aligned} \tag{9.3.3}
$$

or

$$Q = \int_{T_{\text{initial}}, \Phi_{\text{initial}}}^{T_{\text{final}}, \Phi_{\text{final}}} \left(\overline{MS}\langle \overline{C}_p\rangle dT + \Delta H_r d\Phi\right) \tag{9.3.4}$$

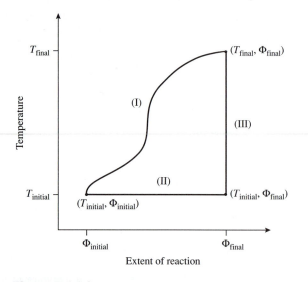

Figure 9.3.2 |
Enthalpy pathways.

where Φ is the extent of reaction (see Chapter 1), ΔH_r is the heat of reaction, $\overline{MS}$ is the total mass of the system, and $\langle \overline{C}_p \rangle$ is an average heat capacity per unit mass for the system. Since enthalpy is a state variable, the solution of the integral in Equation (9.3.4) is path independent. Referring to Figure 9.3.2, a simple way to analyze the integral [pathway (I)] is to allow the reaction to proceed isothermally [pathway (II)] and then evaluate the sensible heat changes using the product composition [pathway (III)]. That is,

$$Q = \Delta H_r |_{T_{\text{initial}}} (\Phi_{\text{final}} - \Phi_{\text{initial}}) + \int_{T_{\text{initial}}}^{T_{\text{final}}} \overline{MS}_{\text{final}} \langle C_{p_{\text{final}}} \rangle dT \qquad (9.3.5)$$

where a positive value for the heat of reaction denotes an endothermic reaction. Since the reaction-rate expressions normally employ moles of species in their evaluation:

$$\int_{T_{\text{initial}}}^{T_{\text{final}}} \overline{MS}_{\text{final}} \langle \overline{C}_{p_{\text{final}}} \rangle dT = \int_{T_{\text{initial}}}^{T_{\text{final}}} \sum_i (n_i C_{p_i}) dT \qquad (9.3.6)$$

where n_i is the moles of species i, and C_{p_i} is the molar heat capacity of species i. Note also that the heat exchange is the integral over time of the heat transfer rate, that is,

$$Q = \int \dot{Q} dt = \int UA_H (T^* - T) dt \qquad (9.3.7)$$

where U is an overall heat transfer coefficient, A_H is the heat transfer area, and T^* is the reference temperature. Combining Equations (9.3.7), (9.3.6), and (9.3.5) and recalling the definition of the extent of reaction in terms of the fractional conversion [Equation (1.2.10)] gives:

$$\int_0^t UA_H\,(T^* - T)\,d\bar{t} = \frac{-\Delta H_r|_{T^0}}{v_\ell}\,n_\ell^0 f_\ell + \sum_i \left(n_i \int_{T_{initial}}^{T_{final}} C_{p_i}\,dT \right) \qquad (9.3.8)$$

or in differential form:

$$UA_H\,(T^* - T) = \frac{-\Delta H_r|_{T^0}}{v_\ell}\,n_\ell^0\,\frac{df_\ell}{dt} + \sum_i (n_i\,C_{p_i})\frac{dT}{dt} \qquad (9.3.9)$$

Notice that [see Equations (1.2.10) and (1.3.2)]:

$$\frac{n_\ell^0}{-v_\ell}\,\frac{df_\ell}{dt} = \frac{d\Phi}{dt} = rV \qquad (9.3.10)$$

Using Equation (9.3.10) in Equation (9.3.9) yields:

$$UA_H\,(T^* - T) = \Delta H_r|_{T^0}rV + \sum_i (n_i\,C_{p_i})\frac{dT}{dt} \qquad (9.3.11)$$

Equations (9.3.11), (9.3.9), and (9.3.8) each define the energy balance for a batch reactor.

EXAMPLE 9.3.1

Show that the general energy balance, Equation (9.3.9), can simplify to an appropriate form for either adiabatic or isothermal reactor operation.

■ Answer

For adiabatic operation there is no heat transfer to the surroundings (i.e., $U = 0$). For this case, Equation (9.3.9) can be written as:

$$0 = \frac{-\Delta H_r|_{T^0}}{v_\ell}\,n_\ell^0\,\frac{df_\ell}{dt} + \sum_i (n_i C_{p_i})\frac{dT}{dt}$$

or when integrated as:

$$T = T^0 + \frac{\Delta H_r|_{T^0}\,n_\ell^0 f_\ell}{v_\ell \sum_i (n_i C_{p_i})} \qquad (9.3.12)$$

If the reactor is operated at isothermal conditions, then no sensible heat effects occur and Equation (9.3.9) becomes:

$$UA_H(T^* - T^0) = \frac{-\Delta H_r|_{T^0}}{v_\ell}\,n_\ell^0\,\frac{df_\ell}{dt} \qquad (9.3.13)$$

The description of the nonisothermal batch reactor then involves Equation (9.3.1) and either Equation (9.3.9) or (9.3.11) for nonisothermal operation or Equation (9.3.12) for adiabatic operation. Notice that for adiabatic conditions, Equation (9.3.12) provides a relationship between T and f_ℓ. Insertion of Equation (9.3.12) into Equation (9.3.1) allows for direct solution of the mass balance.

EXAMPLE 9.3.2 |

The hydration of 1-hexene to 2-hexanol is accomplished in an adiabatic batch reactor:

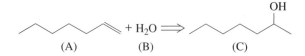

(A)   (B)   (C)

The reactor is charged with 1000 kg of a 10 wt. % H_2SO_4 solution and 200 kg of 1-hexene at 300 K. Assuming that the heat capacities for the reactants and products do not vary with temperature, the heat of reaction does not vary with temperature, and the presence of H_2SO_4 is ignored in the calculation of the heat capacity, determine the time required to achieve 50 percent conversion and the reactor temperature at that point.

Data:

	C_p (cal/(mol-K))	ΔH_f^0 (kcal/mol)
1-hexene	43.8	−10.0
H_2O	16.8	−68.0
2-hexanol	54.0	−82.0

reaction rate constant: $k = 10^4 \exp\left[-10^4/(R_g T)\right]\ (s^{-1})$

■ **Answer**

The material balance on the batch reactor is:

$$\frac{dC_A}{dt} = -kC_A$$

or

$$\frac{df_A}{dt} = k(1 - f_A)$$

since the reaction is first-order (units on k and large excess of water; 1-hexene is the limiting reagent). The energy balance is:

$$T = T^0 + \frac{\Delta H_r|_{T^0}\, n_\ell^0 f_\ell}{\nu_\ell \sum_i (n_i C_{p_i})}$$

where $\nu_\ell = -1$. In order to calculate the heat of reaction, the standard heats of formation can be used as follows:

$$\Delta H_r^0 = -82 + 68 + 10 = -4\ \text{kcal/mol}$$

Thus, the energy balance equation can be written:

$$T = T^0 + \frac{(4000)\, n_A^0\, f_A}{n_A^0\,(1 - f_A)\, C_{p_A} + n_A^0\,(\overline{M} - f_A)\, C_{p_B} + n_A^0 f_A\, C_{p_C}}$$

where

$$\overline{M} = n_B^0 / n_A^0$$

The values of n_A^0 and n_B^0 are:

$$n_A^0 = (2 \times 10^5\,g)(1\,\text{mol}/84\,g) = 2381\,\text{mol}$$
$$n_B^0 = (9 \times 10^5\,g)(1\,\text{mol}/18\,g) = 50000\,\text{mol}$$

so that:

$$n_B^0 / n_A^0 = 21$$

By placing the values for n_A^0, $\overline{M}$, T^0, and the C_p into the energy balance, the result is:

$$T = 300 + \frac{4000\, f_A}{421.8 - 7.8\, f_A}$$

The material balance equation is then solved with $k(T)$ being first converted to $k(f_A)$ by substitution of the energy balance for T:

$$k = 10^4 \exp\left[\dfrac{-10^4}{R_g\left(300 + \dfrac{4000\, f_A}{421.8 - 7.8\, f_A}\right)} \right]$$

The material balance equation must be solved numerically to give $t = 1111$ s or 18.52 min. The reactor temperature at this point is obtained directly from the energy balance with $f_A = 0.5$ to give $T = 304.8$ K.

EXAMPLE 9.3.3

Consider accomplishing the reaction $A + B \Rightarrow C$ in a nonisothermal batch reactor. The reaction occurs in the liquid phase. Find the time necessary to reach 80 percent conversion if the coolant supply is sufficient to maintain the reactor wall at 300 K.

Data:

$C_A^0 = 0.5$ mol/L	$\Delta H_r = -15$ kJ/mol
$C_B^0 = 0.6$ mol/L	$UA_H = 50$ J/(s-K)
$C_{p_A} = C_{p_B} = 65$ J/(mol-K)	$C_{p_C} = 150$ J/(mol-K)
$n_A^0 = 100$ mol	

$$k = 5 \times 10^{-3} \exp\left[\frac{20000\ \text{J/mol}}{R_g}\left(\frac{1}{300} - \frac{1}{T}\right) \right]\ (\text{L/mol/s})$$

■ Answer

The material balance for the batch reactor is:

$$\frac{df_A}{dt} = kC_A^0 (1 - f_A)(1.2 - f_A)$$

The energy balance can be written as:

$$UA_H (T^* - T) = \Delta H_r n_A^0 \frac{df_A}{dt} + [n_A^0 (1 - f_A) C_{P_A} + n_A^0 (1.2 - f_A) C_{P_B} + n_A^0 f_A C_{P_C}] \frac{dT}{dt}$$

The material and energy balance equations must be solved simultaneously. A convenient form for solution by numerical techniques is:

$$\frac{df_A}{dt} = g(T, f_A) = k(T) C_A^0 (1 - f_A)(1.2 - f_A)$$

$$\frac{dT}{dt} = \frac{UA_H(300 - T) - \Delta H_r n_A^0 g(T, f_A)}{n_A^0(1 - f_A) C_{P_A} + n_A^0 (1.2 - f_A) C_{P_B} + n_A^0 f_A C_{P_C}}$$

with $f_A = 0$ and $T = 300$ K at $t = 0$. The results are shown in Figure 9.3.3. From the data given in Figure 9.3.3, it takes 462 s to reach a fractional conversion of 0.8. Additionally, the final temperature is 332 K. Notice that the final temperature is not the maximum temperature achieved during the reaction, in contrast to adiabatic operation.

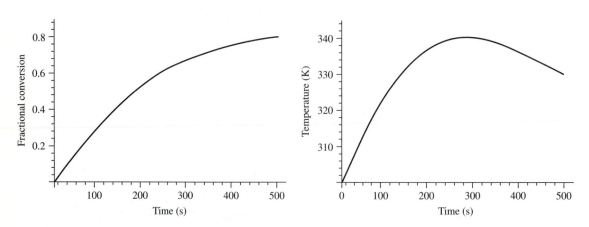

Figure 9.3.3 | Fractional conversion and temperature profiles for the reactor described in Example 9.3.3.

VIGNETTE 9.3.1

M. Kladko [*CHEMTECH*, **1** (1971) 141] presented an interesting case study of performing a liquid-phase, exothermic isomerization reaction. In a 50 gal steam-heated reactor, the reaction of $A \rightarrow B$ was adequately conducted with a 1:1 ratio of A to solvent. However, when scaling-up to a 750 gal glass-lined reactor with a reactant-to-solvent ratio of 2:1, the batch exothermed violently and the reaction ran out of control. The batch erupted through a safety valve and vented. Thus, the effects of the ratio of heat transfer area to reactor volume are nicely demonstrated in this case study. For small vessels the ratio of heat transfer area to reaction volume is higher than for larger vessels and thus the exothermic heat of reaction more efficiently removed.

The analysis of this problem led to the solution illustrated below. A reactor was constructed as a semibatch reactor (solvent initially charged into the reactor and reactant fed over time) and a heating and cooling cycle employed (see Figure 9.3.4). The reactor was operated and the results are shown in Figures 9.3.5–9.3.7.

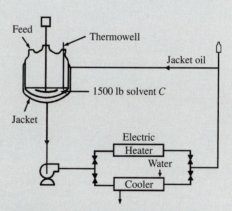

Figure 9.3.4 | Plant installation for "semibatch" isomerization showing heating/cooling circuits. [Adapted from "The Case of a Real Engineering Design Problem" by M. Kladko, *CHEMTECH* (now *Chemical Innovation*), **1** (1971) 141, with permission of the author and the American Chemical Society, copyright 1971.]

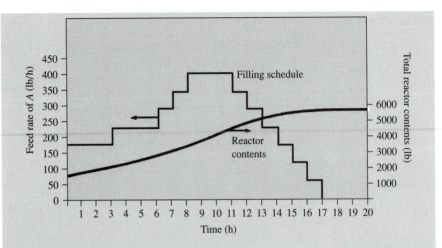

Figure 9.3.5 | Reactor filling schedule and total contents versus time for "semibatch" operation. [Adapted from "The Case of a Real Engineering Design Problem" by M. Kladko, *CHEMTECH* (now *Chemical Innovation*), **1** (1971) 141, with permission of the author and the American Chemical Society, copyright 1971.]

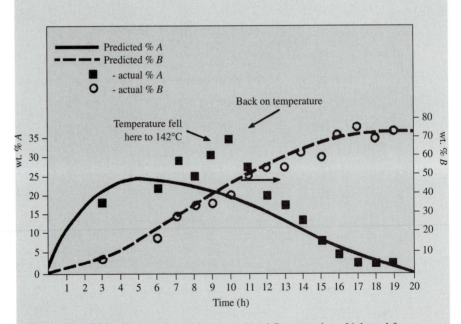

Figure 9.3.6 | Predicted and actual percent *A* and *B* versus time. [Adapted from "The Case of a Real Engineering Design Problem" by M. Kladko, *CHEMTECH* (now *Chemical Innovation*), **1** (1971) 141, with permission of the author and the American Chemical Society, copyright 1971.]

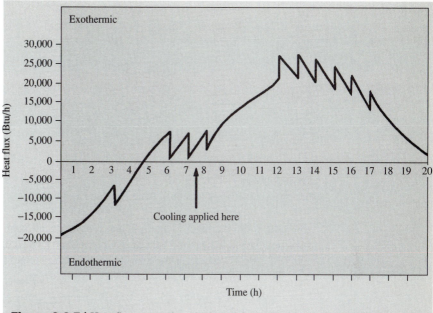

Figure 9.3.7 | Heat flux versus time. [Adapted from "The Case of a Real Engineering Design Problem" by M. Kladko, *CHEMTECH* (now *Chemical Innovation*), **1** (1971) 141, with permission of the author and the American Chemical Society, copyright 1971.]

9.4 | Nonisothermal Plug Flow Reactor

Consider a PFR operating at nonisothermal conditions (refer to Figure 9.4.1). To describe the reactor performance, the material balance, Equation (9.1.1), must be solved simultaneously with the energy balance, Equation (9.2.7). Assuming that the PFR is a tubular reactor of constant cross-sectional area and that T and C_i do not vary over the radial direction of the tube, the heat transfer rate $\dot{Q}$ can be written for a differential section of reactor volume as (see Figure 9.4.1):

$$d\dot{Q} = U(T^* - T)dA_H = U(T^* - T)\frac{4}{d_t}dV_R \tag{9.4.1}$$

since

$$dA_H = \pi d_t dz \quad \text{(differential area for heat transfer)}$$

and

$$dV_R = \frac{\pi d_t^2}{4}dz \quad \text{(differential reactor volume)}$$

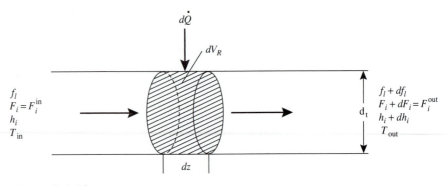

Figure 9.4.1 |
Schematic of differential fluid volume in a nonisothermal PFR.

where d_t is the diameter of the tubular reactor. Recall again that the enthalpy contains both sensible heat and heat of reaction effects. Thus, the energy balance Equation (9.2.7) can be written for the differential fluid element of the PFR as:

$$d\dot{Q} = \sum_i \left(F_i^{\text{out}} \int_{T^0}^{T_{\text{out}}} C_{p_i} dT \right) - \sum_i \left(F_i^{\text{in}} \int_{T^0}^{T_{\text{in}}} C_{p_i} dT \right) - \frac{\Delta H_r|_{T^0}}{v_\ell} F_\ell^0 df_\ell \quad (9.4.2)$$

where the heat of reaction is evaluated at a reference temperature T^0 and C_{p_i} is the molar heat capacity of species i. Normally, T^0 is taken as the reactor entrance temperature.

EXAMPLE 9.4.1

Show that the general energy balance, Equation (9.4.2), can simplify to an appropriate form for either adiabatic or isothermal reactor operation.

■ Answer

For adiabatic operation $d\dot{Q} = 0$. Thus, Equation (9.4.2) simplifies to:

$$\sum_i \left(F_i^{\text{out}} \int_{T^0}^{T_{\text{out}}} C_{p_i} dT \right) - \sum_i \left(F_i^{\text{in}} \int_{T^0}^{T_{\text{in}}} C_{p_i} dT \right) = \frac{\Delta H_r|_{T^0}}{v_\ell} F_\ell^0 df_\ell \quad (9.4.3)$$

If T^0 is the temperature at the reactor entrance and the conversion is zero at this point, then Equation (9.4.3) can be written for any point in the PFR as:

$$\sum_i F_i \int_{T^0}^{T} C_{p_i} dT = \frac{\Delta H_r|_{T^0}}{v_\ell} F_\ell^0 f_\ell \quad (9.4.4)$$

Equation (9.4.4) relates the conversion to the temperature for an adiabatic PFR. If the reactor is operated isothermally, then:

$$d\dot{Q} = -\frac{\Delta H_r|_{T^0}}{v_\ell} F_\ell^0 df_\ell \quad (9.4.5)$$

EXAMPLE 9.4.2

(From C. G. Hill, *An Introduction to Chemical Engineering Kinetics and Reactor Design*, Wiley, 1977, pp. 362–364.)

Butadiene and ethylene can be reacted together to form cyclohexene as follows:

$$CH_2\!=\!CHCH\!=\!CH_2 + CH_2\!=\!CH_2 \Longrightarrow$$

$$(B) \qquad\qquad (E)$$

$$(C)$$

If equimolar butadiene and ethylene at 450°C and 1 atm are fed to a PFR operating adiabatically, what is the space time necessary to reach a fractional conversion of 0.10?

Data:

$$k = 10^{7.5} \exp[-27{,}500/(R_g T)] \text{ L/mol/s}$$
$$\Delta H_r = -30000 \text{ cal/mol}$$
$$C_{p_B} = 36.8 \text{ cal/mol/K}$$
$$C_{p_E} = 20.2 \text{ cal/mol/K}$$
$$C_{p_C} = 59.5 \text{ cal/mol/K}$$

■ **Answer**

Assume that each C_{p_i} is not a strong function of temperature over the temperature range obtained within the PFR (i.e., each is not a function of T). The material and energy balance equations are:

$$\frac{dF_B}{dV_R} = -kC_B^2 = -k(C_B^0)^2 \left[\frac{1 - f_B}{1 + \varepsilon_B f_B}\right]^2 \left(\frac{T^0}{T}\right)^2$$

and

$$\sum_i F_i \int_{T^0}^{T} C_{p_i}\, dT = \frac{\Delta H_r|_{T^0}}{v_B} F_B^0 f_B$$

Since $F_B = F_B^0(1 - f_B) = C_B v$:

$$\frac{dF_B}{dV_R} = -F_B^0 \frac{df_B}{dV_R} = -\frac{F_B^0}{v_0} \frac{df_B}{d(V_R/v_0)} = -C_B^0 \frac{df_B}{d\tau}$$

Thus, the material balance can be written as:

$$\frac{df_B}{d\tau} = kC_B^0 \left[\frac{1 - f_B}{1 - 0.5 f_B}\right]^2 \left(\frac{T^0}{T}\right)^2$$

since $\varepsilon_B = 0.5 \left[\dfrac{1 - 2}{|-1|}\right] = -0.5$. Now for the energy balance,

$$\sum_i F_i \int_{T^0}^{T} C_{p_i}\, dT = \sum_i F_i C_{p_i} \int_{T^0}^{T} dT \quad \text{since } C_{p_i} \neq C_{p_i}(T)$$

Thus,

$$\sum_i F_i C_{p_i} \int_{T^0}^{T} dT = F_B^0 (1 - f_B)(36.8)(T - T^0) + F_B^0 (1 - f_B)(20.2)(T - T^0)$$
$$+ F_B^0 f_B (59.5)(T - T^0)$$

or

$$\sum_i F_i C_{p_i} \int_{T^0}^{T} dT = (57 + 2.5 f_B) F_B^0 (T - T^0)$$

The energy balance then becomes:

$$(57 + 2.5 f_B) F_B^0 (T - T^0) = \frac{(-30,000)}{(-1)} F_B^0 f_B$$

or

$$T = 723 + \frac{(30,000) f_B}{57 + 2.5 f_B}$$

The solution of the material balance equation:

$$\tau = \int_0^{0.10} \frac{df_B}{kC_B^0 \left[\dfrac{1 - f_B}{1 - 0.5 f_B} \right]^2 \left(\dfrac{723}{T} \right)^2}$$

with T from the energy balance gives a value of $\tau = 47.1$ s. Additionally, the exit temperature is 775 K.

VIGNETTE 9.4.1

Over the first year of a child's life in the United States, the child receives vaccines for immunization against hepatitis B, diphtheria, tetanus, pertussis (whooping cough), haemophilus influenzae, polio, measles, mumps, rubella, and chicken pox. Numerous vaccines are based on viruses in some form. The growth of viral-based vaccines typically occurs in bioreactors that are just tanks. However, when the aqueous medium that is used to grow the viral products is to be expelled, it must be "decontaminated." To do so, the fluid is mixed with H_2O_2 and flows through a tubular reactor. The H_2O_2 is used to kill any living material in the tubular reactor. Thus, τ must be fixed to a sufficiently long time to assure complete death of all living matter prior to exiting the reactor. The temperature of this reactor must be maintained below a critical value so that the hydrogen peroxide does not decompose to form dioxygen. Thus, the material balance for the reactor must be solved simultaneously with the energy balance as illustrated in Section 9.4 to properly define the correct τ for complete decontamination.

The energy balance for the PFR can also be written as follows by combining Equations (9.4.1) and (9.4.2):

$$U(T^* - T)\frac{4}{d_t} \, dV_R = \sum_i F_i C_{p_i} \, dT - \frac{\Delta H_r|_{T^0}}{v_\ell} F^0_\ell \, df_\ell \tag{9.4.6}$$

or

$$U(T^* - T)\frac{4}{d_t} = \sum_i F_i C_{p_i} \frac{dT}{dV_R} - \frac{\Delta H_r|_{T^0}}{v_\ell} F^0_\ell \frac{df_\ell}{dV_R} \tag{9.4.7}$$

Using the fact that:

$$F^0_\ell \frac{df_\ell}{dV_R} = (-v_\ell)\mathrm{r}$$

Equation (9.4.7) can be written as:

$$\sum_i F_i C_{p_i} \frac{dT}{dV_R} = (-\Delta H_r|_{T^0})\mathrm{r} - U(T - T^*)\frac{4}{d_t} \tag{9.4.8}$$

Thus, the material and energy balances for the PFR can be written as ($F_\ell = C_\ell \mathrm{v} = C_\ell \pi d_t^2 u/4$; $dV_R = \pi d_t^2 \, dz/4$):

$$\left.\begin{aligned} -u\frac{dC_\ell}{dz} &= (-v_\ell)\mathrm{r} \\[2mm] u\rho \overline{C}_p \frac{dT}{dz} &= (-\Delta H_r|_{T^0})\mathrm{r} - \frac{4U}{d_t}(T - T^*) \end{aligned}\right\} \tag{9.4.9}$$

with $C_\ell = C^0_\ell$ and $T = T^0$ at $z = 0$, where u is the superficial linear velocity, ρ is the average density of the fluid, and $\overline{C}_p$ is the average heat capacity per unit mass. Equation (9.4.9) is only applicable when the density of the fluid is not changing in the reactor and when $\overline{C}_p$ is not a function of temperature, since the following relationship is used to obtain Equation (9.4.9):

$$\sum_i F_i C_{p_i} = \mathrm{v}\rho \overline{C}_p \tag{9.4.10}$$

EXAMPLE 9.4.3

A PFR of dimensions $L = 2$ m and $d_t = 0.2$ m is accomplishing a homogeneous reaction. The inlet concentration of the limiting reactant is $C^0_\ell = 0.3$ kmol/m^3 and the inlet temperature is 700 K. Other data are: $-\Delta H_r|_{T^0} = 10^4$ kJ/kmol, $\overline{C}_p = 1$ kJ/(kg-K), $E = 100$ kJ/mol, $\rho = 1.2$ kg/m^3, $u = 3$ m/s, and $\overline{A} = 5$ s^{-1}. Calculate the dimensionless concentration ($y = C_\ell / C^0_\ell$) and temperature ($\overline{\theta} = T/T^0$) profiles for adiabatic ($U = 0$) and nonisothermal ($U = 70$ J/(m^2-s-K)) operations. (Example adapted from J. Villadsen and M. L. Michelsen, *Solution of Differential Equation Models by Polynomial Approximation,* Prentice-Hall, Englewood Cliffs, 1978, p. 59.)

■ Answer

From the units on $\overline{A}$, it is clear that the reaction rate is first order. Using Equation (9.4.9) with a first-order reaction rate expression gives:

$$-u\frac{dC_\ell}{dz} = \bar{A}\exp[-E/(R_gT)]C_\ell$$

$$u\rho\bar{C}_p\frac{dT}{dz} = (-\Delta H_r)\bar{A}\exp[-E/(R_gT)]C_\ell - \frac{4U}{d_t}(T - T^0)$$

Let $x = z/L$, $y = C_\ell/C_\ell^0$ and $\bar{\theta} = T/T^0$. Using these dimensionless variables in the material and energy balance relationships yields:

$$\frac{dy}{dx} = -(\overline{Da})y\exp\left[\gamma\left(1 - \frac{1}{\bar{\theta}}\right)\right]$$

$$\frac{d\bar{\theta}}{dx} = \beta_T(\overline{Da})y\exp\left[\gamma\left(1 - \frac{1}{\bar{\theta}}\right)\right] - H_w(\bar{\theta} - 1)$$

with $y = 1$ at $x = 0$ and where:

$$\overline{Da} = \frac{L\bar{k}}{u}, \quad \bar{k} = \bar{A}\exp[-E/(R_gT^0)]$$

$$\beta_T = \frac{C_\ell^0(-\Delta H_r)}{\rho\tau T^0}$$

$$\gamma = \frac{E}{R_gT^0}$$

$$H_w = \frac{4U}{d_t}\left(\frac{L}{\rho\bar{C}_pu}\right)$$

Notice that all the groupings $\overline{Da}$, β_T, γ, and H_w are dimensionless. For adiabatic operation, $H_w = 0$. For this case, the mass balance equation can be multiplied by β_T and added to the energy balance to give:

$$\frac{d}{dx}(\bar{\theta} + \beta_T y) = 0$$

Integration of this equation with $y = \bar{\theta} = 1$ at $x = 0$ leads to:

$$\bar{\theta} = 1 + \beta_T(1 - y) \tag{9.4.11}$$

Equation (9.4.11) is the adiabatic relationship between temperature and conversion in dimensionless form. The mass balance can then be written as:

$$\frac{dy}{dx} = -(\overline{Da})y\exp\left[\frac{\gamma\beta_T(1 - y)}{1 + \beta_T(1 - y)}\right]$$

with

$$y = 1 \text{ at } x = 0$$

The solution of this differential equation is straightforward and is shown in Figure 9.4.2. For the nonisothermal case, the material and energy balances must be solved simultaneously by

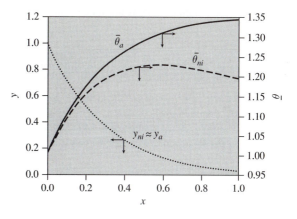

Figure 9.4.2 | Dimensionless concentration and temperature profiles for adiabatic and nonisothermal operation. y_a and $\bar{\theta}_a$ are for adiabatic conditions while y_{ni} and $\bar{\theta}_{ni}$ are for nonisothermal operation.

numerical methods. The nonisothermal results are plotted also in Figure 9.4.2. Notice that there is a maximum in the value of $\bar{\theta}_{ni}$ that occurs at $x = 0.57$ and gives $T = 860$ K. The maximum in the temperature profile from nonisothermal operation is normally denoted as the *hot spot* in the reactor.

9.5 | Temperature Effects in a CSTR

Although the assumption of perfect mixing in the CSTR implies that the reactor contents will be at uniform temperature (and thus the exit stream will be at this temperature), the reactor inlet may not be at the same temperature as the reactor. If this is the case and/or it is necessary to determine the heat transferred to or from the reactor, then an energy balance is required.

The energy balance for a CSTR can be derived from Equation (9.2.7) by again carrying out the reaction isothermally at the inlet temperature and then evaluating sensible heat effects at reactor outlet conditions, that is,

$$\dot{Q} = \frac{F_\ell^0 (\Delta H_r|_{T^0})(f_\ell^f - f_\ell^0)}{(-\upsilon_\ell)} + \sum \left(F_i^f \int_{T^0}^{T^f} C_{p_i} dT \right) \qquad (9.5.1)$$

where the superscript f denotes the final or outlet conditions. For adiabatic operation, $\dot{Q} = 0$.

EXAMPLE 9.5.1

The nitration of aromatic compounds is a highly exothermic reaction that generally uses catalysts that tend to be corrosive (e.g., HNO_3/H_2SO_4). A less corrosive reaction employs N_2O_5

as the nitrating agent as illustrated below:

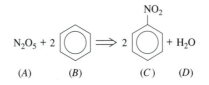

$$N_2O_5 + 2 \quad \Longrightarrow \quad 2 \quad + H_2O$$

$$(A) \qquad\qquad (B) \qquad\qquad (C) \qquad (D)$$

If this reaction is conducted in an adiabatic CSTR, what is the reactor volume and space time necessary to achieve 35 percent conversion of N_2O_5? The reaction rate is first order in A and second order in B.

Data:

$$\Delta H_r = -370.1 \text{ kJ/mol} \qquad\qquad T^0 = 303 \text{ K}$$
$$C_{p_A} = 84.5 \text{ J/(mol-K)} \qquad\qquad F_A^0 = 10 \text{ mol/min}$$
$$C_{p_B} = 137 \text{ J/(mol-K)} \qquad\qquad F_B^0 = 30 \text{ mol/min}$$
$$C_{p_C} = 170 \text{ J/(mol-K)} \qquad\qquad v = 1000 \text{ L/min}$$
$$C_{p_D} = 75 \text{ J/(mol-K)} \qquad\qquad C_A^0 = 0.01 \text{ mol/L}$$

$$k = 0.090 \exp\left[\frac{(40 \text{ kJ/mol})}{R_g}\left(\frac{1}{303} - \frac{1}{T}\right)\right] \qquad (\text{L/mol})^2 (\text{min})^{-1}$$

■ Answer

The reaction occurs in the liquid-phase and the concentrations are dilute so that mole change with reaction does not change the overall density of the reacting fluid. Thus,

$$C_A = C_A^0 (1 - f_A), \qquad F_A = F_A^0 (1 - f_A)$$
$$C_B = C_A^0 (3 - 2f_A), \qquad F_B = F_A^0 (3 - 2f_A)$$
$$F_C = 2F_A^0 f_A, \qquad F_D = F_A^0 f_A$$

The material balance on the CSTR can be written as:

$$V = \frac{F_A^0 f_A}{k(C_A^0)^3 (1 - f_A)(3 - 2f_A)^2}$$

The energy balance for the adiabatic CSTR is:

$$0 = \Delta H_r F_A^0 f_A + F_A^0 (1 - f_A)C_{p_A}(T - T^0) + F_A^0 (3 - 2f_A)C_{p_B}(T - T^0)$$
$$+ 2F_A^0 f_A C_{p_C}(T - T^0) + F_A^0 f_A C_{p_D}(T - T^0)$$

For $f_A = 0.35$, the energy balance yields $T = 554$ K. At 554 K, the value of the rate constant is $k = 119.8$ L^2/(mol^2-min). Using this value of k in the material balance gives $V = 8,500$ L and thus $\tau = 8.5$ min.

EXAMPLE 9.5.2

Consider the aromatic nitration reaction illustrated in Example 9.5.1. Calculate the reactor volume required to reach 35 percent conversion if the reactor is now cooled.

Data:

$$UA_H = 9000 \text{ J/(min-K)}$$
$$T_C^0 = 323 \text{ K (and is constant)}$$
$$v = 100 \text{ L/min}$$
$$C_A^0 = 0.10 \text{ mol/L}$$

All other data are from Example 9.5.1.

■ Answer

The material balance equation remains the same as in Example 9.5.1. The energy balance is now:

$$UA_H\,(T_C^0 - T) = \Delta H_r F_A^0 f_A + F_A^0\,(1 - f_A)C_{p_A}\,(T - T^0) + F_A^0\,(3 - 2f_A)C_{p_B}\,(T - T^0)$$
$$+ 2F_A^0 f_A\, C_{p_C}\,(T - T^0) + F_A^0 f_A\, C_{p_D}\,(T - T^0)$$

when $f_A = 0.35$, the energy balance yields $T = 407$ K. At 407 K, the reaction rate constant is $k = 5.20$ L^2/(mol^2-min). Using this value of k and $C_A^0 = 0.10$ mol/L in the material balance equation gives $V = 196$ L.

9.6 | Stability and Sensitivity of Reactors Accomplishing Exothermic Reactions

Consider the CSTR illustrated in Figure 9.6.1. The reactor is accomplishing an exothermic reaction and therefore must transfer heat to a cooling fluid in order to remain at temperature T. Assume that the heat transfer is sufficiently high to maintain the reactor wall temperature at T_c. Therefore,

$$\dot{Q} = \underbrace{UA_H(T - T_c)}_{\substack{\text{heat removed} \\ \text{from reactor}}} = \underbrace{v_c C_{p_c}(T_c - T_c^0)}_{\substack{\text{sensible heat change} \\ \text{of coolant}}} \qquad (9.6.1)$$

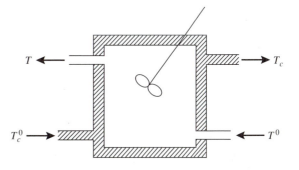

Figure 9.6.1 |
Schematic illustration of a CSTR that is maintained at temperature T by transferring heat to a coolant fluid $(T_c > T_c^0)$.

where v_c is the volumetric flow rate of the coolant and C_{p_c} is the heat capacity of the coolant and is not a function of temperature. Solving for T_c and then substituting the expression back into the heat transfer equation yields:

$$\dot{Q} = \frac{UA_H v_c C_{p_c}}{UA_H + v_c C_{p_c}} (T - T_c^0) = \lambda_\lambda (T - T_c^0) \tag{9.6.2}$$

The energy balance on the CSTR can be written as [from Equation (9.5.1) with a first-order reaction rate expression]:

$$\dot{Q} = \frac{\Delta H_r|_{T^0} F_\ell^0 (f_\ell - f_\ell^0)}{(-v_\ell)} + \sum \left(F_i \int_{T^0}^{T} C_{p_i} dT \right)$$

or since:

$$-v_\ell r V = F_\ell^0 (f_\ell - f_\ell^0) \qquad \text{(material balance)}$$

as:

$$\dot{Q} = (\Delta H_r|_{T^0}) k C_\ell V + v^p \rho C_p (T - T^0) \qquad \text{(first-order reaction rate)} \tag{9.6.3}$$

where v^p is the volumetric flow rate of the product stream and ρ and C_p are the average density and heat capacity of the outlet stream. Rearranging Equation (9.6.3) and substituting Equation (9.6.2) for $\dot{Q}$ gives (note that $\dot{Q}$ is heat removed):

$$\frac{\lambda_\lambda}{v^p \rho C_p} (T_c^0 - T) = \frac{k C_\ell \tau (\Delta H_r|_{T^0})}{\rho C_p} + (T - T^0) \tag{9.6.4}$$

Let:

$$\alpha \alpha_1 = \frac{\lambda_\lambda}{v^p \rho C_p}$$

$$\alpha \alpha_2 = \frac{-k\tau (\Delta H_r|_{T^0})}{\rho C_p}$$

so that Equation (9.6.4) can be written as:

$$(1 + \alpha \alpha_1) T - (T^0 + \alpha \alpha_1 T_c^0) = \alpha \alpha_2 C_\ell \tag{9.6.5}$$

Since:

$$C_\ell = C_\ell^0 / (1 + k\tau)$$

from the solution of the material balance equation, Equation (9.6.5) can be formulated as:

$$(1 + \alpha \alpha_1) T - (T^0 + \alpha \alpha_1 T_c^0) = \frac{\alpha \alpha_2 C_\ell^0}{1 + k\tau} \tag{9.6.6}$$

Q_r (heat removed) $= Q_g$ (heat generated)

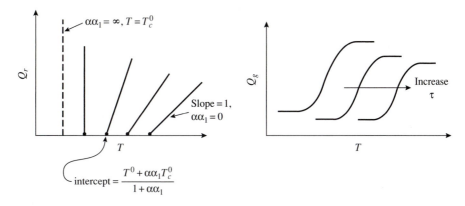

Figure 9.6.2 |
Schematic illustration of Q_r and Q_g as functions of T.

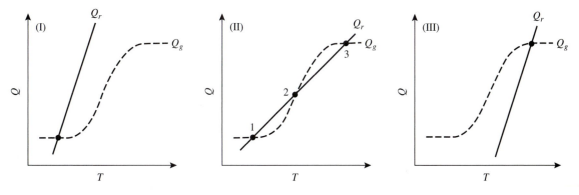

Figure 9.6.3 |
Steady-state solutions to Equation (9.6.3).

If Q_r and Q_g are plotted versus the reaction temperature, T, the results are illustrated in Figure 9.6.2. A solution of Equation (9.6.6) occurs when Q_r equals Q_g, and this can happen as shown in Figure 9.6.3. Notice that for cases (I) and (III) a single solution exists. However, for case (II) three steady-states are possible. An important question of concern with case (II) is whether all three steady-states are stable. This is easy to rationalize as follows. At steady-state 1, if T is increased then $Q_r > Q_g$ so the reactor will return to point 1. Additionally, if T is decreased, $Q_g > Q_r$ so the reactor will also return to point 1 in this case. Thus, steady-state 1 is stable since small perturbations from this position cause the reactor to return to the steady-state. Likewise steady-state 3 is a stable steady-state. However, for steady-state 2, if T is increased, $Q_g > Q_r$ and the reactor will move to position 3. If T is decreased below that of point 2, $Q_r > Q_g$ and the reactor will move to point 1. Therefore, steady-state 2 is unstable. It is important to determine the stability of reactor operation since perturbations from steady-state always occur in a real system. Finally, what determines whether the reactor achieves steady-state 1 or 3 is the start-up of the reactor.

EXAMPLE 9.6.1

Calculate the steady-states for the following reactor configuration. Is there an unstable steady-state?

Data:
$A + B \Rightarrow 2C$ in the liquid phase, $V = 1$ L, $k = 33 \times 10^9 \exp\left[-20{,}000/(R_gT)\right]$ L/(mol · min),
$-\Delta H_r = 20$ kcal/mol, $C_A^0 = 20$ mol/L, $C_B^0 = 3$ mol/L, v = 100 cm³/min, $T^0 = 17$°C, $T_c^0 =$
87°C, $\rho C_p = 650$ cal/(L · °C), $U = 0.1$ cal/(cm² · min · K), and $A_H = 250$ cm².

■ Answer

The reaction rate expression is:

$$(-v_B)\mathbf{r} = kC_AC_B = k(C_B^0)^2\,(1 - f_B)(\overline{M} - f_B), \overline{M} = C_A^0/C_B^0$$

giving the following material balance:

$$0 = C_B^0 f_B - \tau k(C_B^0)^2\,(1 - f_B)(\overline{M} - f_B)$$

Therefore,

$$Q_r = (1 + \alpha\alpha_1)\,T - (T^0 + \alpha\alpha_1\,T_c^0)$$

$$Q_g = \frac{-k(C_B^0)^2\,\tau\,(\Delta H_r)(1 - f_B)(\overline{M} - f_B)}{\rho C_p}$$

First, solve the material balance equation for f_B (gives two values for f_B—one will have physical meaning) at a particular T and next calculate Q_r and Q_g. A sampling of results is provided below (assume $v_cC_{p_c} \gg UA_H$ for the calculation of Q_r):

T (K)	f_B	Q_r	Q_g
310	0.049	0.77	4.55
320	0.124	14.62	11.45
330	0.264	28.46	24.40
340	0.461	42.31	42.50
350	0.658	56.20	60.77
360	0.807	70.00	74.78
370	0.898	83.80	82.85

If these data are plotted, they yield:

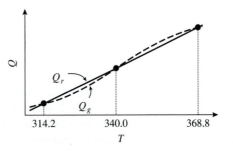

Thus, there are three steady-states and they are:

Steady-state	T (K)	T (°C)	f_B
1	314.2	41.2	0.079
2	340.0	67.0	0.460
3	368.8	95.8	0.888

Steady-state 2 is unstable. Notice the vast differences in f_B for the three steady-states. Thus, it is clear that attempts by a naïve designer to operate at steady-state 2 would not succeed since the reactor would not settle into this steady-state at all.

In addition to knowing the existence of multiple steady-states and that some may be unstable, it is important to assess how stable the reactor operation is to variations in the processing parameters (i.e., the sensitivity). In the above example for the CSTR, it is expected that the stable steady-states have low sensitivity to variations in the processing parameters, while clearly the unstable steady-state would not.

To provide an example of how the sensitivity may be elucidated, consider a tubular reactor accomplishing an exothermic reaction and operating at nonisothermal conditions. As described in Example 9.4.3, hot spots in the reactor temperature pro-

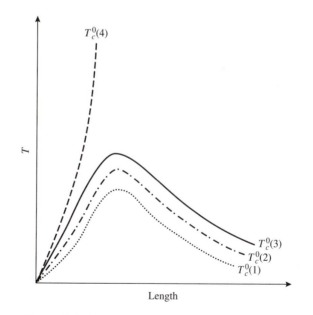

Figure 9.6.4 |

Temperature profiles in a tubular reactor operating nonisothermally and conducting an exothermic reaction. T_c^0 is the temperature of the coolant fluid.

file can occur for this situation. Figure 9.6.4 illustrates what a typical reactor temperature profile would look like. Consider what could happen as the temperature of the coolant fluid, T_c^0, increases. As T_c^0 becomes warmer $T_c^0 (i + 1) > T_c^0 (1)$, $i = 1$, 2, 3, then less heat is removed from the reactor and the temperature at the reactor hot spot increases in value. Eventually, heat is generated at a sufficiently high rate that it cannot be removed [illustrated for $T_c^0 (4)$] such that the hot spot temperature exceeds some physical limit (e.g., phase change of fluid, explosions or fire, the catalyst melts, etc.) and this condition is called *runaway*. Thus, reactor operation close to a runaway point would not be prudent, and determining the sensitivity towards the tendency of runaway a critical factor in the reactor analysis. Several criteria have been developed to assess the sensitivity of reactors; each involves the use of critical assumptions [see, for example, G. F. Froment and K. B. Bischoff, *Chemical Reaction Analysis and Design*, Wiley, New York, 1977, Chapter 1]. Here, an example of how reactor stability can be assessed is illustrated and was provided by J. B. Cropley.

EXAMPLE 9.6.2

A tubular reactor packed with a heterogeneous catalyst is accomplishing an oxidation reaction of an alcohol to an aldehyde. The reactions are:

$$\text{alcohol} + \text{air} \Rightarrow \text{aldehyde} + \text{air} \Rightarrow CO_2 + H_2O$$

In this series reaction pathway, the desired species is the aldehyde. Since both reactions are exothermic (second reaction is highly exothermic), the reactor is operated nonisothermally. The reactor is a shell-and-tube heat exchanger consisting of 2500 tubes of 1 inch diameter. Should the heat exchanger be operated in a cocurrent or countercurrent fashion in order to provide a greater stabilization against thermal runaway?

■ Answer

Since this example comes from the simulation of a real reactor, the amount of data necessary to completely describe it is very high. Thus, only the trends observed will be illustrated in order to conserve the length of presentation.

Schematically, the reactor can be viewed as:

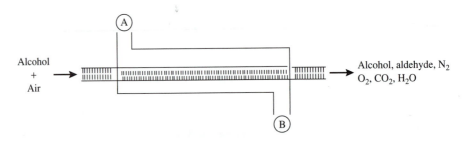

The cooling fluid is fed at point A or at point B for cocurrent and countercurrent operation, respectively. Next, the reactor temperature profiles from the two modes of operation are illustrated. The three profiles in each graph are for different coolant feed temperatures, T_c^0. Notice that a hot spot occurs with countercurrent operation. In order to access the sensitivity,

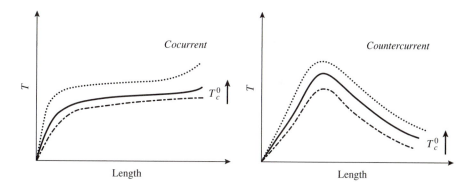

Cropely suggests plotting the maximum temperature in the reactor as a function of the inlet coolant temperature. The results look like:

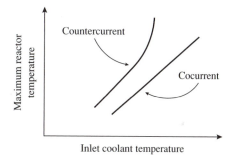

The slope of the line would be an indication of the reactor stability to variations in the inlet coolant temperature. Clearly, cocurrent operation provides better sensitivity and this conclusion is a general one. The reason for this is that by operating in a cocurrent manner the greatest ability to remove heat [largest $\Delta T\,(T_{\text{reactor}} - T_{\text{coolant}})$] can occur in the region of highest heat generation within the reactor.

Exercises for Chapter 9

1. Calculate the final temperature and time required to reach 50 percent conversion in the batch reactor described in Example 9.3.2 if the heat of reaction is now -40 kcal/mol. Do you think that this time is achievable in a large reactor system?

2. Find the final temperature and time required to reach 90 percent conversion in the reactor system of Exercise 1.

3. Plot the fractional conversion and temperature as a function of time for the batch reactor system described in Example 9.3.3 if the reactor is now adiabatic ($U = 0$). Compare your results to those for the nonisothermal situation given in Figure 9.3.3. How much energy is removed from the reactor when it is operated nonisothermally?

4. Consider what happens in the batch reactor given in Example 9.3.3 if the wall temperature does not remain constant. For comparison to the constant wall temperature, calculate the fractional conversion and reactor temperature as a function of time when:

$$T^* = 300 + 0.1t$$

where t is in seconds.

5. Calculate the exit temperature and τ for the PFR described in Example 9.4.2 when mole change with reaction is ignored (i.e., $\varepsilon_B = 0$). How much error is introduced by making this change?

6. Calculate the exit temperature and τ for the PFR described in Example 9.4.2 when the temperature effects on the concentration are ignored. Is this a reasonable simplification or not?

7. An adiabatic PFR can be described by the following set of equations:

$$\frac{dy}{dx} = -4y \exp\left[18\left(1 - \frac{1}{\theta}\right)\right]$$

$$\frac{d\bar{\theta}}{dx} = 0.2y \exp\left[18\left(1 - \frac{1}{\theta}\right)\right]$$

$$y = \bar{\theta} = 1 \text{ at } x = 0$$

Solve these equations and plot y and $\bar{\theta}$ as a function of x for
0 (entrance) $\leq x \leq 1$ (exit). What happens if the heat of reaction is doubled?

8. Ascertain whether the following exothermic reaction:

$$A + A \underset{k_2}{\overset{k_1}{\rightleftharpoons}} P + P \qquad k_1 = 30 \text{ L mol}^{-1}\text{min}^{-1}, k_2 = 0.1 \text{ L mol}^{-1}\text{min}^{-1}$$

(k_1 and k_2 at 80°C) could be carried out in the reactor shown below:

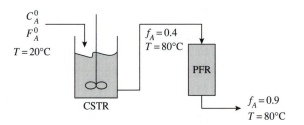

Calculate the volume and heat removed from the CSTR and the PFR. Do the magnitudes of the heat being removed appear feasible? Why or why not?

Data:

$$C_{p_A} = 45 \text{ cal mol}^{-1} \text{ K}^{-1}$$
$$C_{p_P} = 40 \text{ cal mol}^{-1} \text{ K}^{-1}$$
$$-\Delta H_r = 10,000 \text{ cal mol}^{-1}$$
$$C_A^0 = 1.5 \text{ mol L}^{-1}$$
$$F_A^0 = 100 \text{ mol min}^{-1}$$

9. The ester of an organic base is hydrolyzed in a CSTR. The rate of this irreversible reaction is first-order in each reactant. The liquid volume in the vessel is 6500 L. A jacket with coolant at 18°C maintains the reactant mixture at 30°C. Additional data:

Ester feed stream—1 M, 30°C, 20 L/s

Base feed stream—4 M, 30°C, 10 L/s

Rate constant $= 10^{14} \exp(-11{,}000/T) \text{ M}^{-1}\text{s}^{-1}$, T in K

$\Delta H_r = -45 \text{ kcal/mol ester}$

The average heat capacity is approximately constant at 1.0 kcal L^{-1} °C^{-1}.

(a) What is the conversion of ester in the reactor?

(b) Calculate the rate at which energy must be removed to the jacket to maintain 30°C in the reactor. If the heat transfer coefficient is 15 kcal $\text{s}^{-1}\text{m}^{-2}\text{ K}^{-1}$, what is the necessary heat transfer area?

(c) If the coolant supply fails, what would be the maximum temperature the reactor could reach?

10. A reaction is carried out in an adiabatic CSTR with a volume of 10,000 L. The feed solution with reactant A, at a concentration of 5 M, is supplied at 10 L s^{-1}. The reaction is first-order with rate constant:

$$k = 10^{13} \exp(-12500/T) \text{ s}^{-1}, \text{ where } T \text{ is in K}$$

The density is 1000 kg m^{-3} and $C_p = 1.0 \text{ kcal kg}^{-1}\text{K}^{-1}$. The heat of reaction is $\Delta H_r = -70 \text{ kJ mol}^{-1}$.

(a) Calculate the reactor temperature and exit concentration for feed temperatures of 280, 300, and 320 K.

(b) To maintain the reactor temperature below 373 K, a well-mixed cooling jacket at 290 K is used. Show that it is possible to get 90 percent conversion in this reactor with a feed temperature of 320 K. Do you anticipate any start-up problems?

11. The reversible, first-order reaction shown below takes place in a CSTR.

$$A \underset{k_2}{\overset{k_1}{\rightleftarrows}} B$$

The following data are known:

$$k_1 = 10^3 \exp(-2500/T) \text{ s}^{-1}, T \text{ in K}$$
$$\Delta H_r = -10 \text{ kcal mol}^{-1}$$
$$K = 8 \text{ at } 300 \text{ K}$$
$$C_p = 1 \text{ kcal kg}^{-1} \text{K}^{-1}$$
$$\rho = 1 \text{ kg L}^{-1}$$

(a) For a reactor space time of 10 min, what is the conversion for a 300 K operating temperature? What is the conversion at 500 K? (Remember: the equilibrium constant depends on temperature.)

(b) If the feed temperature is 330 K and the feed concentration is 5 M, what is the necessary heat-removal rate per liter of reactor volume to maintain a 300 K operating temperature?

CHAPTER 10

Reactors Accomplishing Heterogeneous Reactions

10.1 | Homogeneous versus Heterogeneous Reactions in Tubular Reactors

In earlier chapters, tubular reactors of several forms have been described (e.g., laminar flow, plug flow, nonideal flow). One of the most widely used industrial reactors is a tubular reactor that is packed with a solid catalyst. This type of reactor is called a *fixed-bed* reactor since the solid catalyst comprises a bed that is in a fixed position. Later in this chapter, reactors that have moving, solid catalysts will be discussed.

A complete and perfect model for a fixed-bed reactor is not technically possible. However, such a model is not necessary. Rather, what is needed is a reasonably good description that accounts for the major effects. In this chapter, the fixed-bed reactor is analyzed at various degrees of sophistication and the applicability of each level of description is discussed.

Consider the PFR illustrated in Figure 10.1.1a. A mass balance on the reactor volume located within dz can be written as:

$$F_i = F_i + dF_i + (-v_i)\mathrm{r}dV \tag{10.1.1}$$

Since $F_i = uA_C C_i$ and $V = A_C z$, Equation (10.1.1) gives:

$$-u\frac{dC_i}{dz} = (-v_i)\mathrm{r} \tag{10.1.2}$$

provided u and A_C are not functions of z. The units of r are moles/(time · volume). Now let the tube be packed with a solid catalyst and write the material balance again (situation depicted in Figure 10.1.1b):

$$F_i = F_i + dF_i + \eta_o \rho_B(-v_i)\mathrm{r}dV \tag{10.1.3}$$

where the units on r (rate of reaction over the solid catalyst) are now mole/(mass of catalyst)/(time) and ρ_B is the bed density (mass of catalyst)/(volume of bed). Equation (10.1.3) can be written as:

$$-u\frac{dC_A}{dz} = \eta_o \rho_B (-v_i) \mathrm{r} \qquad (10.1.4)$$

provided u and A_C are not functions of z. Note the differences between the PFR accomplishing a homogeneous [Equation (10.1.2)] and a solid catalyzed [Equation (10.1.4)] reaction. First of all, the solid-catalyzed reaction rate is per unit mass of catalyst and must therefore be multiplied by the bed density to obtain a reaction rate per unit volume in the mass balance. This is because the amount of catalyst packed into a reactor can vary from packing to packing with even the same solid. Thus, the bed density must be elucidated after each packing of the reactor, while the reaction rate per mass of catalyst need not be. Second, as discussed in Chapter 6, the overall effectiveness factor can be used to relate the reaction rate

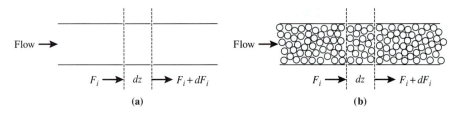

Figure 10.1.1 |
Illustrations of tubular reactors: **(a)** unpacked tube, **(b)** packed tube. $F_i = C_i v = u A_C C_i$.

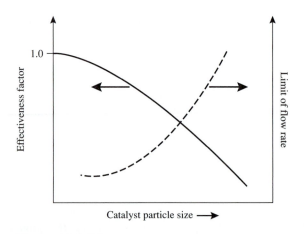

Figure 10.1.2 |
Effects of catalyst particle size on η_o and limits of flow rates.

that occurs within the catalyst particle to the rate that would occur at the local, bulk fluid conditions.

To maximize the reaction rate in the fixed-bed reactor, η_o should be equal to one. In order to do this, smaller particles are necessary (see Chapter 6). As the catalyst particles get smaller, the pressure drop along the reactor increases. (For example, ponder the differences in how difficult it is to have water flow through a column of marbles versus a column of sand.) Thus, a major consideration in the design of a fixed-bed reactor is the trade-off between pressure drop and transport limitations of the rate (illustrated in Figure 10.1.2). A practical design typically involves $\eta_o \neq 1$. Methodologies used to describe fixed-bed reactors given this situation are outlined in the next section.

10.2 | One-Dimensional Models for Fixed-Bed Reactors

Consider a PFR operating at nonisothermal conditions and accomplishing a solid-catalyzed reaction (Figure 10.1.1b). If the fluid properties do not vary over the cross-section of the tube, then only changes along the axial direction need to be considered. If such is the case, then the mass, energy, and momentum balances for a fixed-bed reactor accomplishing a single reaction can be written as:

$$
\left.
\begin{aligned}
-u\frac{dC_i}{dz} &= \eta_o \rho_B (-v_i)\mathrm{r} && \text{(mass)} \\[2mm]
u\rho \overline{C}_p \frac{dT}{dz} &= (-\Delta H_r|_{T^0})\eta_o \rho_B (-v_i)\mathrm{r} - \frac{4U}{d_t}(T - T^*) && \text{(energy)} \\[2mm]
-\frac{dP}{dz} &= \frac{f_f \rho u^2}{g_c d_p} && \text{(momentum)} \\[2mm]
C_i = C_i^0, \quad T = T^0, \quad & P = P^0 \quad \text{at} \quad z = 0
\end{aligned}
\right\}
\tag{10.2.1}
$$

where u and A_C are not functions of z, f_f is a friction factor, and g_c is the gravitational constant. Mass and energy balances of this type were discussed in Chapter 9 for homogeneous reactions [Equation (9.4.9)] and their solutions were illustrated (Example 9.4.3). Here, the only differences are the use of the heterogeneous reaction rate terms and the addition of the momentum balance.

For packed columns with single-phase flow, the Ergun equation can be used for the momentum balance and is:

$$
f_f = \frac{1 - \overline{\varepsilon}_B}{(\overline{\varepsilon}_B)^3}\left[1.75 + 150\frac{(1 - \overline{\varepsilon}_B)}{Re}\right]
\tag{10.2.2}
$$

where $\overline{\varepsilon}_B$ is the void fraction (porosity) of the bed and the Reynolds number is based on the particle size, d_p. The Ergun equation only considers frictional losses because of the packing. However, for $d_t/d_p < 50$, frictional losses from the wall of the tube are also significant. D. Mehta and M. C. Hawley [*Ind. Eng. Chem. Proc.*

Des. Dev., **8** (1969) 280] provided a correction factor to the Ergun equation to consider both the frictional losses from the wall and the packing, and their friction factor expression is:

$$f_f = \left[\frac{1 - \bar{\varepsilon}_B}{(\bar{\varepsilon}_B)^3}\right]\left[1 + \frac{2d_p}{3(1 - \bar{\varepsilon}_B)d_t}\right]^2\left[\frac{1.75}{\left[1 + \dfrac{2d_p}{3(1 - \bar{\varepsilon}_B)d_t}\right]} + 150\frac{(1 - \bar{\varepsilon}_B)}{Re}\right] \quad (10.2.3)$$

Solution of Equation (10.2.1) provides the pressure, temperature, and concentration profiles along the axial dimension of the reactor. The solution of Equation (10.2.1) requires the use of numerical techniques. If the linear velocity is not a function of z [as illustrated in Equation (10.2.1)], then the momentum balance can be solved independently of the mass and energy balances. If such is not the case (e.g., large mole change with reaction), then all three balances must be solved simultaneously.

EXAMPLE 10.2.1

J. P. Kehoe and J. B. Butt [*AIChE J.,* **18** (1972) 347] have studied the kinetics of benzene hydrogenation on a Ni/kieselguhr catalyst. In the presence of excess dihydrogen, the reaction rate is given by:

$$r_B = P_{H_2}k_oK_o\exp\left[\frac{2700\,(\text{cal/mol})}{R_gT}\right]C_B \quad (\text{mol/gcat/s})$$

where

$k_o = 4.22$ mol/(gcat-s-torr)
$K_o = 1.11 \times 10^{-3}$ cm^3/(mol)
$P_{H_2} =$ in torr

T. H. Price and J. B. Butt [*Chem. Eng. Sci.,* **32** (1977) 393] investigated this reaction in an adiabatic, fixed-bed reactor. Using the following data:

$$P_{H_2} = 685 \text{ torr}$$
$$\rho_B = 1.2 \text{ gcat/cm}^3$$
$$L/u = 0.045 \text{ s}$$
$$T^0 = 150°\text{C}$$
$$\bar{C}_p = 1.22 \times 10^5 \text{ J/(kmol-°C)}$$
$$-\Delta H_r|_{T^0} = 2.09 \times 10^8 \text{ J/kmol}$$

calculate the dimensionless concentration of benzene and the dimensionless temperature along the axial reactor dimension assuming $\eta_o = 1$.

■ Answer

The mass and energy balances can be written as (assume u is constant to simplify the solution of the material and energy balances):

$$-u\frac{dC_B}{dz} = \rho_B r_B$$

$$u\rho\overline{C}_p\frac{dT}{dz} = (-\Delta H_r|_{T^0})\rho_B r_B$$

$$C_B = C_B^0, \quad T = T^0 \text{ at } z = 0$$

Let $y = C_B/C_B^0$, $\overline{\theta} = T/T^0$ and $Z = z/L$, so that the mass and energy balances can be expressed as:

$$-\left(\frac{u}{L}\right)\frac{dy}{dZ} = \left(\frac{\rho_B}{C_B^0}\right)r_B$$

$$\rho\overline{C}_p\left(\frac{u}{L}\right)\frac{d\overline{\theta}}{dZ} = (-\Delta H_r|_{T^0})\left(\frac{\rho_B}{T^0}\right)r_B$$

$$y = \overline{\theta} = 1 \text{ at } Z = 0$$

Using the data given, the mass and energy balances reduce to (note that $C_B^0 \simeq 0.1\, C_{\text{total}} = 0.1\rho$):

$$\frac{dy}{dZ} = -0.174\exp\left[\frac{3.21}{\overline{\theta}}\right]y$$

$$\frac{d\overline{\theta}}{dZ} = 0.070\exp\left[\frac{3.21}{\overline{\theta}}\right]y$$

$$y = \overline{\theta} = 1 \text{ at } Z = 0$$

Numerical solution of these equations gives the following results:

Z	y	$\overline{\theta}$
0.0	1.00	1.00
0.1	0.70	1.12
0.2	0.53	1.19
0.3	0.41	1.23
0.4	0.33	1.27
0.5	0.27	1.29
0.6	0.22	1.31
0.7	0.18	1.33
0.8	0.15	1.34
0.9	0.12	1.35
1.0	0.10	1.36

Thus, at the reactor exit $C_B = 0.1C_B^0$ and $T = 302°C$.

Example 10.2.1 illustrates the simultaneous solution of the mass and energy balances for an adiabatic, fixed-bed reactor with no fluid density changes and no transport limitations of the rate, that is, $\eta_o = 1$. Next, situations where these simplifications do not arise are described.

If there is significant mole change with reaction and/or large changes in fluid density because of other factors such as large temperature changes, then Equation (10.2.1) is not appropriate for use. If the fluid density varies along the axial reactor dimension, then the mass and energy balances can be written as:

$$\left.\begin{array}{l} -\dfrac{dF_A}{dz} = A_C\,\eta_o\rho_B(-v_A)\mathrm{r} \\[2mm] \dfrac{d}{dz}\left(\displaystyle\sum_i F_iC_{p_i}T\right) = A_C(-\Delta H_r|_{T^o})\eta_o\rho_B(-v_A)\mathrm{r} - \dfrac{4UA_C}{d_t}(T - T^*) \end{array}\right\} \qquad (10.2.4)$$

Recall that:

$$v = A_Cu$$
$$F_A = F_A^0(1 - f_A)$$
$$F_A = vC_A = v_0(1 + \varepsilon_A f_A)\left[C_A^0\left(\frac{1 - f_A}{1 + \varepsilon_A f_A}\right)\right]\left(\frac{P^0}{P}\right)\left(\frac{T}{T^0}\right)$$
$$\sum_i F_iC_{p_i} = v\rho\overline{C}_p$$

Using these expressions in Equation (10.2.4) gives:

$$\left.\begin{array}{l} -\dfrac{dF_A}{dz} = A_C\,\eta_o\rho_B(-v_A)\mathrm{r} \\[2mm] u\rho\overline{C}_p\dfrac{dT}{dz} = (-\Delta H_r|_{T^o})\eta_o\rho_B(-v_A)\mathrm{r} - \dfrac{4U}{d_t}(T - T^*) \\[2mm] -\dfrac{dP}{dz} = \dfrac{f_f\rho u^2}{g_cd_p} \\[2mm] F_A = F_A^0, \quad T = T^0, \quad P = P^0 \quad \text{at} \quad z = 0 \end{array}\right\} \qquad (10.2.5)$$

assuming $\overline{C}_p$ is not a strong function of T. All three balances in Equation (10.2.5) must be solved simultaneously, since $u = u_o(1 + \varepsilon_A f_A)(P^0/P)(T/T^0)$.

Thus far, the overall effectiveness factor has been used in the mass and energy balances. Since η_o is a function of the local conditions, it must be computed along the length of the reactor. If there is an analytical expression for η_o, for example for an isothermal, first-order reaction rate:

$$\eta_o = \frac{\tanh(\phi)}{\phi\left[1 + \dfrac{\phi \tanh(\phi)}{Bi_m}\right]} \qquad (6.4.34)$$

then the use of η_o is straightforward. As the mass and energy balances are being solved along the axial reactor dimension, η_o can be computed by calculating ϕ and Bi_m at each point. The assumption of an isothermal η_o may be appropriate since it is calculated at each point along the reactor. That is to say that although the temperature varies along the axial direction, at each point along this dimension, the catalyst particle may be isothermal (e.g., at $Z = 0$, $\eta_o = \eta_o(T^0)$ and at $Z \neq 0$, $\eta_o = \eta_o(T) \neq \eta_o(T^0)$).

For most solid-catalyzed reactions, there will not be analytical solutions for η_o. In this case, the effectiveness factor problem must be computed at each point in the reactor. The reactor balances like those given in Equation (10.2.1) that now incorporate the·effectiveness factor description are:

Bulk Fluid Phase:

$$\left.\begin{aligned} -u\frac{dC_{AB}}{dz} &= \bar{k}_c\, a_v\left(C_{AB} - C_{AS}\right) \\[2mm] u\rho\, \bar{C}_p\frac{dT_B}{dz} &= h_t a_v(T_S - T_B) - \frac{4U}{d_t}(T_B - T^*) \\[2mm] -\frac{dP}{dz} &= \frac{f_f\rho u^2}{g_c d_p} \\[2mm] C_{AB} = (C_{AB})^0,\quad T_B &= (T_B)^0,\quad P = P^0 \quad \text{at}\quad Z = 0 \end{aligned}\right\} \tag{10.2.6}$$

Solid Phase (Spherical Catalyst Particle):

$$\left.\begin{aligned} D^e\left(\frac{d^2 C_A}{d\bar{r}^2} + \frac{2}{\bar{r}}\frac{dC_A}{d\bar{r}}\right) &= \rho_p\,(-\nu_A)\mathrm{r}\,(C_A, T) \\[2mm] \lambda^e\left(\frac{d^2 T}{d\bar{r}^2} + \frac{2}{\bar{r}}\frac{dT}{d\bar{r}}\right) &= (-\Delta H_r)\,\rho_p(-\nu_A)\mathrm{r}\,(C_A, T) \end{aligned}\right\} \tag{10.2.7}$$

with

$$\frac{dT}{d\bar{r}} = \frac{dC_A}{d\bar{r}} = 0 \quad \text{at}\quad \bar{r} = 0$$

$$\bar{k}_c\,(C_{AS} - C_{AB}) = -D^e\frac{dC_A}{d\bar{r}} \quad \text{at}\quad \bar{r} = R_p$$

$$h_t(T_S - T_B) = -\lambda^e\frac{dT}{d\bar{r}} \quad \text{at}\quad \bar{r} = R_p$$

where

$$T = T_S \quad \text{at}\quad \bar{r} = R_p$$

$$C_A = C_{AS} \quad \text{at}\quad \bar{r} = R_p$$

and

$$C_{AB}, T_B = \text{bulk fluid phase concentration of } A \text{ and temperature}$$

$$\rho_p = \text{density of the catalyst particle}$$

$$a_v = \text{external catalyst particle surface area per unit reactor volume}$$

Notice how the catalyst particle balances are coupled to the reactor mass and energy balances. Thus, the catalyst particle balances [Equation (10.2.7)] must be solved at each position along the axial reactor dimension when computing the bulk mass and energy balances [Equation (10.2.6)]. Obviously, these solutions are lengthy.

Since most practical problems do not have analytical expressions for the effectiveness factor, the use of Equations (10.2.6) and (10.2.7) are more generally applicable. If Equations (10.2.6) and (10.2.7) are to be solved numerous times (e.g., when performing a reactor design/optimization), then an alternative approach may be applicable. The catalyst particle problem can first be solved for a variety of $C_{AB}, T_B, \phi, Bi_m, Bi_h$ that may be expected to be realized in the reactor design. Second, the η_o values can be fit to a function with $C_{AB}, T_B, \phi, Bi_m, Bi_h$ as variables, that is,

$$\eta_o = \eta_o(C_{AB}, T_B, \phi, Bi_m, Bi_h) \tag{10.2.8}$$

Finally, using the function (normally nonlinear) for η_o in Equation (10.2.1) allows a more efficient numerical solution of the reactor balances.

VIGNETTE 10.2.1

In Example 3.4.3, it was shown that the solution to the PFR mass balance can be obtained by using a series of CSTRs provided the number within the series is sufficiently large. This approach can be used to solve reactor descriptions like Equation (10.2.1) and Equations (10.2.6) and (10.2.7). Since numerical software for solving Equation (10.2.1) and Equations (10.2.6) and (10.2.7) is now readily available it is less likely that the conversion of these differential equations into systems of algebraic equations (series of CSTRs) is necessary. However, there may be situations where this approach is preferred. When doing so it is advisable to always check for the number of stages (CSTRs) that are necessary for solution. As an example of this approach, refer to the reactor situation provided in Vignette 8.5.1. Simulation of the plug flow model of the fixed-film bioreactor was performed by transforming the reactor description into a series of CSTR equations. The solution of these equations is illustrated in Figure 10.2.1 [from Y. Park et al., *Biotech. Bioeng.*, **26** (1984) 457].

Note the differences in the solution for the number of stages (NSTAGE) (i.e., CSTRs) less than 50. Although not shown, all solutions using 50 or greater CSTRs were indistinguishable. Once the number of stages necessary for solution is identified, then the reactor simulations can be performed.

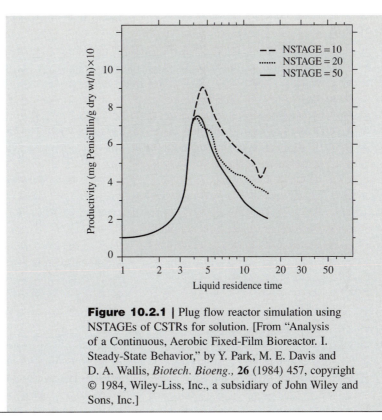

Figure 10.2.1 | Plug flow reactor simulation using NSTAGEs of CSTRs for solution. [From "Analysis of a Continuous, Aerobic Fixed-Film Bioreactor. I. Steady-State Behavior," by Y. Park, M. E. Davis and D. A. Wallis, *Biotech. Bioeng.*, **26** (1984) 457, copyright © 1984, Wiley-Liss, Inc., a subsidiary of John Wiley and Sons, Inc.]

In Chapter 8, axial dispersion in tubular reactors was discussed. Typical industrial reactors have sufficiently high flow rates and reactor lengths so the effects of axial dispersion are minimal and can be neglected. A rule of thumb is that axial dispersion can be neglected if:

$$L/d_p > 50 \qquad \text{isothermal}$$
$$L/d_p > 150 \qquad \text{nonisothermal}$$

Additionally, L. C. Young and B. A. Finlayson [*Ind. Eng. Chem. Fund.*, **12** (1973) 412] showed that if:

$$\left| \frac{\eta_o(-\nu_A)\mathrm{r}\rho_B d_p}{uC_{AB}} \right| \ll \left| \frac{ud_p}{D_a} \right| = Pe_a$$

then axial dispersion can be neglected. However, for laboratory reactors that can have low flow rates and are typically of short axial length, axial dispersion may become important. If such is the case, then the effects of axial dispersion may be needed to accurately describe the reactor.

EXAMPLE 10.2.2

An axially-dispersed, adiabatic tubular reactor can be described by the following mass and energy balances that are in dimensionless form (the reader should verify that these descriptions are correct):

$$\frac{1}{Pe_a}\frac{d^2y}{dZ^2} - \frac{dy}{dZ} - r(y,\bar{\theta}) = 0$$

$$\frac{1}{Bo_a}\frac{d^2\bar{\theta}}{dZ^2} - \frac{d\bar{\theta}}{dZ} - \bar{H}_w r(y,\bar{\theta}) = 0$$

with

$$\frac{1}{Pe_a}\frac{dy}{dZ} = y - 1 \quad \text{at} \quad Z = 0$$

$$\frac{1}{Bo_a}\frac{d\bar{\theta}}{dZ} = \bar{\theta} - 1 \quad \text{at} \quad Z = 0$$

$$\frac{dy}{dZ} = 0 \qquad \text{at} \quad Z = 1$$

$$\frac{d\bar{\theta}}{dZ} = 0 \qquad \text{at} \quad Z = 1$$

If $\bar{H}_w = -0.05$, and $r(y,\bar{\theta}) = 4y\exp\left[18\left(1 - \frac{1}{\bar{\theta}}\right)\right]$, calculate y and $\bar{\theta}$ for $Pe_a = Bo_a = 10$, 20, and 50 (Bo_a is a dimensionless group analogous to the axial Peclet number for the energy balance).

■ **Answer**

The dimensionless system of equations shown above are numerically solved to give the following results.

Z	$Pe_a = 10$		$Pe_a = 20$		$Pe_a = 50$	
	y	$\bar{\theta}$	y	$\bar{\theta}$	y	$\bar{\theta}$
0.0	0.707	1.01	0.824	1.01	0.923	1.00
0.2	0.276	1.03	0.293	1.04	0.309	1.03
0.4	0.092	1.05	0.082	1.05	0.072	1.05
0.6	0.029	1.05	0.021	1.05	0.015	1.05
0.8	0.009	1.05	0.005	1.05	0.003	1.05
1.0	0.004	1.05	0.002	1.05	0.001	1.05

As $Pe_a \to \infty$ the reactor behavior approaches PFR. From the results illustrated, this trend is shown (for PFR, $y = \bar{\theta} = 1$ at $Z = 0$).

In general, a one-dimensional description for fixed-bed reactors can be used to capture the reactor behavior. For nonisothermal reactors, the observation of hot spots (see Example 9.4.3) and reactor stability (see Example 9.6.2) can often be described.

However, there are some situations where the one-dimensional descriptions do not work well. For example, with highly exothermic reactions, a fixed-bed reactor may contain several thousand tubes packed with catalyst particles such that $d_t/d_p \sim 5$ in order to provide a high surface area per reaction volume for heat transfer. Since the heat capacities of gases are small, radial temperature gradients can still exist for highly exothermic gas-phase reactions, and these radial variations in temperature produce large changes in reaction rate across the radius of the tube. In this case, a two-dimensional reactor model is required.

10.3 | Two-Dimensional Models for Fixed-Bed Reactors

Consider a tubular fixed-bed reactor accomplishing a highly exothermic gas-phase reaction. Assuming that axial dispersion can be neglected, the mass and energy balances can be written as follows and allow for radial gradients:

$$
\left.
\begin{aligned}
&\frac{\partial}{\partial z}(uC_A) - \frac{1}{\bar{r}}\frac{\partial}{\partial \bar{r}}\left(\bar{r}D_r\frac{\partial C_A}{\partial \bar{r}}\right) + \eta_o\rho_B(-v_A)\mathrm{r} = 0 \\[2mm]
&\frac{\partial}{\partial z}(u\rho\overline{C}_pT) - \frac{1}{\bar{r}}\frac{\partial}{\partial \bar{r}}\left(\bar{r}\overline{\lambda}_r\frac{\partial T}{\partial \bar{r}}\right) + (-\Delta H_r)\eta_o\rho_B(-v_A)\mathrm{r} = 0
\end{aligned}
\right\} \quad (10.3.1)
$$

where D_r is the radial dispersion coefficient (see Chapter 8) and $\overline{\lambda}_r$ is the effective thermal conductivity in the radial direction. If D_r and $\overline{\lambda}_r$ are not functions of $\bar{r}$ then Equations (10.3.1) can be written as (u not a function of z):

$$
\left.
\begin{aligned}
&u\frac{\partial C_A}{\partial z} = D_r\left[\frac{\partial^2 C_A}{\partial \bar{r}^2} + \frac{1}{\bar{r}}\frac{\partial C_A}{\partial \bar{r}}\right] - \eta_o\rho_B(-v_A)\mathrm{r} \\[2mm]
&u\rho\overline{C}_p\frac{\partial T}{\partial z} = \overline{\lambda}_r\left[\frac{\partial^2 T}{\partial \bar{r}^2} + \frac{1}{\bar{r}}\frac{\partial T}{\partial \bar{r}}\right] - (-\Delta H_r)\eta_o\rho_B(-v_A)\mathrm{r}
\end{aligned}
\right\} \quad (10.3.2)
$$

The conditions necessary to provide a solution to Equation (10.3.2) are:

$$
\left.
\begin{aligned}
&C_A = C_A^0, \qquad T = T^0 \quad \text{at} \quad z = 0 \quad \text{for} \quad 0 \leq \bar{r} \leq d_t/2 \\[2mm]
&\frac{\partial C_A}{\partial \bar{r}} = \frac{\partial T}{\partial \bar{r}} = 0 \quad \text{at} \quad \bar{r} = 0 \quad \text{for} \quad 0 \leq z \\[2mm]
&\frac{\partial C_A}{\partial \bar{r}} = 0 \quad \text{at} \quad \bar{r} = d_t/2 \quad \text{for} \quad 0 \leq z \\[2mm]
&-\overline{\lambda}_r\frac{\partial T}{\partial \bar{r}} = h_t(T - T_w) \quad \text{at} \quad \bar{r} = d_t/2 \quad \text{for} \quad 0 \leq z
\end{aligned}
\right\} \quad (10.3.3)
$$

where h_t is the heat transfer coefficient at the reactor wall and T_w is the wall temperature. Solution of Equations (10.3.2) and (10.3.3) gives $C_A(\bar{r}, z)$ and $T(\bar{r}, z)$.

EXAMPLE 10.3.1

G. F. Froment [*Ind. Eng. Chem.*, **59** (1967) 18] developed a two-dimensional model for a fixed-bed reactor accomplishing the following highly exothermic gas phase reactions:

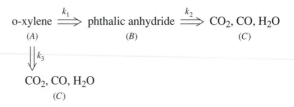

The steady-state mass and energy balances are (in dimensionless form):

$$\frac{\partial y_1}{\partial Z} = Pe_r \left[\frac{\partial^2 y_1}{\partial \overline{R}^2} + \frac{1}{\overline{R}} \frac{\partial y_1}{\partial \overline{R}} \right] + \beta\beta_1 r_1$$

$$\frac{\partial y_2}{\partial Z} = Pe_r \left[\frac{\partial^2 y_2}{\partial \overline{R}^2} + \frac{1}{\overline{R}} \frac{\partial y_2}{\partial \overline{R}} \right] + \beta\beta_1 r_2$$

$$\frac{\partial \overline{\theta}}{\partial Z} = Bo_r \left[\frac{\partial^2 \overline{\theta}}{\partial \overline{R}^2} + \frac{1}{\overline{R}} \frac{\partial \overline{\theta}}{\partial \overline{R}} \right] + \beta\beta_2 r_1 + \beta\beta_3 r_2$$

with

$$y_1 = y_2 = 0 \qquad \text{at} \quad Z = 0 \quad \text{for} \quad 0 \le \overline{R} \le 1$$

$$\overline{\theta} = 1 \qquad \text{at} \quad Z = 0 \quad \text{for} \quad 0 \le \overline{R} \le 1$$

$$\frac{\partial y_1}{\partial \overline{R}} = \frac{\partial y_2}{\partial \overline{R}} = \frac{\partial \overline{\theta}}{\partial \overline{R}} = 0 \quad \text{at} \quad \overline{R} = 0 \quad \text{for} \quad 0 \le Z \le 1$$

$$\frac{\partial y_1}{\partial \overline{R}} = \frac{\partial y_2}{\partial \overline{R}} = 0 \qquad \text{at} \quad \overline{R} = 1 \quad \text{for} \quad 0 \le Z \le 1$$

$$\frac{\partial \overline{\theta}}{\partial \overline{R}} = \overline{H}_w \left(\overline{\theta}_w - \overline{\theta} \right) \qquad \text{at} \quad \overline{R} = 1 \quad \text{for} \quad 0 \le Z \le 1$$

where

$$y_1 = C_B / C_A^0$$
$$y_2 = C_C / C_A^0$$
$\overline{\theta}$ = dimensionless temperature
Z = dimensionless axial coordinate
$\overline{R}$ = dimensionless radial coordinate
$r_1 = k_1(1 - y_1 - y_2) - k_2 y_1$
$r_2 = k_2 y_1 + k_3(1 - y_1 - y_2)$
$\beta\beta_i$ = dimensionless groups
$\overline{\theta}_w$ = dimensionless wall temperature

Using the data given by Froment,

$$Pe_r = 5.706, Bo_r = 10.97, \overline{H}_w = 2.5$$

$$\beta\beta_1 = 5.106, \beta\beta_2 = 3.144, \beta\beta_3 = 11.16$$

Additionally,

$$k_1 = \exp\left[-1.74 + 21.6\left(1 - \frac{1}{\theta}\right)\right]$$

$$k_2 = \exp\left[-4.24 + 25.1\left(1 - \frac{1}{\theta}\right)\right]$$

$$k_3 = \exp\left[-3.89 + 22.9\left(1 - \frac{1}{\theta}\right)\right]$$

If $\overline{\theta}_w = 1$, compute $y_1(Z, \overline{R})$, $y_2(Z, \overline{R})$ and $\overline{\theta}(Z, \overline{R})$.

■ Answer

This reactor description is a coupled set of parabolic partial differential equations that are solved numerically. The results shown here were obtained using the software package PDECOL [N. K. Madsen and R. F. Sincovec, *ACM Toms*, **5** (1979) 326].

   Below are listed the radial profiles for two axial positions within the reactor. Notice that at $Z = 0.2$, there are significant radial gradients while by $Z = 0.6$ these gradients are all but eliminated. If the inlet and coolant temperatures are around 360°C, then the centerline and inner wall temperatures at $Z = 0.2$ are 417°C and 385°C, respectively, giving a radial temperature difference of 32°C.

$\overline{R}$	Z = 0.2		
	y_1	y_2	$\overline{\theta}$
0.0	0.458	0.782	1.09
0.2	0.456	0.777	1.09
0.4	0.451	0.763	1.08
0.6	0.445	0.747	1.07
0.8	0.440	0.734	1.05
1.0	0.438	0.728	1.04

$\overline{R}$	Z = 0.6		
	y_1	y_2	$\overline{\theta}$
0.0	0.613	0.124	1.02
0.2	0.613	0.124	1.02
0.4	0.613	0.124	1.01
0.6	0.612	0.124	1.01
0.8	0.612	0.124	1.01
1.0	0.612	0.124	1.01

   As illustrated in Example 10.3.1, the radial temperature gradient can be significant for highly exothermic, gas-phase reactions. Therefore, one must make a decision for a particular situation under consideration as to whether a 1-D or 2-D analysis is needed.

10.4 | Reactor Configurations

Thus far, fixed-bed reactor descriptions have been presented. Schematic representations of fixed-bed reactors are provided in Figure 10.4.1, and a photograph of a commercial reactor is provided in Figure 10.4.2.

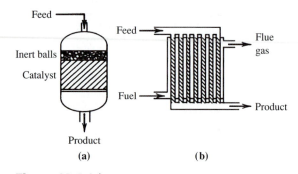

Figure 10.4.1 |

Fixed-bed reactor schematics: **(a)** adiabatic, **(b)** nonadiabatic. [From "Reactor Technology" by B. L. Tarmy, *Kirk-Othmer Encyclopedia of Chemical Technology,* vol. 19, 3rd ed., Wiley (1982). Reprinted by permission of John Wiley and Sons, Inc., copyright © 1982.]

Figure 10.4.2 |

Photograph of a commercial reactor. (Image provided by T. F. Degnan of Exxon Mobil.)

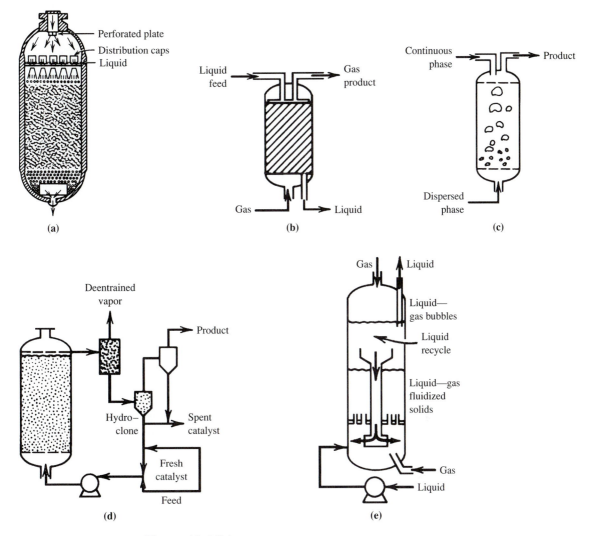

Figure 10.4.3 |

Examples of multiphase reactors: (**a**) trickle-bed reactor, (**b**) countercurrent packed-bed reactor, (**c**) bubble column, (**d**) slurry reactor, and (**e**) a gas-liquid fluidized bed. [From "Reactor Technology" by B. L. Tarmy, *Kirk-Othmer Encyclopedia of Chemical Technology,* vol. 19, 3rd ed., Wiley (1982). Reprinted by permission of John Wiley and Sons, Inc., copyright © 1982.]

The reactor models developed for fixed-bed reactors have been exploited for use in other situations, for example, CVD reactors (see Vignette 6.4.2) for micro-electronics fabrication. These models are applicable to reaction systems involving single fluid phases and nonmoving solids. There are numerous reaction systems that involve more than one fluid phase. Figure 10.4.3 illustrates various types of reactors

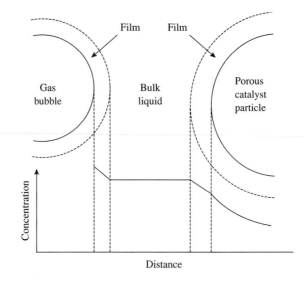

Figure 10.4.4 |
Concentration profile of a gas-phase reactant in a porous
solid-catalyzed reaction occurring in a three phase reactor.

that process multiple fluid phases. Figure 10.4.3a shows a schematic of a trickle-bed reactor. In this configuration, the solid catalyst remains fixed and liquid is sprayed down over the catalyst to create contact. Additionally, gas-phase components can enter either the top (cocurrent) or the bottom (countercurrent; Figure 10.4.3b) of the reactors to create three-phase systems. The three-phase trickle-bed reactor provides large contact between the catalyst and fluid phases. Typical reactions that exploit this type of contact are hydrogenations (H_2—gas, reactant to be hydrogenated—liquid) and halogenations. Bubble (Figure 10.4.3c), slurry (Figure 10.4.3d), and fluidized (Figure 10.4.3e) reactors can also accomplish multiphase reactions (solid catalyst in liquid phase for both cases). As with all multiphase reactors, mass transfer between phases (see Figure 10.4.4) can be very important to their overall performance.

The mathematical analyses of these reactor types will not be presented here. Readers interested in these reactors should consult references specific to these reactor types.

**VIGNETTE
10.4.1**

Fixed-bed reactors are commonly employed by the petrochemicals industries to accomplish a broad spectrum of reactions, including steps to clean up air streams prior to exhausting them into the atmosphere. For example, fixed-bed reactors are used to process air streams containing NO_x by reaction with NH_3 over solid catalysts to give N_2 and H_2O. An interesting application of fixed-bed reactors for environmental cleanup is from workers at Engelhard Corporation, who have developed fixed-bed reactors for treating cooking fumes (U.S. Patent 5,580,535). Exhaust plumes from fast food restaurants contain components that contribute to air pollution. Engelhard has implemented in some fast food restaurants a fixed-bed catalytic reactor to oxidize the exhaust components to CO_2. A second example where fixed beds are used to clean up exhaust streams involves wood stoves. In certain areas of the United States, the chimneys of wood stoves now contain fixed beds of catalysts to assist in the cleanup of wood stove fumes prior to their release into the atmosphere. These two examples show how the fixed-bed reactor is not limited to use in the petrochemical industries.

10.5 | Fluidized Beds with Recirculating Solids

Fluidized beds with recirculating solids developed primarily as reactors for the catalytic cracking of oil to gasoline. Today their uses are much broader. Prior to the early 1940s, catalytic cracking of oil was performed in fixed-bed reactors. After that time, fluidized-bed reactors have been developed for this application. Today, fluidized beds with recirculating zeolite-based catalysts are used for catalytic cracking of oil.

VIGNETTE 10.5.1

In the late 1950s and early 1960s, certain zeolites (Vignette 5.3.1) were found to be excellent catalysts for catalytic cracking. They were observed to improve yields to gasoline, and in the early 1960s the Mobil Corporation introduced the first zeolite-based, fluidized-bed catalytic cracking process. Virtually all refineries in the world use some variation of this theme for catalytic cracking. It has been claimed that the improved efficiency of cracking with zeolite catalysts has provided the United States with a savings of greater than 400 million barrels of oil per year. At $30 a barrel, this savings amounts to more than $12 billion a year.

Figure 10.5.1 shows a schematic of a fluidized bed with recirculating solids. The oil feed enters at the bottom of the riser reactor and is combined with the solid catalyst. The vapor moves upward through the riser at a flow rate sufficient to fluidize the solid catalyst and transport it to the top of the riser in a millisecond timeframe. The catalytic cracking reactions occur during the transportation of the solids to the top of the riser and the catalyst becomes covered with carbonaceous residue (coke) that deactivates it by that point. The solids are separated from the products and enter a fluidized-bed, regeneration reactor. In the regeneration reactor, the carbonaceous residue is removed from the catalyst by combustion with the oxygen in the air. The "clean" catalyst is then returned to the bottom of the riser reactor. Thus, the solid catalyst recirculates between the two reactors. Typical fluidized beds with recirculating solids contain around 250 tons of catalyst of 20–80 microns in particle

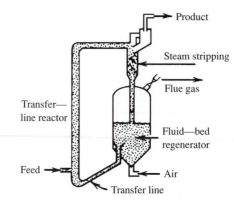

Figure 10.5.1 |
Schematic of fluidized-bed reactor with
recirculating solids. [From "Reactor
Technology" by B. L. Tarmy, *Kirk-Othmer
Encyclopedia of Chemical Technology,* vol.
19, 3rd ed., Wiley (1982). Reprinted by
permission of John Wiley and Sons, Inc.,
copyright © 1982.]

size and have circulation rates of around 25 ton/min. As might be expected, the
solids are continuously being degraded into smaller particles that exit the reactor.
Typical catalyst losses are around 2 ton/day for a catalyst reactor inventory of 250
ton. Thus, catalyst must be continually added.

The fluidized-bed reactor with recirculating solids is an effective reactor configuration for accomplishing reactions that have fast deactivation and regeneration. In addition to catalytic cracking, this configuration is now being used for
another type of reaction. The oxidation of hydrocarbons can be accomplished with
oxide-based catalysts. In the mid-1930s, Mars and van Krevelen postulated that
the oxidation of hydrocarbons by oxide-containing catalysts could occur in two
steps:

$$\textit{Overall:} \quad RH_2 + O_2 \Rightarrow RO + H_2O$$
$$\text{(hydrocarbon)} \quad \text{(oxidized product)}$$

$$\textit{Step 1:} \quad RH_2 + 2O^{-2} \Rightarrow RO + H_2O + 4e^- + 2\square$$
$$\text{(lattice oxygen)} \qquad\qquad \text{(surface oxygen vacancy)}$$

$$\textit{Step 2:} \quad O_2 + 2\square + 4e^- \Rightarrow 2O^{-2}$$

In the first step the hydrocarbon is oxidized by lattice oxygen from the oxide-
containing catalyst to create surface oxygen vacancies in the oxide. The second step
involves filling the vacancies by oxidization of the reduced oxide surface with

dioxygen. This mechanism has been shown to hold for numerous oxide-based catalytic oxidations.

Dupont has recently commercialized the reaction of n-butane to maleic anhydride using a vanadium phosphorus oxide (VPO) catalyst in a recirculating solid reactor system. The overall reaction is:

$$CH_3(CH_2)_2CH_3 \xrightarrow{\text{VPO}}$$

A proposed reaction pathway over the VPO catalyst is as shown below:

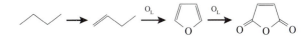

where O_L denotes lattice oxygen. In addition to these reactions, each of the organic compounds can be oxidized to CO_2 and CO. In the absence of gas-phase O_2 the selectivity to CO_2 and CO is minimized. A solids recirculating reactor system like that schematically illustrated in Figure 10.5.1 was developed to accomplish the butane oxidation. In the riser section, oxidized VPO is contacted with butane to produce maleic anhydride. The reduced VPO solid is then re-oxidized in the regeneration reactor. Thus, the VPO is continually being oxidized and reduced (Mars–van Krevelen mechanism). This novel reactor system opens the way to explore the performance of other gas-phase oxidations in solids recirculating reactors.

There are numerous other types of reactors (e.g., membrane reactors) that will not be discussed in this text. Readers interested in reactor configuration should look for references specific to the reactor configuration of interest.

Exercises for Chapter 10

1. Consider the reactor system given in Example 10.2.1.
   (a) Compare the dimensionless concentration profile from the adiabatic case given in the example to that obtained at isothermal conditions.
   (b) What happens if the reactor is operated adiabatically and the heat of reaction is doubled?
   (c) If the reactor in part (b) is now cooled with a constant wall temperature of 150°C, what is the value of $4U/d_t$ that is necessary to bring the outlet temperature to approximately the value given in Example 10.2.1?

2. Plot the dimensionless concentration and temperature profiles for the reactor system given in Example 10.2.2 for $Pe_a = Bo_a = \infty$ and compare them to those obtained when $Pe_a = Bo_a = 15$.

3. Reproduce the results given in Example 10.3.1. What happens as $\overline{H}_w$ is decreased?

4. Design a multitube fixed-bed reactor system to accomplish the reaction of naphthalene (N) with air to produce phthalic anhydride over a catalyst of vanadium pentoxide on silica gel at a temperature of about 610–673 K.

The selective oxidation reaction is carried out in excess air so that it can be considered pseudo-first-order with respect to naphthalene. Analysis of available data indicates that the reaction rate per unit mass of catalyst is represented by:

$$r = k_1 C_N \quad \text{with } k_1 = 1.14 \times 10^{13} \exp(-19,000/T) \text{ (cm}^3 \text{ s}^{-1} \text{ gcat}^{-1})$$

where C_N = the concentration of naphthalene in mol cm^{-3}, and T = absolute temperature in K. (Data for this problem were adapted from C. G. Hill, Jr. *An Introduction to Chemical Engineering Kinetics and Reactor Design,* Wiley, New York, 1977, p. 554.) Assume for this problem that the overall effectiveness factor is unity (although in reality this assumption is likely to be incorrect!). To keep the reactant mixture below the explosion limit of about 1 percent naphthalene in air, consider a feed composition of 0.8 mol % naphthalene vapor in 99.2 percent air at 1.7 atm total pressure. Although there will be some complete oxidation of the naphthalene to CO_2 and H_2O (see part c), initially assume that the only reaction of significance is the selective oxidation to phthalic anhydride as long as the temperature does not exceed 673 K anywhere in the reactor:

$$C_{10}H_8 + 4.5 O_2 \Rightarrow C_8H_4O_3 + 2 H_2O + 2 CO_2$$

The heat of reaction is -429 kcal mol^{-1} of naphthalene reacted. The properties of the reaction mixture may be assumed to be equivalent to those of air ($C_p = 0.255$ cal g^{-1} K^{-1}, $\bar{\mu} = 3.2 \times 10^{-4}$ g cm^{-1} s^{-1}) in the temperature range of interest.

It is desired to determine how large the reactor tubes can be and still not exceed the maximum temperature of 673 K. The reactor will be designed to operate at 95 percent conversion of $C_{10}H_8$ and is expected to have a production rate of 10,000 lb/day of phthalic anhydride. It will be a multitube type reactor with heat transfer salt circulated through its jacket at a temperature equivalent to the feed temperature.

The catalyst consists of 0.32 cm diameter spheres that have a bulk density of 0.84 g cm^{-3} and pack into a fixed bed with a void fraction of 0.4. The mass velocity of the gas through each tube will be 0.075 g cm^{-2} s^{-1}, which corresponds to an overall heat transfer coefficient between the tube wall and the reacting fluid of about 10^{-3} cal cm^{-2} s^{-1} K^{-1}.

(a) Show the chemical structures of the reactants and products. Give the relevant material, energy, and momentum balances for a plug flow reactor. Define all terms and symbols carefully.

(b) Determine the temperature, pressure, and naphthalene concentration in the reactor as a function of catalyst bed depth using a feed temperature of 610 K and tubes of various inside diameters, so that a maximum temperature of 673 K is never exceeded. Present results for several different tube diameters. What is the largest diameter tube that can be used without exceeding 673 K? How many tubes will be required, and how long must they be to attain the desired conversion and production rates? For this optimal tube diameter, vary the feed temperature (615, 620 K) and naphthalene mole fraction (0.9 percent and 1 percent) to examine the effects of these parameters on reactor behavior.

(c) Combustion can often occur as an undesirable side reaction. Consider the subsequent oxidation of phthalic anhydride (PA) that also occurs in the reactor according to:

$$C_8H_4O_3 + 7.5\,O_2 \Rightarrow 2\,H_2O + 8\,CO_2$$

with rate: $r = k_2 C_{PA}$ where $k_2 = 4.35 \times 10^4 \exp(-10,000/T)$ cm^3 s^{-1} gcat^{-1} and $\Delta H_r = -760$ kcal mol^{-1}. Redo part (b) and account for the subsequent oxidation of PA in the reactor. Include the selectivity to phthalic anhydride in your analysis.

5. When exothermic reactions are carried out in fixed-bed reactors, hot spots can develop. Investigate the stability of the fixed-bed reactor described below for: (a) cocurrent coolant flow and (b) countercurrent coolant flow when the inlet coolant temperature (T_C^0) is 350 K and higher. Calculate the sensitivity by plotting the maximum temperature in the reactor versus the inlet coolant temperature (the slope of this line is the sensitivity). Which mode of cooling minimizes the sensitivity? (This problem is adapted from material provided by Jean Cropley.)

Data:
Reaction system (solid-catalyzed):

$$\text{Dammitol} + \frac{1}{2}O_2 \overset{r_1}{\Rightarrow} \text{Valualdehyde} + H_2O,$$
$$\text{(DT)} \qquad\qquad\qquad \text{(VA)}$$

$$\text{Valualdehyde} + \frac{5}{2}O_2 \overset{r_2}{\Rightarrow} 2CO_2 + 2H_2O,$$
$$\text{(VA)}$$

with $\Delta H_{r_1} = -74.32$ kBTU/lbmol, $\Delta H_{r_2} = -474.57$ kBTU/lbmol. The rate expressions for these reactions are:

$$RT = \left(\frac{1}{T} - \frac{1}{373}\right)\Big/ 1.987$$

$$P_{O_2} = X_{O_2} P_T$$
$$P_{DT} = X_{DT} P_T$$
$$P_{VA} = X_{VA} P_T$$

$$r_1 = \frac{k_1 \exp[-k_2 RT] P_{O_2}^{k_3} P_{DT}^{k_4}}{1 + k_5 P_{O_2}^{k_6} + k_7 P_{DT}^{k_8} + k_9 P_{VA}^{k_{10}}} \quad \text{lbmol/(lbcat-h)}$$

$$r_2 = \frac{k_{11} \exp[-k_{12} RT] P_{O_2}^{k_{13}} P_{VA}^{k_{14}}}{1 + k_5 P_{O_2}^{k_6} + k_7 P_{DT}^{k_8} + k_9 P_{VA}^{k_{10}}} \quad \text{lbmol/(lbcat-h)}$$

where P_T is the total pressure and:

$$
\begin{array}{ll}
k_1 = 1.771 \times 10^{-3} & k_8 = 1.0 \\
k_2 = 23295.0 & k_9 = 1.25 \\
k_3 = 0.5 & k_{10} = 2.0 \\
k_4 = 1.0 & k_{11} = 2.795 \times 10^{-4} \\
k_5 = 0.8184 & k_{12} = 33000.0 \\
k_6 = 0.5 & k_{13} = 0.5 \\
k_7 = 0.2314 & k_{14} = 2.0
\end{array}
$$

The governing differential equations are (the reader should derive these equations):

mass balances:

$$\frac{dX_i}{dz} = \frac{(NT) R_i \rho_B A_C \cdot \overline{MW}}{W} + \frac{X_i}{\overline{MW}} \frac{d\overline{MW}}{dz}, \quad i = 1(\text{DT}), 2(\text{VA}), 3(\text{O}_2), 4(\text{CO}_2), 5(\text{H}_2\text{O})$$

energy balance on reacting fluid:

$$\frac{dT_R}{dz} = \frac{-(NT) A_C (r_1 \Delta H_{r_1} + r_2 \Delta H_{r_2}) \rho_B - \pi (NT) d_t U(T_R - T_C)}{W \cdot \overline{C}_p}$$

energy balance on coolant:

$$\frac{dT_C}{dz} = \frac{(\text{mode}) \pi (NT) d_t U(T_R - T_C)}{F_C C_{pC}}$$

mole change due to reactions—reactions in the presence of inert N_2:

$$\frac{d\overline{MW}}{dz} = \sum_{i=1}^{NC} M_i \frac{dX_i}{dz} - 28 \sum_{i=1}^{NC} \frac{dX_i}{dz}$$

momentum balance [alternative form from Equation (10.2.3) that accounts for the losses from both the wall and catalyst particles]:

$$\frac{dP_T}{dz} = \frac{-\alpha^2}{(32.2)(144)} \left[\frac{150 \overline{\mu} u}{d_p^2} \frac{(1 - \overline{\varepsilon}_B)^2}{\overline{\varepsilon}_B^3} + \frac{1.75 \overline{MW} u^2}{\alpha R_g T_R d_p} \frac{(1 - \overline{\varepsilon}_B)}{\overline{\varepsilon}_B^3} P_T \right]$$

where

$$u = \frac{(359)(14.7)W \cdot T}{(273)(3600)(NT) \cdot A_C \cdot \overline{MW} \cdot P_T}$$

(conversion of mass flow rate to superficial velocity)

$$\alpha = 1 + \frac{2d_p}{3(1 - \overline{\varepsilon}_B)d_t}$$

X_i = mole fraction of component i

u = superficial velocity (ft/s)

T_R = reacting fluid temperature (K)(T_R^0 = inlet value)

T_C = coolant temperature (K) (T_C^0 = inlet value)

$\overline{MW}$ = mean molecular weight of the reacting gas

P_T = total pressure (psia) (P_T^0 = inlet value)

W = total mass flow rate (lb/h)

NT = number of reactor tubes

A_C = cross-sectional area of a reactor tube

d_t = diameter of tube

d_p = catalyst pellet diameter

ρ_B = bed density (lbcat/ft^3)

z = axial coordinate (ft)

R_i = net reaction rate of component i (lbmol/lbcat/h), for example, $R_1 = -r_1$, $R_2 = r_1 - r_2$, $R_3 = -0.5r_1 - 2.5r_2$: assumes effectiveness factors are equal to one

$\overline{\varepsilon}_B$ = porosity of bed

U = overall heat transfer coefficient (BTU/h/ft^2/°C)

mode = (+1) for cocurrent coolant flow, (−1) for countercurrent coolant flow

$F_C C_{pC}$ = coolant rate (BTU/h/°C)

$\overline{\mu}$ = reacting fluid viscosity (lb/ft/h)

$\overline{C}_p$ = gas heat capacity (BTU/lb/°C)

R_g = universal gas constant (psia-ft^3/lbmol/K)

NC = number of components = 5

i	Component	M_i
1	Dammitol	46
2	Valuadehyde	44
3	O_2	32
4	CO_2	44
5	H_2O	18

reactor	2500 tubes	$\frac{1}{12}$ ft (ID) $\times$ 20 ft (length)
W	100,000	lb/h
X_1	0.1	
X_3	0.07	
X_5	0.02	
remainder is N_2 (O_2 comes from air)		
$F_C C_{pC}$	10^6	BTU/(h-°C)
P_T^0	114.7	psia
T_R^0	353	K
d_p	0.25	inches (spherical)
ρ_B	100	lb/ft^3
$\bar{\varepsilon}_B$	0.5	—
$\bar{\mu}$	0.048	lb/(ft-h)
$\bar{C}_p$	0.50	BTU/(lb-°C)
U	120.5	BTU/(h-ft^2-°C)

6. Refer to the reactor system described in Exercise 5. Using the mode of cooling that minimizes the sensitivity, what does the effect of changing $\bar{\varepsilon}_B$ from 0.5 to 0.4 (a realistic change) have on the reactor performance?

7. If the effectiveness factors are now not equal to one, write down the additional equations necessary to solve Exercise 5. Solve the reactor equations without the restriction of the effectiveness factors being unity.

Review of Chemical Equilibria

A.1 | Basic Criteria for Chemical Equilibrium of Reacting Systems

The basic criterion for equilibrium with a single reaction is:

$$\Delta G = \sum_{i=1}^{NCOMP} v_i \mu_i = 0 \tag{A.1.1}$$

where ΔG is the Gibbs function, $NCOMP$ is the number of components in the system, v_i is the stoichiometric coefficient of species i, and μ_i is the chemical potential of species i. The chemical potential is:

$$\mu_i = \mu_i^0 + R_g T \ln a_i \tag{A.1.2}$$

where R_g is the universal gas constant, μ_i^0 is the standard chemical potential of species i in a reference state such that $a_i = 1$, and a_i is the activity of species i. The reference states are: (1) for gases (i.e., $\bar{f}^0 = 1$) (ideal gas, $P = 1$ atm) where $\bar{f}$ is the fugacity, (2) for liquids, the pure liquid at T and one atmosphere, and (3) for solids, the pure solid at T and one atmosphere. If multiple reactions are occurring in a network, then Equation (A.1.1) can be extended to give:

$$\Delta G_j = \sum_{i=1}^{NCOMP} v_{i,j} \mu_i = 0, \quad j = 1, \cdots, NRXN \tag{A.1.3}$$

where $NRXN$ is the number of independent reactions in the network.

In general it is not true that the change in the standard Gibbs function, ΔG^0, is zero. Thus,

$$\Delta G^0 = \sum_{i=1}^{NCOMP} v_i \mu_i^0 \neq 0 \tag{A.1.4}$$

Therefore,

$$\Delta G - \Delta G^0 = \sum_{i=1}^{NCOMP} v_i(\mu_i - \mu_i^0) \tag{A.1.5}$$

or by using Equation (A.1.2):

$$\Delta G - \Delta G^0 = R_g T \sum_{i=1}^{NCOMP} v_i \ln a_i = R_g T \ln\left(\prod_i a_i^{v_i}\right) \tag{A.1.6}$$

Now consider the general reaction:

$$\bar{a}A + \bar{b}B + \cdots = \bar{w}W + \bar{s}S + \cdots \tag{A.1.7}$$

Application of Equation (A.1.6) to Equation (A.1.7) and recalling that $\Delta G = 0$ at equilibrium gives:

$$\Delta G^0 = -R_g T \ln\left[\frac{a_W^{\bar{w}} a_S^{\bar{s}} \cdots}{a_A^{\bar{a}} a_B^{\bar{b}} \cdots}\right] = -R_g T \ln K_a \tag{A.1.8}$$

Thus, the equilibrium constant K_a is defined as:

$$K_a = \prod_{i=1}^{NCOMP} a_i^{v_i} \tag{A.1.9}$$

Differentiation of Equation (A.1.8) with respect to T yields:

$$\left[\frac{\partial(\Delta G^0/T)}{\partial T}\right]\bigg|_P = -R_g\left[\frac{\partial(\ln K_a)}{\partial T}\right]\bigg|_P \tag{A.1.10}$$

Note that $\Delta G^0 = \Delta H^0 - T\Delta S^0$, where ΔH^0 and ΔS^0 are the standard enthalpy and entropy, respectively, and differentiation of this expression with respect to T gives:

$$\left[\frac{\partial(\Delta G^0/T)}{\partial T}\right]\bigg|_P = -\frac{\Delta H^0}{T^2} \tag{A.1.11}$$

Equating Equations (A.1.10) and (A.1.11) provides the functional form for the temperature dependence of the equilibrium constant:

$$\left[\frac{\partial(\ln K_a)}{\partial T}\right]\bigg|_P = \frac{\Delta H^0}{R_g T^2} \tag{A.1.12}$$

or after integration (assume ΔH^0 is independent of T):

$$K_a = \bar{K} \exp[-\Delta H^0/(R_g T)] \tag{A.1.13}$$

Notice that when the reaction is exothermic (ΔH^0 is negative), K_a increases with decreasing T. For endothermic reactions the opposite is true. From Equation (A.1.8):

$$\frac{-\Delta G^0}{R_g T} = \ln K_a \qquad (A.1.14)$$

and

$$K_a = \exp\left[-\Delta G^0/(R_g T)\right] \qquad (A.1.15)$$

Since ΔG^0 is not a function of pressure, it is clear that pressure has no influence on K_a.

A.2 | Determination of Equilibrium Compositions

Consider a gas-phase reaction. If the Lewis and Randall mixing rules are used (simplest form of mixing rules—more sophisticated relationships could be applied if deemed necessary) to give for the fugacity of species i, $\bar{f}_i$:

$$a = \bar{f}_i/\bar{f}_i^0 \qquad (A.2.1)$$

where

$$\bar{f}_i = X_i \bar{\phi}_i P$$

and $\bar{\phi}_i$ = fugacity coefficient of pure i at T and P of system for any mole fraction X_i. Substituting the above expression into Equation (A.1.9) for the reaction given in Equation (A.1.7) yields (let all $\bar{f}_i^0 = 1$):

$$K_a = \left[\frac{X_W^{\bar{w}} X_S^{\bar{s}} \cdots}{X_A^{\bar{a}} X_B^{\bar{b}} \cdots}\right]\left[\frac{\bar{\phi}_W^{\bar{w}} \bar{\phi}_S^{\bar{s}} \cdots}{\bar{\phi}_A^{\bar{a}} \bar{\phi}_B^{\bar{b}} \cdots}\right] P^{\bar{w}+\bar{s}+\cdots-\bar{a}-\bar{b}\cdots} \qquad (A.2.2)$$

or

$$K_a = K_x K_{\bar{\phi}} P^{\bar{w}+\bar{s}+\cdots-\bar{a}-\bar{b}\cdots}$$

where

$$X_i = \frac{n_i}{n_{\text{inert}} + \sum_j n_j}$$

Equation (A.2.2) can be written in terms of moles as:

$$K_a = K_{\bar{\phi}}\left[\frac{n_W^{\bar{w}} n_S^{\bar{s}} \cdots}{n_A^{\bar{a}} n_B^{\bar{b}} \cdots}\right]\left[\frac{P}{n_{\text{inert}} + \sum_j n_j}\right]^{\bar{w}+\bar{s}+\cdots-\bar{a}-\bar{b}\cdots} \qquad (A.2.3)$$

Note that K_a is not a function of pressure and that $K_{\bar{\phi}}$ is only weakly dependent on pressure. Thus, if:

$$\bar{w} + \bar{s} + \cdots - \bar{a} - \bar{b}\cdots = \begin{cases} - & \text{then} \quad P{\uparrow}, K_x{\uparrow} \\ + & \text{then} \quad P{\uparrow}, K_x{\downarrow} \\ 0 & \text{then} \quad \text{no effect} \end{cases}$$

and the effect of inert species are:

$$\overline{w} + \overline{s} + \cdots - \overline{a} - \overline{b} \cdots = \begin{cases} - & \text{then add inert,} \quad K_x \downarrow \\ + & \text{then add inert,} \quad K_x \uparrow \\ 0 & \text{then no effect} \end{cases}$$

Finally, just to state clearly once again, a catalyst has no effect on equilibrium yields.

APPENDIX B

Regression Analysis

B.1 | Method of Least Squares

Below is illustrated the method of least squares to fit a straight line to a set of data points (y_i, x_i). Extensions to nonlinear least squares fits are discussed in Section B.4.

Consider the problem of fitting a set of data (y_i, x_i) where y and x are the dependent and independent variables, respectively, to an equation of the form:

$$y = \bar{\alpha}_1 + \bar{\alpha}_2 x \tag{B.1.1}$$

by determining the coefficients $\bar{\alpha}_1$ and $\bar{\alpha}_2$ so that the differences between y_i and $y_i = \bar{\alpha}_1 + \bar{\alpha}_2 x_i$ are minimized. Given $\bar{\alpha}_1$ and $\bar{\alpha}_2$, the deviations Δy_i can be calculated as:

$$\Delta y_i = y_i - \bar{\alpha}_1 - \bar{\alpha}_2 x_i \tag{B.1.2}$$

For any value of $x = x_i$, the probability PP_i for making the observed measurement y_i with a Gaussian distribution and a standard deviation σ_i for the observations about the actual value $y(x_i)$ is (P. R. Bevington, *Data Reduction and Error Analysis for the Physical Sciences,* McGraw-Hill, New York, 1969, p. 101):

$$PP_i = \frac{1}{\sigma_i \sqrt{2\pi}} \exp\left[-\frac{1}{2}\left[\frac{y_i - y(x_i)}{\sigma_i} \right]^2 \right] \tag{B.1.3}$$

The probability of making the observed data set of measurements of the N values of y_i is the product of the individual PP_i or:

$$PP(\bar{\alpha}_1, \bar{\alpha}_2) = \prod_i^N PP_i = \prod_i^N \left(\frac{1}{\sigma_i \sqrt{2\pi}} \right) \exp\left[-\frac{1}{2} \sum_i^N \left[\frac{y_i - y(x_i)}{\sigma_i} \right]^2 \right] \tag{B.1.4}$$

The best estimates for $\bar{\alpha}_1$ and $\bar{\alpha}_2$ are the values that maximize $PP(\bar{\alpha}_1, \bar{\alpha}_2)$ (method of maximum likelihood). Define:

$$\overline{X}^2 = \sum_i^N \left[\frac{y_i - y(x_i)}{\sigma_i} \right]^2 = \sum_i^N \left[\frac{1}{\sigma_i^2} (y_i - \overline{\alpha}_1 - \overline{\alpha}_2 x_i)^2 \right] \tag{B.1.5}$$

Note that in order to maximize $PP(\overline{\alpha}_1, \overline{\alpha}_2)$, $\overline{X}^2$ is minimized. Thus, the method to find the optimum fit to the data is to minimize the sum of the squares of the deviations (i.e., least-squares fit).

As an example of how to calculate $\overline{\alpha}_1$ and $\overline{\alpha}_2$, consider here the case where $\sigma_i = \sigma = $ constant. To minimize $\overline{X}^2$, the partial derivatives of $\overline{X}^2$ with respect to $\overline{\alpha}_1$ and $\overline{\alpha}_2$ must be set equal to zero:

$$\frac{\partial \overline{X}^2}{\partial \overline{\alpha}_1} = 0 = \frac{\partial}{\partial \overline{\alpha}_1} \left[\frac{1}{\sigma^2} \sum (y_i - \overline{\alpha}_1 - \overline{\alpha}_2 x_i)^2 \right] = \frac{-2}{\sigma^2} \sum (y_i - \overline{\alpha}_1 - \overline{\alpha}_2 x_i)$$

$$\frac{\partial \overline{X}^2}{\partial \overline{\alpha}_2} = 0 = \frac{\partial}{\partial \overline{\alpha}_2} \left[\frac{1}{\sigma^2} \sum (y_i - \overline{\alpha}_1 - \overline{\alpha}_2 x_i)^2 \right] = \frac{-2}{\sigma^2} \sum x_i (y_i - \overline{\alpha}_1 - \overline{\alpha}_2 x_i)$$

These equations can be rearranged to give:

$$\sum y_i = \sum \overline{\alpha}_1 + \sum \overline{\alpha}_2 x_i = \overline{\alpha}_1 N + \overline{\alpha}_2 \sum x_i$$

$$\sum x_i y_i = \sum \overline{\alpha}_1 x_i + \sum \overline{\alpha}_2 x_i^2 = \overline{\alpha}_1 \sum x_i + \overline{\alpha}_2 \sum x_i^2$$

The solutions to these equations yield $\overline{\alpha}_1$ and $\overline{\alpha}_2$ in the following manner:

$$\left. \begin{array}{l} \overline{\alpha}_1 = \dfrac{\begin{vmatrix} \sum y_i & \sum x_i \\ \sum x_i y_i & \sum x_i^2 \end{vmatrix}}{\begin{vmatrix} N & \sum x_i \\ \sum x_i & \sum x_i^2 \end{vmatrix}} = \dfrac{\sum y_i \sum x_i^2 - \sum x_i \sum x_i y_i}{N \sum x_i^2 - (\sum x_i)^2} \\[30pt] \overline{\alpha}_2 = \dfrac{\begin{vmatrix} N & \sum y_i \\ \sum x_i & \sum x_i y_i \end{vmatrix}}{\begin{vmatrix} N & \sum x_i \\ \sum x_i & \sum x_i^2 \end{vmatrix}} = \dfrac{N \sum x_i y_i - \sum x_i \sum y_i}{N \sum x_i^2 - (\sum x_i)^2} \end{array} \right\} \tag{B.1.6}$$

The calculation is straightforward. First compute $\sum x_i$, $\sum y_i$, $\sum x_i^2$, and $\sum x_i y_i$. Second use the summed values in Equation (B.1.6) to obtain $\overline{\alpha}_1$ and $\overline{\alpha}_2$.

B.2 | Linear Correlation Coefficient

Referring to Equations (B.1.1) and (B.1.6), if there is no correlation between x and y, then there are no trends for y to either increase or decrease with increasing x. Therefore, the least-squares fit must yield $\overline{\alpha}_2 = 0$. Now, consider the question of whether the data correspond to a straight line of the form:

$$x = \overline{\alpha}_1' + \overline{\alpha}_2' y \tag{B.2.1}$$

The solution for $\overline{\alpha}_2'$ is:

$$\overline{\alpha}_2' = \frac{N \sum x_i y_i - \sum x_i \sum y_i}{N \sum y_i^2 - (\sum y_i)^2} \tag{B.2.2}$$

Again, if there is no correlation between x and y, then $\overline{\alpha}_2' = 0$. At the other extreme, if there is complete correlation between x and y, then there is a relationship between $\overline{\alpha}_1$, $\overline{\alpha}_2$, $\overline{\alpha}_1'$, and $\overline{\alpha}_2'$ that is:

$$y = -\frac{\overline{\alpha}_1'}{\overline{\alpha}_2'} + \frac{1}{\overline{\alpha}_2'}x = \overline{\alpha}_1 + \overline{\alpha}_2 x$$

$$\overline{\alpha}_1 = -\frac{\overline{\alpha}_1'}{\overline{\alpha}_2'}$$

$$\overline{\alpha}_2 = \frac{1}{\overline{\alpha}_2'}$$

Thus, a perfect correlation gives $\overline{\alpha}_2\overline{\alpha}_2' = 1$ and no correlation yields $\overline{\alpha}_2\overline{\alpha}_2' = 0$ since both $\overline{\alpha}_2$ and $\overline{\alpha}_2'$ are zero for this condition. The linear correlation coefficient is therefore defined as:

$$\overline{R}_{cc} = \sqrt{\overline{\alpha}_2\overline{\alpha}_2'} = \frac{N\sum x_i y_i - \sum x_i \sum y_i}{\left[N\sum x_i^2 - (\sum x_i)^2\right]^{1/2}\left[N\sum y_i^2 - (\sum y_i)^2\right]^{1/2}} \qquad (B.2.3)$$

The values of $\overline{R}_{cc}$ are $-1 \leq \overline{R}_{cc} \leq 1$ with $\overline{R}_{cc} = 1$ and $\overline{R}_{cc} = 0$ defining perfect and no correlations, respectively. Although the linear correlation coefficient is commonly quoted as a measure of "goodness of fit," it is really not appropriate as a direct measure of the degree of correlation. If the data can be represented in a manner such that the fit should result in a y-intercept equal to zero, then a simple method can be used to determine the "goodness of fit."

B.3 | Correlation Probability with a Zero Y-Intercept

Numerous kinetic expressions can be placed into a form that would yield a zero y-intercept when using the linear least-squares method. A survey of a few of these models is provided in Table B.3.1. Given that the y-intercept is a known value (i.e., zero), if a perfect correlation could be achieved, the hypothesis that the true value of the parameter, $\overline{\alpha}_1$, is equal to the specified value, $\overline{\alpha}_1^*$, could be tested by referring the quantity:

$$\bar{t}^* = \frac{\overline{\alpha}_1 - \overline{\alpha}_1^*}{SE(\overline{\alpha}_1)} \qquad (B.3.1)$$

to the table of Student's $\bar{t}^*$ values with N-2 degrees of freedom. The standard error, SE, can be calculated as follows. The standard deviation σ_z of the determination of a parameter z is via the chain rule:

$$\sigma_z^2 = \sum\left[\sigma_i^2\left(\frac{\partial z}{\partial y_i}\right)^2\right] \qquad (B.3.2)$$

where σ_i is the standard deviation of each datum point i. If $\sigma_i = \sigma = $ constant, then σ_z^2 is approximately equal to the sample variance, which is (P. R. Bevington, *Data*

Table B.3.1 | Examples of kinetic relationships yielding zero intercepts.

Kinetics	Reactions	Relationship
1. 0th-order, irreversible (one-way), (constant volume)	$A \xrightarrow{k} \bar{P}$	$C_A^0 - C_A = kt$
2. First-order, irreversible (one-way)	$A \xrightarrow{k} \bar{P}$	$\ln\left[\dfrac{1}{1-f_A}\right] = kt$
3. First-order, reversible (two-way) • no product present at $t = 0$	$A \underset{k_2}{\overset{k_1}{\rightleftharpoons}} \bar{P}$	$\ln\left[\dfrac{1}{1-(f_A/f_A^{eq})}\right] = \left(\dfrac{k_1}{f_A^{eq}}\right)t$
• product present at $t = 0$, C_P^0		$\ln\left[\dfrac{1}{1-(f_A/f_A^{eq})}\right] = \left[\dfrac{k_1\left(\frac{C_P^0}{C_A^0}+1\right)}{\left[\frac{C_P^0}{C_A^0}+f_A^{eq}\right]}\right]t$
4. Second-order, irreversible (one-way), (constant volume) • $C_A^0 = C_B^0$ • $C_A^0 \neq C_B^0$	$A + B \xrightarrow{k} \bar{P}$	$\dfrac{1}{C_A} - \dfrac{1}{C_A^0} = kt$ $\ln\left[\dfrac{C_B C_A^0}{C_A C_B^0}\right] = (C_B^0 - C_A^0)kt$
5. Second-order, reversible (two-way) $C_A^0 = C_B^0,\ C_C^0 = C_D^0 = 0$	$A + B \underset{k_2}{\overset{k_1}{\rightleftharpoons}} C + D$	$\ln\left[\dfrac{f_A^{eq} - (2f_A^{eq} - 1)f_A}{(f_A^{eq} - f_A)}\right] = 2k_1\left[\dfrac{1}{f_A^{eq}} - 1\right]C_A^0 t$
6. Third-order, irreversible (one-way), (constant volume)	$A + 2B \xrightarrow{k} \bar{P}$	$\ln\left[\dfrac{C_B C_A^0}{C_A C_B^0}\right] - \dfrac{2\left[\frac{C_B^0}{C_A^0} - 2\right](C_A^0 - C_A)}{(C_B^0/C_A^0)C_B} = (C_A^0)^2\left[\dfrac{C_B^0}{C_A^0} - 2\right]kt$
	$A + B + C \xrightarrow{k} \bar{P}$	$\ln\left[\dfrac{C_A^0}{C_A}\right] + \left[\dfrac{C_C^0/C_A^0 - 1}{\left(\frac{C_B^0}{C_A^0} - \frac{C_C^0}{C_A^0}\right)}\ln\left(\dfrac{C_B^0}{C_B}\right)\right] + \left[\dfrac{C_B^0/C_A^0 - 1}{\left(\frac{C_B^0}{C_A^0} - \frac{C_C^0}{C_A^0}\right)}\ln\left(\dfrac{C_C^0}{C_C}\right)\right] =$ $(C_A^0)^2\left[\dfrac{C_B^0}{C_A^0} - 1\right]\left[\dfrac{C_C^0}{C_A^0} - 1\right]kt$
7. nth-order, irreversible (one-way), (constant volume)	$A \xrightarrow{k} \bar{P}$	$(C_A)^{(1-n)} - (C_A^0)^{(1-n)} = (n-1)kt$

Reduction and Error Analysis for the Physical Sciences, McGraw-Hill, New York, 1969, p. 114):

$$\sigma^2 \cong \frac{1}{N-2} \sum \left(y_i - \overline{\alpha}_1 - \overline{\alpha}_2 x_i\right)^2 \tag{B.3.3}$$

for the linear equation (B.1.1). (The sample variance is the sum of squares of the residuals divided by the number of data points minus the number of parameters fitted.) Now Equation (B.3.2) written for the linear equation (B.1.1) gives:

$$\left.\begin{aligned}
\sigma_{\overline{\alpha}_1}{}^2 &= \sigma^2 \sum \left(\frac{\partial \overline{\alpha}_1}{\partial y_i}\right)^2 \\
\sigma_{\overline{\alpha}_2}{}^2 &= \sigma^2 \sum \left(\frac{\partial \overline{\alpha}_2}{\partial y_i}\right)^2
\end{aligned}\right\} \tag{B.3.4}$$

Using Equation (B.1.6) to calculate the partial derivatives in Equation (B.3.4) yields:

$$\sigma_{\overline{\alpha}_1}{}^2 = \frac{\sigma^2}{\Delta} \left(\sum x_i^2\right) \tag{B.3.5}$$

$$\sigma_{\overline{\alpha}_2}{}^2 = \frac{N\sigma^2}{\Delta} \tag{B.3.6}$$

where

$$\Delta = N \sum x_i^2 - \left(\sum x_i\right)^2$$

Thus, $SE(\overline{\alpha}_1) = \sigma_{\overline{\alpha}_1}$ and $SE(\overline{\alpha}_2) = \sigma_{\overline{\alpha}_2}$. Now returning to Equation (B.3.1), $\overline{t}^*$ is the experimental deviation over the standard error and if this value is larger than the value in a Student's $\overline{t}^*$-distribution table (see any text on statistics for this table) for a given degree of confidence, for example, 95 percent ($\overline{t}_{\exp}^* =$ expected deviation/standard error), then the hypothesis is rejected, that is, the y-intercept is significantly different than zero. If $\overline{t}^* < \overline{t}_{\exp}^*$ then the hypothesis is accepted and $\overline{\alpha}_2$ can be reported as:

$$\overline{\alpha}_2 \pm \sigma_{\overline{\alpha}_2} \tag{B.3.7}$$

B.4 | Nonlinear Regression

There are numerous methods for performing nonlinear regression. Here, a simple analysis is presented in order to provide the reader the general concepts used in performing a nonlinear regression analysis.

To begin a nonlinear regression analysis, the model function must be known. Let:

$$y = f(x, a) \tag{B.4.1}$$

where the function f is nonlinear in the dependent variable x and unknown parameters designated by the set $a = [a_1, a_2, \ldots, a_n]$. A least-squares fit of the observed measurements y_i to the function shown in Equation (B.4.1) can be performed as follows. First, define $\overline{X}^2$ [for linear regression see Equation (B.1.5)] as:

$$\overline{X}^2 = \sum_{i=1}^{N} \left[\frac{y_i - y(x_i)}{\sigma_i} \right] = \sum_{i=1}^{N} \left\{ \frac{1}{\sigma^2} [y_i - f(x_i, a)]^2 \right\} \qquad (B.4.2)$$

As with linear least squares analysis, $\overline{X}^2$ is minimized as follows. The partial derivatives of $\overline{X}^2$ with respect to the parameters of a are set equal to zero, for example, with a_1:

$$0 = \frac{\partial \overline{X}^2}{\partial a_1} = \frac{\partial}{\partial a_1} \left\{ \frac{1}{\sigma^2} \sum_{i=1}^{N} [y_i - f(x_i, a)]^2 \right\} = \frac{-2}{\sigma^2} \sum_{i=1}^{N} [y_i - f(x_i, a)] \frac{\partial f(x_i, a)}{\partial a_1} \qquad (B.4.3)$$

Thus, there will be n equations containing the n parameters of a. These equations involve the function $f(x_i, a)$ and the partial derivatives of the function, that is,

$$\frac{\partial f(x_i, a)}{\partial a_j}, \quad \text{for } j = 1, \ldots, n$$

The set of n equations of the type shown in Equation (B.4.3) needs to be solved. This set of equations is nonlinear if $f(x_i, a)$ is nonlinear. Thus, the solution of this set of equations requires a nonlinear algebraic equation solver. These are readily available. For information on the type of solution, consult any text on numerical analysis. Since the solution involves a set nonlinear algebraic equation, it is performed by an iterative process. That is, initial guesses for the parameters a are required. Often, the solution will terminate at local minimum rather than the global minimum. Thus, numerous initial guesses should be used to assure that the final solution is independent of the initial guess.

The issue of "goodness-of-fit" with nonlinear regression is not straightforward. Numerous methods can be used to explore the "goodness-of-fit" of the model to the data (e.g., residual analysis, variance analysis, and Chi-squared analysis). It is always a good idea to inspect the plot of the predicted $[y(x_i)]$ versus observed y_i values to watch for systematic deviations. Additionally, some analytical measure for "goodness-of-fit" should also be employed.

APPENDIX C

Transport in Porous Media

C.1 | Derivation of Flux Relationships in One-Dimension

Consider a tube filled with an isobaric binary gas mixture of components A and B. When component A moves it exerts a force on B. This frictional force, ff_{AB}, can be described as:

ff_{AB} = (proportionality constant) (concentration of A) (concentration of B) (relative velocity of A to B)

$\underbrace{\qquad\qquad\qquad\qquad\qquad\qquad}_{\text{number of collisions}}$ $\underbrace{\qquad\qquad\qquad\qquad\qquad}_{\substack{\text{momentum exchange} \\ \text{per collision}}}$

or

$$ff_{AB} = (\text{const})_{AB}\, C_A C_B (V_A - V_B) \qquad (\text{C.1.1})$$

where V_i is the molecular velocity of species i. The total frictional losses have to equal the driving force. Thus, for the section dz,

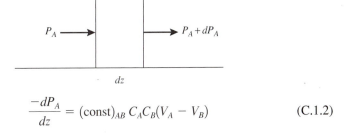

$$\frac{-dP_A}{dz} = (\text{const})_{AB}\, C_A C_B (V_A - V_B) \qquad (\text{C.1.2})$$

Multiply Equation (C.1.2) by C (total concentration) and rearrange to give:

$$-CP\frac{dX_A}{dz} = (\text{const})_{AB}\, C C_A C_B (V_A - V_B)$$

$$\frac{-CP}{(\text{const})_{AB}\, C_A C_B} \frac{dX_A}{dz} = C(V_A - V_B)$$

$$\frac{-R_g T}{(\text{const})_{AB}\, X_A X_B} \frac{dX_A}{dz} = C(V_A - V_B) \tag{C.1.3}$$

If the proportionality factor is defined as:

$$(\text{const})_{AB} = \frac{R_g T}{CD_{AB}} \tag{C.1.4}$$

then Equation (C.1.3) can be written as follows:

$$\frac{-CD_{AB}}{X_A X_B} \frac{dX_A}{dz} = C(V_A - V_B) \tag{C.1.5}$$

Equation (C.1.5) is Fick's First Law. To see this, rearrange Equation (C.1.5) as shown below:

$$C\, X_A X_B (V_A - V_B) = -CD_{AB} \frac{dX_A}{dz}$$

$$C_A (X_B V_A - X_B V_B) = -CD_{AB} \frac{dX_A}{dz} \tag{C.1.6}$$

Recall that Fick's First Law is:

$$\bar{J}_A = -CD_{AB} \frac{dX_A}{dz} = C_A (V_A - V_{\text{total}}) \tag{C.1.7}$$

where $\bar{J}_A$ is the flux of A with respect to a coordinate system that is moving at V_{total} and:

$$V_{\text{total}} = X_A V_A + X_B V_B$$

Further arrangements of Equation (C.1.6) can be made as shown below:

$$C_A (X_B V_A - X_B V_B + X_A V_A - X_A V_A) = -CD_{AB} \frac{dX_A}{dz}$$

$$C_A (-X_A V_A - X_B V_B + X_A V_A + X_B V_A) = -CD_{AB} \frac{dX_A}{dz}$$

$$C_A (V_A - V_{\text{total}}) = -CD_{AB} \frac{dX_A}{dz}$$

which is Equation (C.1.7).

Now consider a multicomponent system. For a multicomponent mixture of *NCOMP* species:

$$ff_{ij} = \frac{R_g T \, C_i C_j}{C D_{ij}} (V_i - V_j) \tag{C.1.8}$$

by analogy to the frictional force for a binary mixture [Equation (C.1.1) with Equation (C.1.4)]. Using Equation (C.1.8) gives a total force balance of:

$$\frac{-dX_i}{dz} = \sum_{\substack{j=1 \\ i \neq j}}^{NCOMP} \frac{X_i X_j (V_i - V_j)}{D_{ij}} \tag{C.1.9}$$

that is the Stefan-Maxwell equations.

The Stefan-Maxwell equations are normally written in terms of fluxes, N_i. Since $N_i = C_i V_i$, Equation (C.1.9) can be expressed in terms of fluxes as:

$$\frac{-dX_i}{dz} = \sum_{\substack{j=1 \\ i \neq j}}^{NCOMP} \left[\frac{X_j N_i - X_i N_j}{C D_{ij}} \right] \tag{C.1.10}$$

C.2 | Flux Relationships in Porous Media

Consider the movement of a binary gas mixture in a pore:

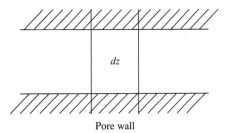

Pore wall

The loss of momentum in dz due to molecule-wall collisions is:

$$(dP_A)_K (g_c A_C^P) = N_A \frac{R_g T}{D_{KA}} dz (g_c A_C^P) \tag{C.2.1}$$

where

$$g_c = \text{gravitational constant}$$
$$A_C^P = \text{cross-sectional area of the pore}$$
$$D_{KA} = \text{Knudsen diffusion coefficient of } A$$
$$(dP_A)_K = \text{change in pressure from molecule–wall collisions}$$

Fick's First Law is (momentum losses due to molecule-molecule collisions):

$$N_A = \bar{J}_A + X_A(N_A + N_B) = -\frac{D_{AB}}{R_g T}\frac{dX_A}{dz} + X_A(N_A + N_B) \qquad \text{(C.2.2)}$$

(diffusive flux) + (bulk flow)

Let:

$$FR = 1 + \frac{N_B}{N_A} \qquad \text{(C.2.3)}$$

so that Equation (C.2.2) can be written as:

$$N_A = \frac{-D_{AB}}{\left[1 - (FR)X_A\right]R_g T}\frac{dX_A}{dz} \qquad \text{(C.2.3)}$$

or

$$-(dP_A)_{mm}\, g_c A_C^P = N_A \frac{R_g T}{D_{AB}}\left[1 - (FR)X_A\right]dz(g_c A_C^P) \qquad \text{(C.2.4)}$$

where $(dP_A)_{mm}$ is the change in pressure from molecule-molecule collisions. Now the total momentum loss due to molecule-wall and molecule-molecule collisions is the sum of Equation (C.2.1) and Equation (C.2.4) $[(dP_A) = (dP_A)_K + (dP_A)_{mm}]$:

$$-(dP_A)g_c A_C^P = N_A \frac{R_g T}{D_{KA}}\, dz\, g_c A_C^P + N_A \frac{R_g T}{D_{AB}}\left[1 - (FR)X_A\right]dz(g_c A_C^P)$$

or after rearrangement:

$$N_A = \frac{-1}{\left[\dfrac{1 - (FR)X_A}{D_{AB}} + \dfrac{1}{D_{KA}}\right]}\frac{1}{R_g T}\frac{dP_A}{dz} \qquad \text{(C.2.5)}$$

If there is equimolar counterdiffusion ($N_A = -N_B$) and/or if X_A is small, Equation (C.2.5) reduces to:

$$N_A = \frac{-1}{\dfrac{1}{D_{AB}} + \dfrac{1}{D_{KA}}}\frac{1}{R_g T}\frac{dP_A}{dz} = \frac{-D_{TA}}{R_g T}\frac{dP_A}{dz} \qquad \text{(C.2.6)}$$

where

$$\frac{1}{D_{TA}} = \frac{1}{D_{AB}} + \frac{1}{D_{KA}} \qquad \text{(C.2.7)}$$

Equation (C.2.7) is called the Bosanquet equation.

A momentum balance for multicomponent mixtures can be formulated in a manner analogous to that used to derive Equation (C.2.4) using molecule-wall and molecule-molecule (Stefan-Maxwell) relationships to give:

$$-\left(dP_i\right)g_c A_C^P = \frac{N_i R_g T}{D_{Ki}}\,dz\,g_c A_C^P + \sum_{\substack{j=1 \\ i \neq j}}^{NCOMP} \frac{R_g T}{D_{ij}}\left(X_j N_i - X_i N_j\right)dz\,g_c A_C^P$$

or

$$\frac{-1}{R_g T}\frac{dP_i}{dz} = \sum_{\substack{j=1 \\ i \neq j}}^{NCOMP}\left[\frac{X_j N_i - X_i N_j}{D_{ij}}\right] + \frac{N_i}{D_{Ki}} \tag{C.2.8}$$

INDEX